Advances in Near-Earth Space and Atmospheric Physics from Ground-Based and Satellite Observations

Advances in Near-Earth Space and Atmospheric Physics from Ground-Based and Satellite Observations

Guest Editors

Veronika Barta
Christina Arras
Jaroslav Urbar

MDPI

Basel • Beijing • Wuhan • Barcelona • Belgrade • Novi Sad • Cluj • Manchester

Guest Editors

Veronika Barta
HUN-REN Institute of Earth Physics and Space Science
Sopron
Hungary

Christina Arras
Institute of Geodesy
Technische Universität Berlin
Berlin
Germany

Jaroslav Urbar
Institute of Atmospheric Physics of Czech Academy of Sciences
Prague
Czech Republic

Editorial Office

MDPI AG
Grosspeteranlage 5
4052 Basel, Switzerland

This is a reprint of the Special Issue, published open access by the journal *Remote Sensing* (ISSN 2072-4292), freely accessible at: https://www.mdpi.com/journal/remotesensing/special_issues/M31FRT69P8.

For citation purposes, cite each article independently as indicated on the article page online and as indicated below:

Lastname, A.A.; Lastname, B.B. Article Title. *Journal Name* **Year**, *Volume Number*, Page Range.

ISBN 978-3-7258-6980-0 (Hbk)
ISBN 978-3-7258-6981-7 (PDF)
https://doi.org/10.3390/books978-3-7258-6981-7

Contents

Ercha Aa, Nicholas M. Pedatella and Guiping Liu
Impacts of the Sudden Stratospheric Warming on Equatorial Plasma Bubbles: Suppression of EPBs and Quasi-6-Day Oscillations
Reprinted from: *Remote Sensing* **2024**, *16*, 1469, https://doi.org/10.3390/rs16081469 . . . . **1**

Veronika Barta, Tamás Bozóki, Dávid Péter Süle, Daniel Kouba, Jens Mielich, Tero Raita and Attila Buzás
Ionospheric Absorption Variation Based on Ionosonde and Riometer Data and the NOAA D-RAP Model over Europe During Intense Solar Flares in September 2017
Reprinted from: *Remote Sensing* **2024**, *16*, 3975, https://doi.org/10.3390/rs16213975 . . . . **17**

Kitti Alexandra Berényi, Veronika Barta, Csilla Szárnya, Attila Buzás and Balázs Heilig
Subauroral and Auroral Conditions in the Mid- and Low-Midlatitude Ionosphere over Europe During the May 2024 Mother's Day Superstorm
Reprinted from: *Remote Sensing* **2025**, *17*, 2492, https://doi.org/10.3390/rs17142492 . . . . **39**

Mingyang Liu, Bingqiang Sun and Rui Lu
Radiative Transfer Simulation in the Near-Space Region for a Point Source at High Temperature Based on a Monte Carlo Method
Reprinted from: *Remote Sensing* **2025**, *17*, 2769, https://doi.org/10.3390/rs17162769 . . . . **66**

Yujiao Jin, Xianguo Zhang, Maosheng He, Yongping Li, Xiangguang Meng, Jiangzhao Ai, et al.
Comparisons and Analyses of Thermospheric Mass Densities Derived from Global Navigation Satellite System–Precise Orbit Determination and an Ionization Gauge–Orbital Neutral Atmospheric Detector Onboard a Spherical Satellite at 520 km Altitude
Reprinted from: *Remote Sensing* **2025**, *17*, 98, https://doi.org/10.3390/rs17010098 . . . . **91**

Yufen Ma, Kang Zeng, Ailiyaer Aihaiti, Junjian Liu, Zonghui Liu and Ali Mamtimin
Diurnal Variations of Infrared Land Surface Emissivity in the Taklimakan Desert: An Observational Analysis
Reprinted from: *Remote Sensing* **2025**, *17*, 1276, https://doi.org/10.3390/rs17071276 . . . . **109**

Anna Morozova, Luca Spogli, Teresa Barata, Rayan Imam, Emanuele Pica, Juan Andrés Cahuasquí, et al.
Scintillations in Southern Europe During the Geomagnetic Storm of June 2015
Reprinted from: *Remote Sensing* **2025**, *17*, 535, https://doi.org/10.3390/rs17030535 . . . . **128**

Csilla Szárnya, Zbyšek Mošna, Antal Igaz, Daniel Kouba, Tobias G. W. Verhulst, Petra Koucká Knížová, et al.
Technical Possibilities and Limitations of the DPS-4D Type of Digisonde in Individual Meteor Detections
Reprinted from: *Remote Sensing* **2024**, *16*, 2658, https://doi.org/10.3390/rs16142658 . . . . **151**

Xin Tian, Mingsheng Chen, Pinhua Xie, Jin Xu, Ang Li, Bo Ren, et al.
Evaluation of MAX-DOAS Profile Retrievals under Different Vertical Resolutions of Aerosol and NO_2 Profiles and Elevation Angles
Reprinted from: *Remote Sensing* **2023**, *15*, 5431, https://doi.org/10.3390/rs15225431 . . . . **175**

Qian Wang, Peiqi Hao and Han Pan
Automatic Identification and Segmentation of Diffuse Aurora from Untrimmed All-Sky Auroral Videos
Reprinted from: *Remote Sensing* **2026**, *18*, 402, https://doi.org/10.3390/rs18030402 . . . . **203**

remote sensing

MDPI

Article

Impacts of the Sudden Stratospheric Warming on Equatorial Plasma Bubbles: Suppression of EPBs and Quasi-6-Day Oscillations

Ercha Aa [1,*], Nicholas M. Pedatella [2] and Guiping Liu [3]

1 Haystack Observatory, Massachusetts Institute of Technology, Westford, MA 01886, USA
2 High Altitude Observatory, NSF National Center for Atmospheric Research, Boulder, CO 80301, USA
3 ITM Physics Laboratory, Heliophysics Division, NASA Goddard Space Flight Center, Greenbelt, MD 20771, USA
* Correspondence: aercha@mit.edu

Abstract: This study investigates the day-to-day variability of equatorial plasma bubbles (EPBs) over the Atlantic–American region and their connections to atmospheric planetary waves during the sudden stratospheric warming (SSW) event of 2021. The investigation is conducted on the basis of the GOLD (Global Observations of the Limb and Disk) observations, the ICON (Ionospheric Connection Explorer) neutral wind dataset, ionosonde measurements, and simulations from the WACCM-X (Whole Atmosphere Community Climate Model with thermosphere–ionosphere eXtension). We found that the intensity of EPBs was notably reduced by 35% during the SSW compared with the non-SSW period. Furthermore, GOLD observations and ionosonde data show that significant quasi-6-day oscillation (Q6DO) was observed in both the intensity of EPBs and the localized growth rate of Rayleigh–Taylor (R-T) instability during the 2021 SSW event. The analysis of WACCM-X simulations and ICON neutral winds reveals that the Q6DO pattern coincided with an amplification of the quasi-6-day wave (Q6DW) in WACCM-X simulations and noticeable ~6-day periodicity in ICON zonal winds. The combination of these multi-instrument observations and numerical simulations demonstrates that certain planetary waves like the Q6DW can significantly influence the day-to-day variability of EPBs, especially during the SSW period, through modulating the strength of prereversal enhancement and the growth rate of R-T instability via the wind-driven dynamo. These findings provide novel insights into the connection between atmospheric planetary waves and ionospheric EPBs.

Keywords: equatorial plasma bubbles; atmospheric planetary waves; sudden stratospheric warming; quasi-6-day oscillations; Rayleigh–Taylor instability

1. Introduction

A significant portion of the ionospheric variability can be driven by lower atmospheric forcing via vertical coupling processes [1,2]. Sudden stratospheric warming (SSW) is one of the most dramatic meteorological events of vertical coupling that occurs in the winter polar stratosphere, driven by the growth and dissipation of upward-propagating planetary waves from the troposphere, as well as their nonlinear interaction with the zonal mean flow [3]. SSW is characterized by a sudden rise in high-latitude stratospheric temperatures by several tens of Kelvin within merely a few days, a displacement or breakdown of the polar vortex, and the weakening (minor warming) or even reversal (major warming) of the zonal-mean zonal wind at the location of 60° latitude and 10 hPa (~30 km) [4]. Numerous studies have indicated that SSW can not only change the chemistry and dynamics of the stratosphere and mesosphere but also influence the whole atmosphere by adjusting the dynamics, composition, and electrodynamics of the coupled thermosphere–ionosphere system [5,6]. SSW-related wind and temperature changes can alter the amplitudes and phases of tidal waves, which is particularly pronounced in causing the amplification of the lunar semidiurnal tide through resonance, attributed to the atmospheric Pekeris mode, and the solar semidiurnal tide, influenced by the interaction with stationary planetary waves [7, 8]. The enhanced tidal perturbations can propagate into the altitudes of the ionospheric E-region and modulate the electric fields through the ionospheric wind dynamo, leading to changes in the E $\times$ B plasma drifts and subsequent electron density variation in the ionospheric F-region [9]. Thus, significant variability with diverse temporal scales is observed in the global ionosphere during the SSW period, particularly the enhanced semidiurnal variations over the low-latitude and equatorial regions, such as shown in the vertical plasma drift, e.g., [10–16], equatorial electrojet, e.g., [17–21], and ionospheric electron density (Ne) and total electron content (TEC), e.g., [5,13,22–28].

Moreover, notable ionospheric variations during the SSW period with longer temporal scales of multiple days (e.g., $\sim$2, $\sim$3, $\sim$6, $\sim$10, and $\sim$16 days) over the low-latitude and equatorial ionosphere have also been extensively documented, e.g., [29–35]. These multiday oscillations are related to the modulation of planetary waves, including the westward-propagating Rossby waves and the eastward-propagating equatorial Kelvin waves, either through direction propagation into the ionospheric E-region height or via their interactions with and modulation of upward-propagating tides [36–38]. Among various planetary waves, the westward-propagating quasi-6-day wave (Q6DW) of the Rossby normal mode, characterized by a zonal wavenumber 1 and a period of 5–8 days, is one of the most prominent and recurrent oscillations within the MLT (mesosphere and lower thermosphere) region, e.g., [39–42]. Several studies have indicated that the interaction between Q6DW and tides and the subsequent modulation of the equatorial E-region dynamo because of wind disturbances could modify the plasma drift, thereby leading to the generation of quasi-6-day oscillations (Q6DOs) in the ionospheric F-region electron density and TEC over the equatorial ionization anomaly (EIA) regions, e.g., [34,43–47]. Moreover, some studies suggest that the dissipation of planetary waves in the lower thermosphere may induce an additional meridional flow, which could amplify the mixing of thermospheric composition and cause an overall reduction in the thermospheric O/N_2 column ratio and ionospheric TEC, e.g., [35,48].

While substantial progress has been achieved in understanding the low-latitude and equatorial ionospheric variation during the SSW period, the potential impact of SSW on equatorial plasma bubbles (EPBs) is an important issue that has not been thoroughly analyzed and explained yet. EPBs refer to irregular ionospheric structures characterized by plasma density depletion with scale sizes ranging from tens to hundreds of kilometers, which usually occur in the equatorial and low-latitude ionosphere after local sunset, spanning from the bottom-side F-region to 1000–2000 km [49–51]. EPBs represent some of the most prominent ionospheric plasma disturbances that constitute a noteworthy space weather phenomenon, capable of adversely impacting radio wave propagation and posing severe challenges to navigation systems. Thus, the morphological features and the dynamic evolution of EPBs have been extensively studied using multi-instrument observations and numerical simulations, e.g., [52–62]. It is known that EPBs are generated by the generalized Rayleigh–Taylor (R-T) instability when the bottom-side F-region becomes unstable in the evening with a steep vertical density gradient after the decay of the E-region [63–65]. The prereversal enhancement (PRE), characterized by an increase in the eastward electric field and upward plasma drift at the equatorial evening terminator, creates favorable conditions to amplify the growth rate of R-T instability and is considered one of the pivotal factors controlling the development of EPBs [66–68]. Although we currently possess a reasonably sound knowledge of the climatological characteristics of EPBS, such as their seasonal and longitudinal variations, e.g., [69–74], understanding the complicated day-to-day variability of EPBs and their relationship with lower atmospheric forcing, especially during an SSW, remains a challenging issue for the equatorial space weather community [75–78].

Some studies have conducted preliminary analyses on the relationship between SSW events and the occurrence rate of equatorial spread-F/scintillation. For example, ref. [79] reported a prolonged weakening of the ionospheric scintillation at a ground-based Global

Navigation Satellite System (GNSS) station located in Brazil during three SSW events. Yu et al. [80] and Ye et al. [81] also observed the inhibition of scintillation over the American sector in COSMIC satellite measurements during certain SSW events. Jose et al. [82] presented that the timing of the equatorial spread-F occurrence showed a quasi-16-day oscillation during the SSW years. These studies suggested that the modification of plasma vertical drift during PRE and the altered meridional wind pattern are potential factors causing variations in the equatorial spread-F/scintillation. Moreover, recent pioneering studies have shown that the modulation of tidal winds by planetary waves could have a large impact on the strength of the PRE, thereby leading to planetary-wave-scale oscillations in the intensity of PRE that may potentially influence the day-to-day variability of EPBs [54,83–87]. Nevertheless, there is currently limited observational evidence to demonstrate that signatures of planetary-wave-scale oscillations can be detected in EPBs or the growth rate of R-T instability, and the day-to-day variability of EPBs during SSW deserved a further examination with improved observational and modeling capabilities [86,88].

In the current study, we use multi-instrumental observations and a whole atmospheric model to collectively investigate the day-to-day variation in EPBs during the 2021 SSW, as well as to examine the possible connection between EPBs and planetary waves through the modulation of the R-T instability growth rate. Specifically, the ultraviolet imaging data from the Global-scale Observations of the Limb and Disk (GOLD) is used to characterize the day-to-day variation in EPBs throughout SSW, and equatorial ionosonde data are used to compute the local growth rate of R-T instability as a proxy for EPBs' occurrence. We found that the intensity of EPBs was notably reduced by 35% during the SSW period compared with the non-SSW period. Moreover, a significant quasi-6-day oscillation (Q6DO) pattern was recorded in both the EPBs and R-T instability growth rate during the SSW. The Ionospheric Connection Explorer (ICON) neutral winds data, combined with simulation results from the Whole Atmosphere Community Climate Model with thermosphere–ionosphere eXtension (WACCM-X), are utilized to analyze the corresponding Q6DW activities in the MLT region and middle atmosphere, respectively. The results show that lower atmospheric waves could play an essential role in controlling the day-to-day variation in EPBs during the SSW period by modulating the strength of the PRE and the growth rate of R-T instability.

2. Observations and Simulations

The GOLD instrument is a UV imaging spectrograph that operates from the geostationary orbit at a longitude of 47.5°W, which observes airglow emissions of Earth within the wavelength range of 134 to 160 nm through the disk, limb, and stellar occultation technique [89,90]. In the evening, the disk measurements of the OI 135.6 nm emission by the GOLD instrument offer continuous and crucial information on the low-latitude and equatorial ionosphere from West Africa to South America, which is particularly valuable for revealing the two-dimensional spatial morphology of EPBs, e.g., [57,88,91–94]. To characterize the day-to-day variation in EPBs, Aa et al. [88] has developed a new GOLD Bubble Index capable of quantifying the intensity of the bubbles in two dimensions on any given night within the field-of-view of GOLD measurements. In essence, the Bubble Index is computed based on calculating the standard deviation values of the normalized residual radiance at 135.6 nm, which can effectively quantify the short-term (e.g., day-to-day) and long-term (e.g., seasonal and solar cycle) variations in 2D EPBs. To get more information about the GOLD Bubble Index, readers are encouraged to refer to Aa et al. [88]. Moreover, the ionosonde observations from the São Luís station (2.6°S, 44.2°W), located in the equatorial ionosphere, will be used in the study to derive the local growth rate of R-T instability.

The ICON is a low-Earth orbiting satellite that operated within an altitude range of 580–610 km and had an inclination angle of 27° for studying the low-latitude and midlatitude ionosphere and thermosphere [95]. The Michelson interferometer for Global High-resolution Thermospheric Imaging (MIGHTI) is one of the four science instruments of ICON, which can measure the thermospheric wind velocities derived from the Doppler shift of the atomic oxygen emissions at red (630.0 nm) and green (557.7 nm) lines [96]. MIGHTI has two identical sensors to orthogonally measure the horizontal line-of-sight wind at two angles of 45° and 135° relative to the ICON orbit to collectively derive the horizontal wind vector on the northern limb (∼13°) of the orbit. In this study, the ICON-MIGHTI wind data from the green line measurements is utilized to analyze the signatures of planetary waves in the lower thermosphere [97].

WACCM-X is a whole atmosphere community climate model that extends from the surface of Earth to the upper thermosphere (500–700 km), which integrates a thorough treatment of physics, chemistry, electrodynamics, and thermodynamics spanning from the troposphere to the thermosphere–ionosphere [98]. In the current study, to accurately simulate the atmospheric condition during the 2021 SSW event, the WACCM-X simulation uses the specified dynamics to nudge the model toward the meteorological reanalysis data [99], where the lower atmosphere meteorology up to 50 km is constrained to the NASA Modern Era Retrospective Analysis for Research and Applications version-2 (MERRA-2, Gelaro et al. [100]). The F10.7 solar radio flux is utilized to represent the solar irradiance variations, and the Kp index is employed to drive the high-latitude convection and auroral precipitation [101]. This simulation is referred to as LA + S/G (Lower Atmosphere + Solar/Geomagnetic). Besides using the realistic F10.7 and Kp indices, WACCM-X was also run with troposphere–stratosphere being constrained to MERRA-2 but using constant F10.7 (75 solar flux unit) and Kp (0+) values to isolate forcing from the lower atmosphere (LA only). For more content on the WACCM-X simulation, readers could refer to Pedatella et al. [86].

3. Results

Figure 1 depicts the meteorological and solar/geomagnetic activities between December 2020 and March 2021. Figure 1a displays the temporal variation in the stratospheric polar temperature at 60°–90°N, 10 hPa (red) and zonal mean zonal wind at 60°N, 10 hPa (blue). The stratospheric polar temperature showed a rapid increase of more than 30 K in just a few days at the beginning of January 2021. The zonal mean zonal wind showed a significant decreasing trend from late December of 2020 and reversed its direction on 5 January 2021. Following the initial reversal, the zonal mean zonal wind remained weak for ∼30 days and experienced two more reversals around days 12–22 and 31–34, associated with small temperature increases. These conditions correspond to a major SSW event, and some early studies have reported enhanced semidiurnal tidal perturbations in the ionosphere [16,26,99,102]. Figure 1b displays the amplitudes of the planetary waves of zonal wavenumber 1 (Z1) and wavenumber 2 (Z2) at 60°N, 10 hPa. As is evident, the occurrence of the 2021 SSW was mainly associated with a considerable enhancement of the Z1 component. Moreover, Figure 1c,d shows the variation in F10.7 cm solar radio flux and Kp indices. The F10.7 solar flux decreased from ∼100 solar flux units (sfu, 1 sfu = 10^{-22} W/m^2/Hz) at the beginning of December 2020 to 72 sfu on 5 January 2021, where it maintained a constantly low level of around 70 sfu for the remainder of the time period. The Kp index did not exhibit strong geomagnetic activity during this time period, although it occasionally reached four, indicating some minor to moderate perturbations. Furthermore, Figure 1e illustrates the temporal variation in the Q6DW amplitudes in the WACCM-X simulated temperature at 60°S and 0.001 hPa (∼95 km). The amplitude was derived by fitting the temperature to a westward propagating wavenumber-1 wave with a 6.7-day period within a moving window of 19 days. For more detailed descriptions of the derivation of Q6DW, readers could refer to Pedatella et al. [86]. The amplitude of the simulated Q6DW exhibited a large enhancement, beginning in mid-December 2020 and persisting through the SSW till early February 2021.

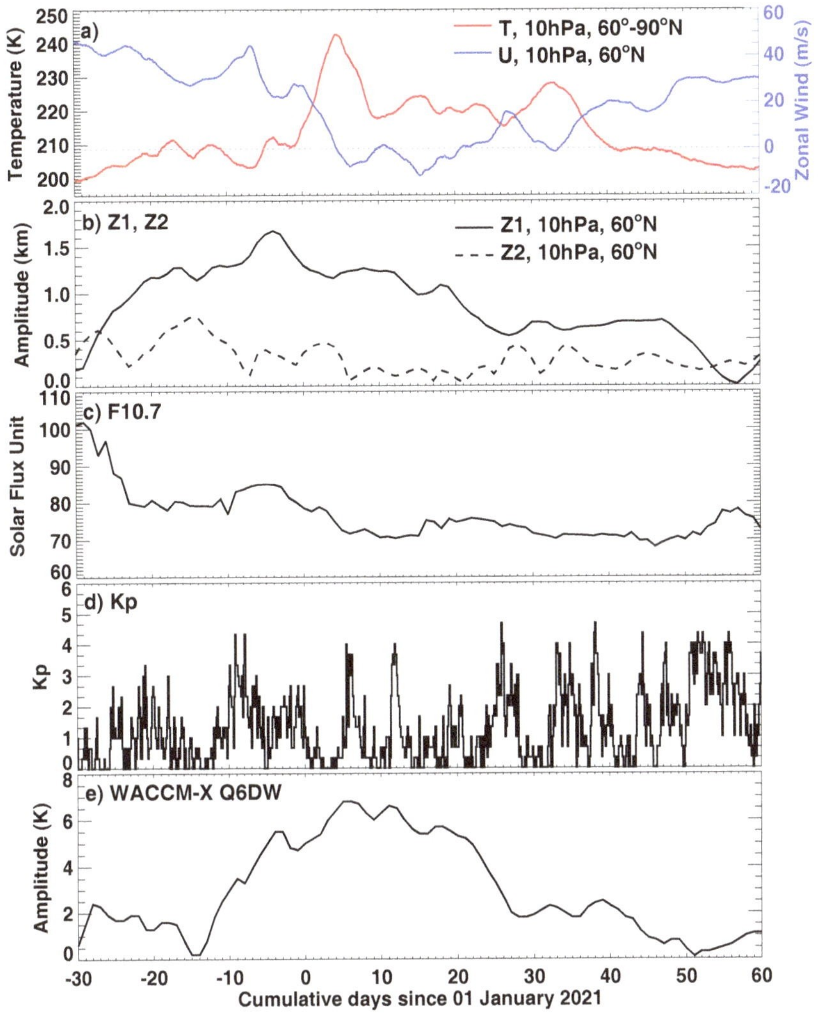

Figure 1. Temporal variations in (**a**) stratospheric polar temperature at 60°–90°N, 10 hPa (red) and zonal mean zonal wind at 60°N, 10 hPa (blue), (**b**) the amplitude of planetary waves with zonal wavenumber 1 (solid) and 2 (dashed) at 60°N, 10 hPa, (**c**) F10.7 cm solar radio flux, (**d**) Kp index, and (**e**) the amplitude of the westward wavenumber-1 Q6DW in temperature given by WACCM-X simulation (LA + S/G) with respect to the cumulative days from 1 January 2021.

Subsequently, we investigate the day-to-day variability of EPBs observed by the GOLD instrument before and after the beginning of the SSW event. For example, Figure 2a,b shows two GOLD postsunset partial disk images of OI 135.6 nm radiance in geomagnetic coordinates at 23:40 UT on 28 December 2020, a day before the SSW onset, and 8 January 2021, a day during the SSW period. As can be seen, the intensity of EPBs over the American–Atlantic region exhibited a significant difference between these two days: On 28 December 2020, before the onset of the SSW, at least five strong EPBs were noticeable as parallel dark streaks in the equatorial region, elongated in the geomagnetic meridional orientation. However, at the same time on 8 January, after the beginning of the SSW, there were almost no discernible EPBs within the GOLD field of view. To specify the intensity of observed bubbles before and during the SSW for further analysis, Figure 2c,d displays the corresponding longitudinal variation in the normalized differential radiance (ΔR), after removing the latitudinal-integrated reference radiance values within the equatorial rectangle box shown in the top panels. On 28 December (Figure 2c), the curve showed significant fluctuations with five distinct valleys, which effectively delineated the longitudinal variation in EPBs-induced bite-outs in radiance. In contrast, the curve on 8 January (Figure 2d) exhibited very small fluctuations indicating weak/no EPBs. The GOLD Bubble Index is determined as the standard deviation values of the normalized ΔR, which were 0.30 on December 28 and only 0.06 on 8 January, indicating a significant reduction in the intensity of EPBs during the SSW period. For a more detailed mathematical description of the GOLD Bubble Index calculation, readers may refer to [88], which has demonstrated the advantage of this Bubble Index in quantifying the multiday variability of the 2D EPBs' magnitude in GOLD observations. Given the fact that solar activity was at quite low levels (F10.7 = 84 and 73) and geomagnetic conditions were extremely quiet ($Kp \leq 2$) on these two days, external forcing from solar and geomagnetic activities were thus unlikely the causes of such a significant reduction in the intensity of EPBs during the SSW period. Instead, forcing from the lower atmosphere could play an important role.

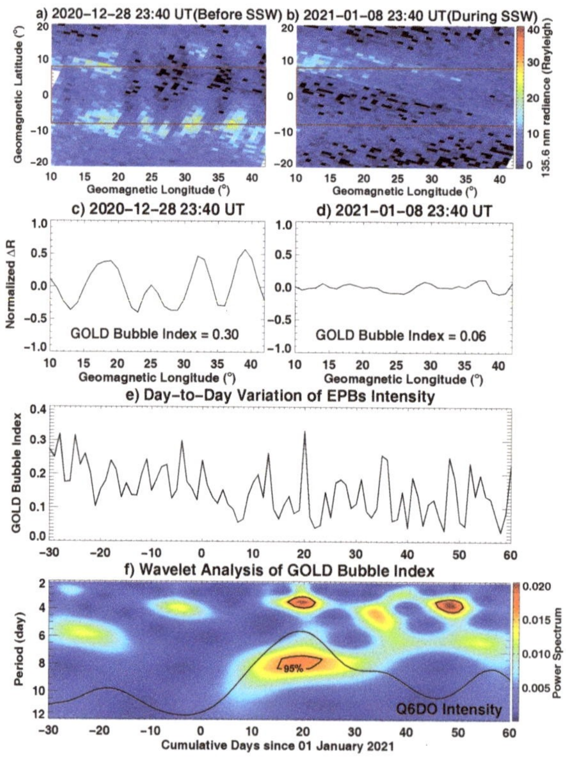

Figure 2. (**a**,**b**) Disk images of GOLD nighttime measurements of OI 135.6 nm emission at 23:40 UT on 28 December 2020 (before SSW) and 8 January 2021 (during SSW) in geomagnetic coordinates. The red lines denote the chosen equatorial region between ±8° MLAT. (**c**,**d**) The corresponding longitudinal fluctuation of the normalized differential emission radiance (ΔR) after removing the MLAT-integrated background reference. (**e**,**f**) Temporal variation in the GOLD Bubble Index and associated wavelet power spectrum since 1 January 2021. Circles mark the 95% significance level. The black line in panel f represents the temporal variation in the averaged power spectrum with a periodicity of 5.5–8 days (i.e., the Q6DO intensity).

Moreover, Figure 2e depicts the temporal fluctuation of the GOLD Bubble Index with respect to the cumulative days since 1 January 2021. Following the onset of the SSW, the Bubble Index exhibits a substantial reduction during the period of 5–10 January, significantly lower than that in December 2020. More specifically, the average Bubble Index during the 35 days before the SSW period (i.e., 1 December 2020–4 January 2021) is 0.196, while the average Bubble Index during the 35 days within the SSW period (i.e., 5 January 2021–10 February 2021) is 0.125. This indicates that the intensity of EPBs was notably reduced by approximately 35% during the SSW period. Given that December–January is typically the season with a high occurrence of EPBs over the American–Atlantic sector, it is unlikely that such a significant decrease was solely due to seasonal variations. Moreover, it is known that the occurrence of EPBs can be impacted by solar activity, with a typically higher occurrence rate during solar maximum than during solar minimum [69]. However, solar activity remained at low levels between December 2020 and February 2021, with the average F10.7 index during the pre-SSW and SSW period being 83 and 73 solar flux units, respectively. The significant suppression of observed EPBs is unlikely to be merely caused by small changes in solar flux during this period.

Besides the effects due to seasonal variation and changes in solar flux, such a substantial inhibition of EPBs could be related to the SSW through the following three mechanisms: (1) Suppression of gravity waves seeding. Gravity waves propagating upward could induce wavelike modulations in the electron density and/or polarization electric field at the bottom-side F-region, serving as an important seeding source to initiate the development of EPBs. Several studies have reported that the amplitudes of gravity waves may experience slight enhancements before the reversal of zonal winds at 10 hPa and 60°N but tend to undergo considerable reductions after the onset of SSW, e.g., [103–105]. Consequently, the reduction in seeding perturbations caused by gravity waves would effectively suppress the development of EPBs. (2) Changes in the dusk equatorial vertical plasma drifts and in the growth rate of the R-T instability. As previously mentioned in the Introduction, the dusktime equatorial vertical plasma drifts are one of the most critical factors influencing the growth rate of the R-T instability and thus governing the generation of EPBs. Several studies have reported that a phase shift of the migrating semidiurnal tide would cause a reduction in equatorial $E \times B$ drifts in the late afternoon during SSW events, e.g., [14,17,79,106], which is considered to be related to the suppression of dusktime PRE. In this study, we further verify this mechanism by examining the variation in equatorial plasma drifts and changes in the growth rate of the R-T instability, as is discussed in the following paragraph. (3) The modification of the flux-tube-integrated conductivity due to variations in meridional neutral winds. It is known that the poleward (equatorward) neutral winds can decrease (increase) the height of the ionospheric layer so as to enhance (reduce) the field-line-integrated Pedersen conductivity, thereby leading to an inhibition (enhancement) of the growth rate of the R-T instability [107]. In particular, Zhang et al. [16] reported poleward disturbances in the field-aligned plasma drift were primarily attributed to the poleward disturbances of meridional winds during the 2021 SSW event. Thus, the poleward neutral wind disturbance tends to reduce the growth rate of the R-T instability, partially contributing to the suppression of EPBs after the onset of SSW.

For an in-depth investigation of the day-to-day variability of EPBs during the SSW period, Figure 2f depicts the wavelet power spectrum of the GOLD Bubble Index with respect to the cumulative days since 1 January 2021. The power spectrum revealed some prominent spectral peaks with a periodicity of 6–8 days and ~3 days, as circled by the 95% significance lines, as well as some moderate peaks with a quasi-6-day period. Notably, these oscillation components were most noticeable after the onset of SSW from early January through February 2021. In particular, the quasi-6-day spectrum peak observed around 15–20 January corresponds to the period when the zonal winds undergo their most significant reversal at 10 hPa and 60°N, as shown in Figure 1a. Moreover, the black line in Figure 2f represents the temporal variation in the averaged power spectrum with a periodicity of 5.5–8 days (i.e., the Q6DO intensity). As observed, the temporal variation in the Q6DO intensity in EPBs showed a generally consistent trend with that of the Q6DW in the WACCM-X simulation (Figure 1e), both demonstrating a small peak in mid-December and a pronounced enhancement in January with a correlation coefficient of 0.67. This suggests that the amplified planetary wave during this period potentially played a crucial role in modulating the day-to-day variability of EPBs as observed by GOLD.

To explore the potential factors contributing to the significant variability from one day to another day in EPBs during the SSW period, as well as to further validate the connection between planetary waves and ionospheric irregularities, we analyzed the variation in the growth rate of the R-T instability. In order to maximize the utilization of real measurements, we calculated the equatorial localized growth rate of R-T instability at the bottom-side F-region using the equatorial Sao Luis ionosonde measurements on the basis of the following Equation in Sultan [65]:

$$\gamma = (\frac{E}{B} - \frac{g}{v_{\text{in}}})\frac{1}{n_0}\frac{\partial n_0}{\partial z}, \tag{1}$$

where γ is the growth rate of R-T instability; E and B are the magnitude of the zonal electric field, as well as the geomagnetic field, respectively; g represents the gravitational term, and v_{in} denotes the collision frequency between ion and neutral; n_0 is the ionospheric electron density in the F-region; and z is the altitudinal term. It has been demonstrated that this equation is an effective approximation for estimating the localized growth rate of the Rayleigh–Taylor (R-T) instability near the geomagnetic equator, e.g., [63,108–110], although its value may be somewhat larger compared with those obtained utilizing flux-tube-integrated values. We note that the term of meridional neutral wind is omitted in the simplified version, partly due to magnetic meridional wind not significantly altering the equatorial conductivity because of the small magnetic inclination angles in that region. The term recombination damping, and other flux-tube-integrated terms, are omitted in this equation due to their either ineffective nature or being less likely to undergo large day-to-day variations [111].

We next employ a similar approach as introduced by [108,109] to calculate the aforementioned parameters by utilizing realistic measurements: the speed factor term (E/B) is substituted with vertical plasma drift data obtained from the Sao Luis digisonde drift mode observations. The inversed vertical gradient scale length of Ne, $1/n_0(\partial n_0/\partial z)$, is determined based on the Sao Luis digisonde bottom-side electron density profiles. The neutral-ion collision frequency v_{in} is calculated according to the methodology outlined in Kelley [63] as follows:

$$v_{\text{in}} = 2.6 \times 10^{-9}(n_n + n_i)A^{-\frac{1}{2}}, \tag{2}$$

where n_n and n_i represent neutral and ion density, respectively, while the variable A is the averaged neutral molecular mass in atomic mass units. These items are computed utilizing the NRLMSISE-00 model [112]. For a more detailed description of this equation, readers could refer to Kelley [63].

Figure 3a,b shows an example of electron density profiles, F2-layer peak height (hmF2), and vertical plasma drift observed by the Sao Luis ionosonde during 4–10 January 2021. Figure 3c illustrates the associated localized growth rate of the R-T instability at the bottom-side F-region that is derived using Sao Luis ionosonde measurements based on Equation (1). The daily peak values of the growth rate were observed after local sunset, attributed to the aforementioned equatorial PRE effect and steep density gradient at the bottom-side F-region in the postsunset hours. Evidently, the postsunset peak values of both the vertical plasma drift and the R-T instability growth rate exhibit a considerable decrease following the onset of the SSW on 5 January. For example, the postsunset peak value of the R-T instability growth rate was about 36 $\times$ 10^{-4}/s on 4 January, but largely reduced to around 15–20 $\times$ 10^{-4}/s on 7–8 January, and further decreased to 5–10 $\times$ 10^{-4}/s on 9–10 January. Such a decreasing trend is consistent with the suppression of EPBs, as shown in Figure 2e. This demonstrates that changes in the amplitudes of PRE during the SSW period play an essential role in modulating the growth rate of the R-T instability, thereby contributing dominantly to the observed inhibition of the intensity of EPBs.

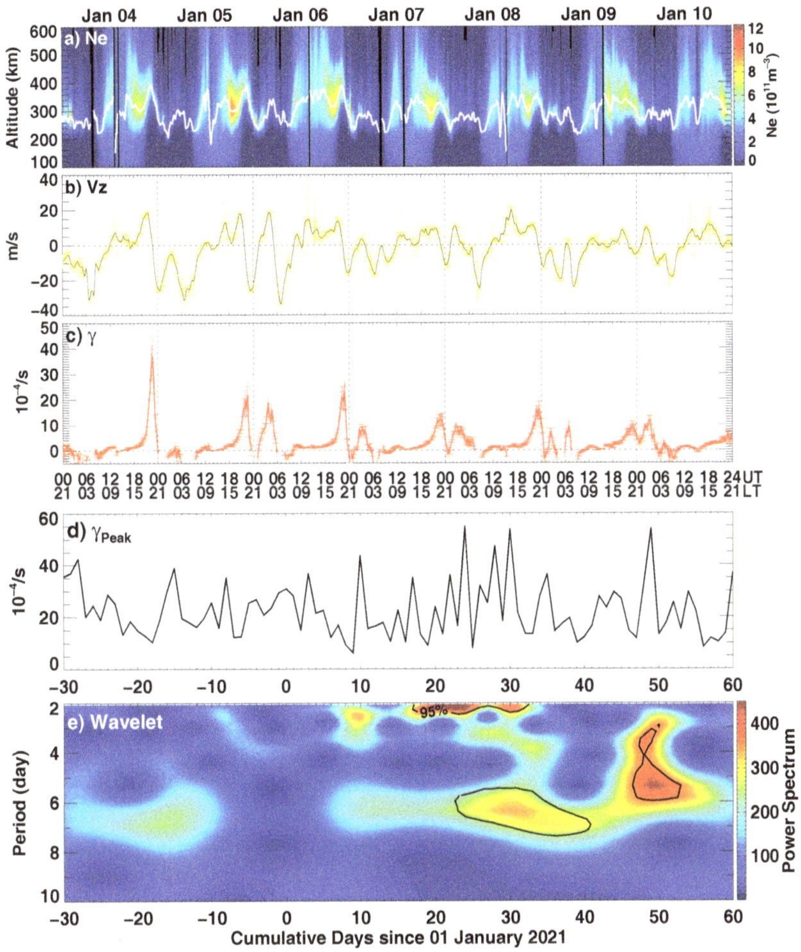

Figure 3. Ionosonde observations: of (**a**) Ne profiles and hmF2 (F2-region peak height). (**b**) F-region vertical plasma drift at Sao Luis during 4–10 January 2021. (**c**) Temporal fluctuation of the derived linear growth rate of the R-T instability. (**d**,**e**) Day-to-day variation in postsunset maximum values of R-T instability growth rate and corresponding wavelet analysis with respect to cumulative days since 1 January 2021.

Moreover, Figure 3d,e shows the day-by-day variation in postsunset maximum values of R-T instability growth rate and the corresponding wavelet analysis as a function of the accumulated days since 1 January 2021. As is evident, the wavelet power spectrum exhibits dominant peaks with a periodicity of ∼2–3 days and ∼6 days, particularly noticeable in January through mid-February following the onset of SSW. Additionally, a modest peak with a periodicity of ∼6 days is observed in mid-December. In general, the presence of the Q6DO in the wavelet power spectrum is coincident with the enhancement of Q6DW in the WACCM-X simulation (Figure 1e) and the observed ∼6-day periodicity in the GOLD Bubble Index (Figure 2f); all results exhibit similar predominant spectral peaks in January. This suggests that the amplified Q6DW in the MLT region during the SSW serves as the source of strong ∼6-day oscillations in the R-T instability growth rate and in the day-to-day variation in EPBs. We note that the dominant peak in the GOLD bubble index has a periodicity of 6–8 days, slightly larger than that observed in the R-T instability growth rate. This is because the R-T instability is not equivalent to EPBs, although it is the most pivotal factor determining the development of EPBs. The variability of EPBs is also influenced by potential seeding factors, such as gravity waves and TIDs, as well as forcing from above, such as geomagnetic activities. These factors will further complicate the multiday variability of EPBs, resulting in a slight offset from the dominant 6-day peaks driven by planetary waves.

To further examine the coupling mechanism between the ionosphere and atmosphere, Figure 4 displays the periodogram plot derived from ICON-MIGHTI zonal winds measurements at 100 km altitude from 15 to 30 January 2021. Evident spectrum peaks, especially a quasi-6-day periodicity, could be discerned, suggesting the existence of such periodicity in the ionospheric altitudes, which were coincident with the noticeable Q6DW in the MLT region, as simulated by WACCM-X and consistent with the observed Q6DO signatures in the growth rate of R-T instability and in the intensity of EPBs. These coordinated periodic variations indicate that the observed enhancement of Q6DO in the EPBs during the SSW can be attributed to neutral wind modulations by planetary waves of Q6DW via the ionospheric wind-driven dynamo mechanism.

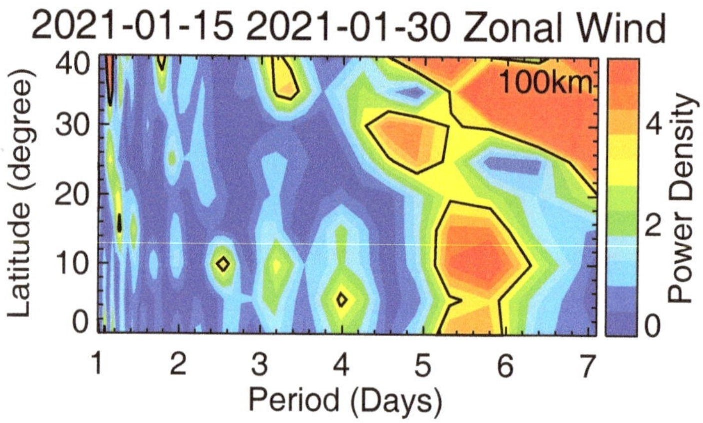

Figure 4. Periodograms of the ICON/MIGHTI zonal winds measurements at 100 km during 15–30 January 2021.

Figure 5a shows the normalized daily values of the flux-tube-integrated growth rate of R-T instability and the PRE averaged between 20–120°W longitudes, derived from the WACCM-X simulations (LA only). Figure 5b depicts the cross-correlation analysis of the PRE and R-T instability. As observed, the two parameters exhibit significant common powers manifested as encircled spectral peaks in mid-December 2020 and January 2021, generally coinciding with that of the enhanced Q6DW in the MLT region shown in Figure 1e, although the connection appears to be nonlinear. The arrows within the common power region mainly point to the right, indicating an in-phase variation between PRE and R-T instability, as expected. The spectral peak region in January exhibits a dominant periodicity of 5–7 days, consistent with that of the ICON wind measurements as shown in Figure 4 and the localized R-T instability growth rate, as shown in Figure 3e. Pedatella et al. [86] has shown that the amplitude of semidiurnal migrating tides exhibits a distinct spectral peak, with a periodicity of approximately 6 days after the onset of the SSW, consistent with the timing of enhanced Q6DW in the MLT region. Therefore, we deduce that the Q6DW in the MLT region acted to induce periodic oscillations in the zonal winds through interaction with semidiurnal migrating tides, which subsequently modulated the strength of PRE via the E-region wind-driven dynamo at a similar periodicity, leading to the observed Q6DO in the growth rate of the R-T instability and ultimately modulating the intensity of EPBs.

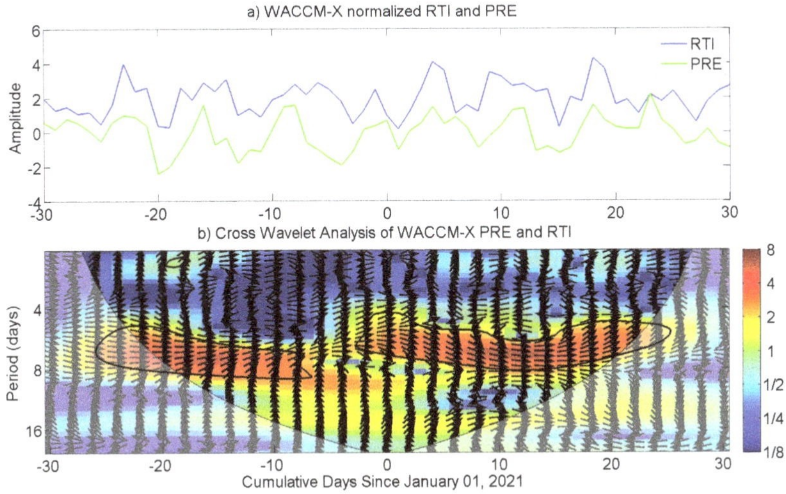

Figure 5. (**a**) WACCM-X simulations (LA only) of the normalized growth rate of the R-T instability (blue) and normalized PRE (green) with respect to the cumulative days from 1 January 2021. (**b**) Cross-wavelet analysis of the WACCM-X normalized R-T instability and PRE time series. The relative phase is represented by black arrows, pointing right for in-phase and left for antiphase.

4. Conclusions

This study uses multi-instrument observations and the WACCM-X simulation to investigate the day-to-day variability of EPBs and their relationship to atmospheric planetary waves during the 2021 SSW event. The main findings are as follows:

1. We found that the intensity of EPBs was notably reduced by 35% during the SSW period compared with the non-SSW period. Such a significant inhibition of EPBs during the SSW period could be collectively attributed to the suppression of gravity waves seeding, changes in the dusk equatorial vertical plasma drifts and in the growth rate of the R-T instability, and the modification of Pedersen conductivity due to the variations in neutral winds. In addition to SSW, there could also be some effects due to seasonal variations and changes in solar flux.
2. We found that significant Q6DO signatures were observed in both the intensity of EPBs and the associated growth rate of R-T instability during the SSW, which were coincident with the amplification of the Q6DW in the WACCM-X simulation and noticeable ∼6-day periodicity in ICON-MIGHTI zonal winds. These results demonstrate that certain planetary waves like the Q6DW can play a crucial role in controlling the day-to-day variability of EPBs, especially during the SSW event. This influence is exerted through the modulation of zonal winds and the ionospheric E-region dynamo via the interaction between planetary waves and tides, which led to periodic oscillations in the PRE strength and the growth rate of the R-T instability, ultimately resulting in the Q6DO in the intensity of EPBs. These findings provide new insights into the day-to-day variability of ionospheric irregularities and their potential correlation with atmospheric planetary waves.

Author Contributions: Conceptualization, E.A. and N.M.P.; methodology, E.A. and N.M.P.; software, N.M.P.; validation, E.A., N.M.P. and G.L.; formal analysis, E.A. and N.M.P.; investigation, E.A. and N.M.P.; resources, N.M.P.; writing—original draft preparation, E.A.; writing—review and editing,

N.M.P. and G.L.; visualization, E.A.; supervision, N.M.P.; project administration, E.A.; funding acquisition, E.A. and N.M.P. All authors have read and agreed to the published version of the manuscript.

Funding: This work is supported by NSF awards AGS-1952737, AGS-2033787, PHY-2028125, and NSFC41974184, NASA support 80NSSC22K0171, 80NSSC21K1310, 80GSFC22CA011, NNH220B82A, and NNH17ZDA001N07.

Data Availability Statement: GOLD data can be accessed at (https://gold.cs.ucf.edu/data/search/ (accessed on 1 January 2024)). The ionosonde data are provided by the University of Massachusetts Lowell DIDB database of Global Ionospheric Radio Observatory (https://giro.uml.edu/didbase/scaled.php (accessed on 1 January 2024)). The NCEP stratosphere reanalysis data can be accessed at NASA Atmospheric Chemistry and Dynamics Laboratory (https://acd-ext.gsfc.nasa.gov/Data_services/met/ann_data.html (accessed on 1 January 2024)).

Acknowledgments: WACCM-X is part of the Community Earth System Model (CESM) and the source code is available at (http://www.cesm.ucar.edu (accessed on 1 January 2024)).

Conflicts of Interest: The authors declare no conflicts of interest.

References

1. Forbes, J.M.; Palo, S.E.; Zhang, X. Variability of the ionosphere. *J. Atmos. Sol. Terr. Phys.* **2000**, *62*, 685–693. [CrossRef]
2. Liu, H.L. Variability and predictability of the space environment as related to lower atmosphere forcing. *Space Weather* **2016**, *14*, 634–658. [CrossRef]
3. Matsuno, T. A Dynamical Model of the Stratospheric Sudden Warming. *J. Atmos. Sci.* **1971**, *28*, 1479–1494. [CrossRef]
4. Baldwin, M.P.; Ayarzagüena, B.; Birner, T.; Butchart, N.; Butler, A.H.; Charlton-Perez, A.J.; Domeisen, D.I.V.; Garfinkel, C.I.; Garny, H.; Gerber, E.P.; et al. Sudden Stratospheric Warmings. *Rev. Geophys.* **2021**, *59*, e2020RG000708. [CrossRef]
5. Goncharenko, L.P.; Harvey, V.L.; Liu, H.; Pedatella, N.M. Sudden Stratospheric Warming Impacts on the Ionosphere-Thermosphere System: A Review of Recent Progress. In *Proceedings of the Ionosphere Dynamics and Applications*; Huang, C., Lu, G., Eds.; AGU Publications: Washington, DC, USA, 2021; Volume 3, p. 369. [CrossRef]
6. Pedatella, N.M.; Chau, J.; Schmidt, H.; Goncharenko, L.; Stolle, C.; Hocke, K.; Harvey, V.; Funke, B.; Siddiqui, T. How sudden stratospheric warming affects the whole atmosphere. *EOS* **2018**, *99*, 35–38. [CrossRef]
7. Forbes, J.M.; Zhang, X. Lunar tide amplification during the January 2009 stratosphere warming event: Observations and theory. *J. Geophys. Res. Space Phys.* **2012**, *117*, A12312. [CrossRef]
8. Zhang, X.; Forbes, J.M. Lunar tide in the thermosphere and weakening of the northern polar vortex. *Geophys. Res. Lett.* **2014**, *41*, 8201–8207. [CrossRef]
9. Chau, J.L.; Goncharenko, L.P.; Fejer, B.G.; Liu, H.L. Equatorial and Low Latitude Ionospheric Effects During Sudden Stratospheric Warming Events. Ionospheric Effects During SSW Events. *Space Sci. Rev.* **2012**, *168*, 385–417. [CrossRef]
10. Chau, J.L.; Fejer, B.G.; Goncharenko, L.P. Quiet variability of equatorial E × B drifts during a sudden stratospheric warming event. *Geophys. Res. Lett.* **2009**, *36*, L05101. [CrossRef]
11. Fang, T.W.; Fuller-Rowell, T.; Wang, H.; Akmaev, R.; Wu, F. Ionospheric response to sudden stratospheric warming events at low and high solar activity. *J. Geophys. Res. Space Phys.* **2014**, *119*, 7858–7869. [CrossRef]
12. Fejer, B.G.; Tracy, B.D.; Olson, M.E.; Chau, J.L. Enhanced lunar semidiurnal equatorial vertical plasma drifts during sudden stratospheric warmings. *Geophys. Res. Lett.* **2011**, *38*, L21104. [CrossRef]
13. Goncharenko, L.P.; Chau, J.L.; Liu, H.L.; Coster, A.J. Unexpected connections between the stratosphere and ionosphere. *Geophys. Res. Lett.* **2010**, *37*, L10101. [CrossRef]
14. Maute, A.; Hagan, M.E.; Richmond, A.D.; Roble, R.G. TIME-GCM study of the ionospheric equatorial vertical drift changes during the 2006 stratospheric sudden warming. *J. Geophys. Res. Space Phys.* **2014**, *119*, 1287–1305. [CrossRef]
15. Maute, A.; Fejer, B.G.; Forbes, J.M.; Zhang, X.; Yudin, V. Equatorial vertical drift modulation by the lunar and solar semidiurnal tides during the 2013 sudden stratospheric warming. *J. Geophys. Res. Space Phys.* **2016**, *121*, 1658–1668. [CrossRef]
16. Zhang, R.; Liu, L.; Ma, H.; Chen, Y.; Le, H. ICON Observations of Equatorial Ionospheric Vertical ExB and Field-Aligned Plasma Drifts During the 2020-2021 SSW. *Geophys. Res. Lett.* **2022**, *49*, e99238. [CrossRef]
17. Fejer, B.G.; Olson, M.E.; Chau, J.L.; Stolle, C.; Lühr, H.; Goncharenko, L.P.; Yumoto, K.; Nagatsuma, T. Lunar-dependent equatorial ionospheric electrodynamic effects during sudden stratospheric warmings. *J. Geophys. Res. Space Phys.* **2010**, *115*, A00G03. [CrossRef]
18. Park, J.; Lühr, H.; Kunze, M.; Fejer, B.G.; Min, K.W. Effect of sudden stratospheric warming on lunar tidal modulation of the equatorial electrojet. *J. Geophys. Res. Space Phys.* **2012**, *117*, A03306. [CrossRef]
19. Patra, A.K.; Pavan Chaitanya, P.; Sripathi, S.; Alex, S. Ionospheric variability over Indian low latitude linked with the 2009 sudden stratospheric warming. *J. Geophys. Res. Space Phys.* **2014**, *119*, 4044–4061. [CrossRef]
20. Siddiqui, T.A.; Maute, A.; Pedatella, N.; Yamazaki, Y.; Lühr, H.; Stolle, C. On the variability of the semidiurnal solar and lunar tides of the equatorial electrojet during sudden stratospheric warmings. *Ann. Geophys.* **2018**, *36*, 1545–1562. [CrossRef]

21. Yamazaki, Y.; Yumoto, K.; McNamara, D.; Hirooka, T.; Uozumi, T.; Kitamura, K.; Abe, S.; Ikeda, A. Ionospheric current system during sudden stratospheric warming events. *J. Geophys. Res. Space Phys.* **2012**, *117*, A03334. [CrossRef]
22. Fagundes, P.R.; Goncharenko, L.P.; Abreu, A.J.; Venkatesh, K.; Pezzopane, M.; Jesus, R.; Gende, M.; Coster, A.J.; Pillat, V.G. Ionospheric response to the 2009 sudden stratospheric warming over the equatorial, low, and middle latitudes in the South American sector. *J. Geophys. Res. Space Phys.* **2015**, *120*, 7889–7902. [CrossRef]
23. Jin, H.; Miyoshi, Y.; Pancheva, D.; Mukhtarov, P.; Fujiwara, H.; Shinagawa, H. Response of migrating tides to the stratospheric sudden warming in 2009 and their effects on the ionosphere studied by a whole atmosphere-ionosphere model GAIA with COSMIC and TIMED/SABER observations. *J. Geophys. Res. Space Phys.* **2012**, *117*, A10323. [CrossRef]
24. Lin, C.H.; Lin, J.T.; Chang, L.C.; Liu, J.Y.; Chen, C.H.; Chen, W.H.; Huang, H.H.; Liu, C.H. Observations of global ionospheric responses to the 2009 stratospheric sudden warming event by FORMOSAT-3/COSMIC. *J. Geophys. Res. Space Phys.* **2012**, *117*, A06323. [CrossRef]
25. Liu, H.; Yamamoto, M.; Tulasi Ram, S.; Tsugawa, T.; Otsuka, Y.; Stolle, C.; Doornbos, E.; Yumoto, K.; Nagatsuma, T. Equatorial electrodynamics and neutral background in the Asian sector during the 2009 stratospheric sudden warming. *J. Geophys. Res. Space Phys.* **2011**, *116*, A08308. [CrossRef]
26. Oberheide, J. Day-to-Day Variability of the Semidiurnal Tide in the F-Region Ionosphere During the January 2021 SSW From COSMIC-2 and ICON. *Geophys. Res. Lett.* **2022**, *49*, e00369. [CrossRef]
27. Pancheva, D.; Mukhtarov, P. Stratospheric warmings: The atmosphere-ionosphere coupling paradigm. *J. Atmos.-Sol.-Terr. Phys.* **2011**, *73*, 1697–1702. [CrossRef]
28. Pedatella, N.M.; Forbes, J.M. Evidence for stratosphere sudden warming-ionosphere coupling due to vertically propagating tides. *Geophys. Res. Lett.* **2010**, *37*, L11104. [CrossRef]
29. Gan, Q.; Eastes, R.W.; Burns, A.G.; Wang, W.; Qian, L.; Solomon, S.C.; Codrescu, M.V.; McClintock, W.E. New Observations of Large-Scale Waves Coupling With the Ionosphere Made by the GOLD Mission: Quasi-16-Day Wave Signatures in the F-Region OI 135.6-nm Nightglow During Sudden Stratospheric Warmings. *J. Geophys. Res. Space Phys.* **2020**, *125*, e27880. [CrossRef]
30. Gu, S.Y.; Li, T.; Dou, X.; Wu, Q.; Mlynczak, M.G.; Russell, J.M. Observations of Quasi-Two-Day wave by TIMED/SABER and TIMED/TIDI. *J. Geophys. Res. Atmos.* **2013**, *118*, 1624–1639. [CrossRef]
31. Liu, G.; England, S.L.; Janches, D. Quasi Two-, Three-, and Six-Day Planetary-Scale Wave Oscillations in the Upper Atmosphere Observed by TIMED/SABER Over 17 Years During 2002-2018. *J. Geophys. Res. Space Phys.* **2019**, *124*, 9462–9474. [CrossRef]
32. Mo, X.; Zhang, D. Quasi-10 d wave modulation of an equatorial ionization anomaly during the Southern Hemisphere stratospheric warming of 2002. *Ann. Geophys.* **2020**, *38*, 9–16. [CrossRef]
33. Vineeth, C.; Pant, T.K.; Sridharan, R. Equatorial counter electrojets and polar stratospheric sudden warmings—A classical example of high latitude-low latitude coupling? *Ann. Geophys.* **2009**, *27*, 3147–3153. [CrossRef]
34. Yamazaki, Y.; Matthias, V.; Miyoshi, Y.; Stolle, C.; Siddiqui, T.; Kervalishvili, G.; Laštovička, J.; Kozubek, M.; Ward, W.; Themens, D.R.; et al. September 2019 Antarctic Sudden Stratospheric Warming: Quasi-6-Day Wave Burst and Ionospheric Effects. *Geophys. Res. Lett.* **2020**, *47*, e86577. [CrossRef]
35. Yue, J.; Wang, W.; Ruan, H.; Chang, L.C.; Lei, J. Impact of the interaction between the quasi-2 day wave and tides on the ionosphere and thermosphere. *J. Geophys. Res. Space Phys.* **2016**, *121*, 3555–3563. [CrossRef]
36. Forbes, J.M.; Maute, A.; Zhang, X.; Hagan, M.E. Oscillation of the Ionosphere at Planetary-Wave Periods. *J. Geophys. Res. Space Phys.* **2018**, *123*, 7634–7649. [CrossRef]
37. Laštovička, J. Forcing of the ionosphere by waves from below. *J. Atmos.-Sol.-Terr. Phys.* **2006**, *68*, 479–497. [CrossRef]
38. Pancheva, D.; Merzlyakov, E.; Mitchell, N.J.; Portnyagin, Y.; Manson, A.H.; Jacobi, C.; Meek, C.E.; Luo, Y.; Clark, R.R.; Hocking, W.K.; et al. Global-scale tidal variability during the PSMOS campaign of June–August 1999: Interaction with planetary waves. *J. Atmos.-Sol.-Terr. Phys.* **2002**, *64*, 1865–1896. [CrossRef]
39. Lieberman, R.S.; Riggin, D.M.; Franke, S.J.; Manson, A.H.; Meek, C.; Nakamura, T.; Tsuda, T.; Vincent, R.A.; Reid, I. The 6.5-day wave in the mesosphere and lower thermosphere: Evidence for baroclinic/barotropic instability. *J. Geophys. Res. Atmos.* **2003**, *108*, 4640. [CrossRef]
40. Liu, H.L.; Talaat, E.R.; Roble, R.G.; Lieberman, R.S.; Riggin, D.M.; Yee, J.H. The 6.5-day wave and its seasonal variability in the middle and upper atmosphere. *J. Geophys. Res. Atmos.* **2004**, *109*, D21112. [CrossRef]
41. Talaat, E.R.; Yee, J.H.; Zhu, X. The 6.5-day wave in the tropical stratosphere and mesosphere. *J. Geophys. Res. Atmos.* **2002**, *107*, 4133. [CrossRef]
42. Wu, D.L.; Hays, P.B.; Skinner, W.R. Observations of the 5-day wave in the mesosphere and lower thermosphere. *Geophys. Res. Lett.* **1994**, *21*, 2733–2736. [CrossRef]
43. Gan, Q.; Oberheide, J.; Pedatella, N.M. Sources, Sinks, and Propagation Characteristics of the Quasi 6-Day Wave and Its Impact on the Residual Mean Circulation. *J. Geophys. Res. Atmos.* **2018**, *123*, 9152–9170. [CrossRef]
44. Gu, S.Y.; Ruan, H.; Yang, C.Y.; Gan, Q.; Dou, X.; Wang, N. The Morphology of the 6-Day Wave in Both the Neutral Atmosphere and F Region Ionosphere Under Solar Minimum Conditions. *J. Geophys. Res. Space Phys.* **2018**, *123*, 4232–4240. [CrossRef]
45. Lin, J.T.; Lin, C.H.; Rajesh, P.K.; Yue, J.; Lin, C.Y.; Matsuo, T. Local-Time and Vertical Characteristics of Quasi-6-Day Oscillation in the Ionosphere During the 2019 Antarctic Sudden Stratospheric Warming. *Geophys. Res. Lett.* **2020**, *47*, e90345. [CrossRef]
46. Qin, Y.; Gu, S.Y.; Dou, X.; Gong, Y.; Chen, G.; Zhang, S.; Wu, Q. Climatology of the Quasi-6-Day Wave in the Mesopause Region and Its Modulations on Total Electron Content During 2003-2017. *J. Geophys. Res. Space Phys.* **2019**, *124*, 573–583. [CrossRef]

47. Yamazaki, Y. Quasi-6-Day Wave Effects on the Equatorial Ionization Anomaly Over a Solar Cycle. *J. Geophys. Res. Space Phys.* **2018**, *123*, 9881–9892. [CrossRef]
48. Chang, L.C.; Yue, J.; Wang, W.; Wu, Q.; Meier, R.R. Quasi two day wave-related variability in the background dynamics and composition of the mesosphere/thermosphere and the ionosphere. *J. Geophys. Res. Space Phys.* **2014**, *119*, 4786–4804. [CrossRef] [PubMed]
49. Kil, H. The Morphology of Equatorial Plasma Bubbles—A review. *J. Astron. Space Sci.* **2015**, *32*, 13–19. [CrossRef]
50. Woodman, R.F.; La Hoz, C. Radar observations of F region equatorial irregularities. *J. Geophys. Res.* **1976**, *81*, 5447–5466. [CrossRef]
51. Hysell, D.L. An overview and synthesis of plasma irregularities in equatorial spread/F. *J. Atmos. Sol. Terr. Phys.* **2000**, *62*, 1037–1056. [CrossRef]
52. Aa, E.; Huang, W.; Liu, S.; Ridley, A.; Zou, S.; Shi, L.; Chen, Y.; Shen, H.; Yuan, T.; Li, J.; et al. Midlatitude Plasma Bubbles Over China and Adjacent Areas During a Magnetic Storm on 8 September 2017. *Space Weather* **2018**, *16*, 321–331. [CrossRef]
53. Aa, E.; Zou, S.; Ridley, A.; Zhang, S.; Coster, A.J.; Erickson, P.J.; Liu, S.; Ren, J. Merging of Storm Time Midlatitude Traveling Ionospheric Disturbances and Equatorial Plasma Bubbles. *Space Weather* **2019**, *17*, 285–298. [CrossRef]
54. Abdu, M.A.; Batista, P.P.; Batista, I.S.; Brum, C.G.M.; Carrasco, A.J.; Reinisch, B.W. Planetary wave oscillations in mesospheric winds, equatorial evening prereversal electric field and spread F. *Geophys. Res. Lett.* **2006**, *33*, L07107. [CrossRef]
55. Huang, C.S.; Hairston, M.R. The postsunset vertical plasma drift and its effects on the generation of equatorial plasma bubbles observed by the C/NOFS satellite. *J. Geophys. Res. Space Phys.* **2015**, *120*, 2263–2275. [CrossRef]
56. Huba, J.D.; Joyce, G. Global modeling of equatorial plasma bubbles. *Geophys. Res. Lett.* **2010**, *37*, L17104. [CrossRef]
57. Karan, D.K.; Eastes, R.W.; Daniell, R.E.; Martinis, C.R.; McClintock, W.E. GOLD Mission's Observation About the Geomagnetic Storm Effects on the Nighttime Equatorial Ionization Anomaly (EIA) and Equatorial Plasma Bubbles (EPB) During a Solar Minimum Equinox. *Space Weather* **2023**, *21*, e2022SW003321. [CrossRef]
58. Klenzing, J.; Halford, A.J.; Liu, G.; Smith, J.M.; Zhang, Y.; Zawdie, K.; Maruyama, N.; Pfaff, R.; Bishop, R.L. A system science perspective of the drivers of equatorial plasma bubbles. *Front. Astron. Space Sci.* **2023**, *9*, 420. [CrossRef]
59. Li, G.; Ning, B.; Liu, L.; Wan, W.; Liu, J.Y. Effect of magnetic activity on plasma bubbles over equatorial and low-latitude regions in East Asia. *Ann. Geophys.* **2009**, *27*, 303–312. [CrossRef]
60. Makela, J.J.; Vadas, S.L.; Muryanto, R.; Duly, T.; Crowley, G. Periodic spacing between consecutive equatorial plasma bubbles. *Geophys. Res. Lett.* **2010**, *37*, L14103. [CrossRef]
61. Otsuka, Y.; Shiokawa, K.; Ogawa, T.; Wilkinson, P. Geomagnetic conjugate observations of equatorial airglow depletions. *Geophys. Res. Lett.* **2002**, *29*, 1753. [CrossRef]
62. Yokoyama, T.; Shinagawa, H.; Jin, H. Nonlinear growth, bifurcation, and pinching of equatorial plasma bubble simulated by three-dimensional high-resolution bubble model. *J. Geophys. Res. Space Phys.* **2014**, *119*, 10474–10482. [CrossRef]
63. Kelley, M.C. Equatorial Plasma Instabilities. In *The Earth's Ionosphere: Plasma Physics and Electrodynamics*; Kelley, M.C., Ed.; Academic Press: Cambridge, MA, USA, 1989; pp. 113–185. [CrossRef]
64. Ott, E. Theory of Rayleigh-Taylor bubbles in the equatorial ionosphere. *J. Geophys. Res.* **1978**, *83*, 2066–2070. [CrossRef]
65. Sultan, P.J. Linear theory and modeling of the Rayleigh-Taylor instability leading to the occurrence of equatorial spread F. *J. Geophys. Res.* **1996**, *101*, 26875–26892. [CrossRef]
66. Abadi, P.; Otsuka, Y.; Tsugawa, T. Effects of pre-reversal enhancement of E × B drift on the latitudinal extension of plasma bubble in Southeast Asia. *Earth Planets Space* **2015**, *67*, 74. [CrossRef]
67. Fejer, B.G.; Scherliess, L.; de Paula, E.R. Effects of the vertical plasma drift velocity on the generation and evolution of equatorial spread F. *J. Geophys. Res.* **1999**, *104*, 19859–19870. [CrossRef]
68. Huang, C.S. Effects of the postsunset vertical plasma drift on the generation of equatorial spread F. *Prog. Earth Planet. Sci.* **2018**, *5*, 3. [CrossRef]
69. Aa, E.; Zou, S.; Liu, S. Statistical Analysis of Equatorial Plasma Irregularities Retrieved From Swarm 2013-2019 Observations. *J. Geophys. Res. Space Phys.* **2020**, *125*, e27022. [CrossRef]
70. Burke, W.J.; Gentile, L.C.; Huang, C.Y.; Valladares, C.E.; Su, S.Y. Longitudinal variability of equatorial plasma bubbles observed by DMSP and ROCSAT-1. *J. Geophys. Res. Space Phys.* **2004**, *109*, A12301. [CrossRef]
71. Gentile, L.C.; Burke, W.J.; Rich, F.J. A global climatology for equatorial plasma bubbles in the topside ionosphere. *Ann. Geophys.* **2006**, *24*, 163–172. [CrossRef]
72. Smith, J.; Heelis, R.A. Equatorial plasma bubbles: Variations of occurrence and spatial scale in local time, longitude, season, and solar activity. *J. Geophys. Res. Space Phys.* **2017**, *122*, 5743–5755. [CrossRef]
73. Su, S.Y.; Chao, C.K.; Liu, C.H. On monthly/seasonal/longitudinal variations of equatorial irregularity occurrences and their relationship with the postsunset vertical drift velocities. *J. Geophys. Res. Space Phys.* **2008**, *113*, A05307. [CrossRef]
74. Wan, X.; Xiong, C.; Rodriguez-Zuluaga, J.; Kervalishvili, G.N.; Stolle, C.; Wang, H. Climatology of the Occurrence Rate and Amplitudes of Local Time Distinguished Equatorial Plasma Depletions Observed by Swarm Satellite. *J. Geophys. Res. Space Phys.* **2018**, *123*, 3014–3026. [CrossRef]
75. Abdu, M.A. Day-to-day and short-term variabilities in the equatorial plasma bubble/spread F irregularity seeding and development. *Prog. Earth Planet. Sci.* **2019**, *6*, 11. [CrossRef]
76. Retterer, J.M.; Roddy, P. Faith in a seed: On the origins of equatorial plasma bubbles. *Ann. Geophys.* **2014**, *32*, 485–498. [CrossRef]

77. Tsunoda, R.T. On the enigma of day-to-day variability in equatorial spread F. *Geophys. Res. Lett.* **2005**, *32*, L08103. [CrossRef]
78. Tsunoda, R.T. Day-to-day variability in equatorial spread F: Is there some physics missing? *Geophys. Res. Lett.* **2006**, *33*, L16106. [CrossRef]
79. de Paula, E.R.; Jonah, O.F.; Moraes, A.O.; Kherani, E.A.; Fejer, B.G.; Abdu, M.A.; Muella, M.T.A.H.; Batista, I.S.; Dutra, S.L.G.; Paes, R.R. Low-latitude scintillation weakening during sudden stratospheric warming events. *J. Geophys. Res. Space Phys.* **2015**, *120*, 2212–2221. [CrossRef]
80. Yu, T.; Ye, H.; Liu, H.; Xia, C.; Zuo, X.; Yan, X.; Yang, N.; Sun, Y.; Zhao, B. Ionospheric F Layer Scintillation Weakening as Observed by COSMIC/FORMOSAT-3 During the Major Sudden Stratospheric Warming in January 2013. *J. Geophys. Res. Space Phys.* **2020**, *125*, e27721. [CrossRef]
81. Ye, H.; Xue, X.; Yu, T.; Sun, Y.Y.; Yi, W.; Long, C.; Zhang, W.; Dou, X. Ionospheric F-Layer Scintillation Variabilities Over the American Sector During Sudden Stratospheric Warming Events. *Space Weather* **2021**, *19*, e2020SW002703. [CrossRef]
82. Jose, L.; Vineeth, C.; Pant, T.K. Impact of Stratospheric Sudden Warming on the Occurrence of the Equatorial Spread-F. *J. Geophys. Res. Space Phys.* **2017**, *122*, 12544–12555. [CrossRef]
83. Abdu, M.A.; Brum, C.G.; Batista, P.P.; Gurubaran, S.; Pancheva, D.; Bageston, J.V.; Batista, I.S.; Takahashi, H. Fast and ultrafast Kelvin wave modulations of the equatorial evening F region vertical drift and spread F development. *Earth Planets Space* **2015**, *67*, 1. [CrossRef]
84. Ghosh, P.; Otsuka, Y.; Mani, S.; Shinagawa, H. Day-to-day variation of pre-reversal enhancement in the equatorial ionosphere based on GAIA model simulations. *Earth Planets Space* **2020**, *72*, 93. [CrossRef]
85. Liu, H.L. Day-to-Day Variability of Prereversal Enhancement in the Vertical Ion Drift in Response to Large-Scale Forcing From the Lower Atmosphere. *Space Weather* **2020**, *18*, e02334. [CrossRef]
86. Pedatella, N.M.; Aa, E.; Maute, A. Quasi 6-Day Planetary Wave Oscillations in Equatorial Plasma Irregularities. *J. Geophys. Res. Space Phys.* **2024**, *129*, e2023JA032312. [CrossRef]
87. Yamazaki, Y.; Diéval, C. Modeling of Planetary Wave Influences on the Pre reversal Enhancement of the Equatorial F Region Vertical Plasma Drift. *Space Weather* **2021**, *19*, e02685. [CrossRef]
88. Aa, E.; Zhang, S.R.; Liu, G.; Eastes, R.W.; Wang, W.; Karan, D.K.; Qian, L.; Coster, A.J.; Erickson, P.J.; Derghazarian, S. Statistical Analysis of Equatorial Plasma Bubbles Climatology and Multi-Day Periodicity Using GOLD Observations. *Geophys. Res. Lett.* **2023**, *50*, e2023GL103510. [CrossRef]
89. Eastes, R.W.; McClintock, W.E.; Burns, A.G.; Anderson, D.N.; Andersson, L.; Codrescu, M.; Correira, J.T.; Daniell, R.E.; England, S.L.; Evans, J.S.; et al. The Global-Scale Observations of the Limb and Disk (GOLD) Mission. *Space Sci. Rev.* **2017**, *212*, 383–408. [CrossRef]
90. Eastes, R.W.; Solomon, S.C.; Daniell, R.E.; Anderson, D.N.; Burns, A.G.; England, S.L.; Martinis, C.R.; McClintock, W.E. Global-Scale Observations of the Equatorial Ionization Anomaly. *Geophys. Res. Lett.* **2019**, *46*, 9318–9326. [CrossRef]
91. Aa, E.; Zou, S.; Eastes, R.; Karan, D.K.; Zhang, S.R.; Erickson, P.J.; Coster, A.J. Coordinated Ground-Based and Space-Based Observations of Equatorial Plasma Bubbles. *J. Geophys. Res. Space Phys.* **2020**, *125*, e27569. [CrossRef]
92. Aa, E.; Zhang, S.R.; Erickson, P.J.; Wang, W.; Qian, L.; Cai, X.; Coster, A.J.; Goncharenko, L.P. Significant Mid- and Low-Latitude Ionospheric Disturbances Characterized by Dynamic EIA, EPBs, and SED Variations During the 13–14 March 2022 Geomagnetic Storm. *J. Geophys. Res. Space Phys.* **2023**, *128*, e2023JA031375. [CrossRef]
93. Eastes, R.W.; McClintock, W.E.; Burns, A.G.; Anderson, D.N.; Andersson, L.; Aryal, S.; Budzien, S.A.; Cai, X.; Codrescu, M.V.; Correira, J.T.; et al. Initial Observations by the GOLD Mission. *J. Geophys. Res. Space Phys.* **2020**, *125*, e27823. [CrossRef]
94. Martinis, C.; Daniell, R.; Eastes, R.; Norrell, J.; Smith, J.; Klenzing, J.; Solomon, S.; Burns, A. Longitudinal Variation of Postsunset Plasma Depletions From the Global Scale Observations of the Limb and Disk (GOLD) Mission. *J. Geophys. Res. Space Phys.* **2021**, *126*, e28510. [CrossRef]
95. Immel, T.J.; England, S.L.; Mende, S.B.; Heelis, R.A.; Englert, C.R.; Edelstein, J.; Frey, H.U.; Korpela, E.J.; Taylor, E.R.; Craig, W.W.; et al. The Ionospheric Connection Explorer Mission: Mission Goals and Design. *Space Sci. Rev.* **2018**, *214*, 13. [CrossRef] [PubMed]
96. Harding, B.J.; Makela, J.J.; Englert, C.R.; Marr, K.D.; Harlander, J.M.; England, S.L.; Immel, T.J. The MIGHTI Wind Retrieval Algorithm: Description and Verification. *Space Sci. Rev.* **2017**, *212*, 585–600. [CrossRef] [PubMed]
97. Liu, G.; England, S.L.; Lin, C.S.; Pedatella, N.M.; Klenzing, J.H.; Englert, C.R.; Harding, B.J.; Immel, T.J.; Rowland, D.E. Evaluation of Atmospheric 3-Day Waves as a Source of Day-to-Day Variation of the Ionospheric Longitudinal Structure. *Geophys. Res. Lett.* **2021**, *48*, e94877. [CrossRef]
98. Liu, H.L.; Bardeen, C.G.; Foster, B.T.; Lauritzen, P.; Liu, J.; Lu, G.; Marsh, D.R.; Maute, A.; McInerney, J.M.; Pedatella, N.M.; et al. Development and Validation of the Whole Atmosphere Community Climate Model with Thermosphere and Ionosphere Extension (WACCM-X 2.0). *J. Adv. Model. Earth Syst.* **2018**, *10*, 381–402. [CrossRef]
99. Pedatella, N.M. Ionospheric Variability during the 2020-2021 SSW: COSMIC-2 Observations and WACCM-X Simulations. *Atmosphere* **2022**, *13*, 368. [CrossRef]
100. Gelaro, R.; McCarty, W.; Suárez, M.J.; Todling, R.; Molod, A.; Takacs, L.; Randles, C.A.; Darmenov, A.; Bosilovich, M.G.; Reichle, R.; et al. The Modern-Era Retrospective Analysis for Research and Applications, Version 2 (MERRA-2). *J. Clim.* **2017**, *30*, 5419–5454. [CrossRef] [PubMed]

101. Heelis, R.A.; Lowell, J.K.; Spiro, R.W. A model of the high-latitude ionospheric convection pattern. *J. Geophys. Res.* **1982**, *87*, 6339–6345. [CrossRef]
102. Liu, J.; Zhang, D.; Sun, S.; Hao, Y.; Xiao, Z. Ionospheric Semidiurnal Lunitidal Perturbations During the 2021 Sudden Stratospheric Warming Event: Latitudinal and Inter-Hemispheric Variations in the American, Asian-Australian, and African-European Sectors. *J. Geophys. Res. Space Phys.* **2022**, *127*, e2022JA030313. [CrossRef]
103. Nayak, C.; Yiğit, E. Variation of Small-Scale Gravity Wave Activity in the Ionosphere During the Major Sudden Stratospheric Warming Event of 2009. *J. Geophys. Res. Space Phys.* **2019**, *124*, 470–488. [CrossRef]
104. Thurairajah, B.; Bailey, S.M.; Cullens, C.Y.; Hervig, M.E.; Russell, J.M. Gravity wave activity during recent stratospheric sudden warming events from SOFIE temperature measurements. *J. Geophys. Res. Atmos.* **2014**, *119*, 8091–8103. [CrossRef]
105. Yamashita, C.; Liu, H.L.; Chu, X. Gravity wave variations during the 2009 stratospheric sudden warming as revealed by ECMWF-T799 and observations. *Geophys. Res. Lett.* **2010**, *37*, L22806. [CrossRef]
106. Pedatella, N.M.; Liu, H.L. The influence of atmospheric tide and planetary wave variability during sudden stratosphere warmings on the low latitude ionosphere. *J. Geophys. Res. Space Phys.* **2013**, *118*, 5333–5347. [CrossRef]
107. Maruyama, T. A diagnostic model for equatorial spread F. 1. Model description and application to electric field and neutral wind effects. *J. Geophys. Res.* **1988**, *93*, 14611–14622. [CrossRef]
108. Aa, E.; Zhang, S.R.; Coster, A.J.; Erickson, P.J.; Rideout, W. Multi-instrumental analysis of the day-to-day variability of equatorial plasma bubbles. *Front. Astron. Space Sci.* **2023**, *10*, 1167245. [CrossRef]
109. Das, S.K.; Patra, A.K.; Niranjan, K. On the Assessment of Day To Day Occurrence of Equatorial Plasma Bubble. *J. Geophys. Res. Space Phys.* **2021**, *126*, e29129. [CrossRef]
110. Otsuka, Y. Review of the generation mechanisms of post-midnight irregularities in the equatorial and low-latitude ionosphere. *Prog. Earth Planet. Sci.* **2018**, *5*, 57. [CrossRef]
111. Huba, J.D.; Bernhardt, P.A.; Ossakow, S.L.; Zalesak, S.T. The Rayleigh-Taylor instability is not damped by recombination in the F region. *J. Geophys. Res.* **1996**, *101*, 24553–24556. [CrossRef]
112. Picone, J.M.; Hedin, A.E.; Drob, D.P.; Aikin, A.C. NRLMSISE-00 empirical model of the atmosphere: Statistical comparisons and scientific issues. *J. Geophys. Res. Space Phys.* **2002**, *107*, 1468. [CrossRef]

remote sensing

MDPI

Article

Ionospheric Absorption Variation Based on Ionosonde and Riometer Data and the NOAA D-RAP Model over Europe During Intense Solar Flares in September 2017

Veronika Barta [1,*], Tamás Bozóki [1,2], Dávid Péter Süle [1], Daniel Kouba [3], Jens Mielich [4], Tero Raita [5] and Attila Buzás [1,6]

1 HUN-REN Institute of Earth Physics and Space Science, 9400 Sopron, Hungary
2 Department of Geophysics and Space Science, Institute of Geography and Earth Sciences, ELTE Eötvös Loránd University, 1117 Budapest, Hungary
3 Institute of Atmospheric Physics, Czech Academy of Sciences, 141 31 Prague, Czech Republic
4 Leibniz Institute of Atmospheric Physics at the University of Rostock, 18225 Kuehlungsborn, Germany; mielich@iap-kborn.de
5 Sodankylä Geophysical Observatory, University of Oulu, 99600 Sodankylä, Finland
6 HUN-REN-ELTE Space Research Group, 1117 Budapest, Hungary
* Correspondence: barta.veronika@epss.hun-ren.hu

Abstract: A novel method was developed based on the amplitude data of the EM waves measured by Digisondes to calculate and investigate the relative ionospheric absorption changes. The effect of 13 solar flares (>C8) that occurred from 4 to 10 September 2017 were studied at three European Digisonde stations (Juliusruh (54.63°N, 13.37°E), Průhonice (49.98°N, 14.55°E) and San Vito (40.6°N, 17.8°E)). The present study compares the results of the amplitude method with the absorption changes measured by the Finnish Riometer Network and determined by the NOAA D-RAP model during the same events. The X-class flares caused 1.5–2.5 dB of attenuation at 30–32.5 MHz based on the riometer data, while the absorption changes were between 10 and 15 dB in the 2.5–4.5 MHz frequency range according to the amplitude data. The impact caused by energetic particles after the solar flares are clearly seen in the riometer data, while among the Digisonde stations it can be observed only at Juliusruh in some certain cases. Comparing the results of the amplitude method with the D-RAP model it seems evident that the observed absorption values almost always exceed the values given by the model both at 2.5 MHz and at 4 MHz during the investigated period. According to the comparison between the riometer data with the D-RAP, generally, the model underestimates the absorption values obtained from the riometers during solar flares except at the highest latitude stations, while D-RAP overestimates the impact during the particle events.

Keywords: space weather; solar flare; ionosphere; ionospheric absorption; ionosonde data; HF radio wave absorption; radio fade-out; riometer; D-RAP model

1. Introduction

The ionosphere is a dynamic interface between the plasma space above and the neutral atmosphere below it. Its state is affected both from above, from the direction of outer space (mainly related to solar activity), and from below, from the neutral atmosphere (e.g., by tropospheric events, sudden stratospheric warmings). Radio and satellite communication and navigation highly depend on the state of the ionosphere, since the electromagnetic (EM) waves propagate through and/or reflect from it. The reduction of the electromagnetic waves traveling through the ionosphere is most pronounced in the high frequency (HF) range (3–30 MHz), which is crucial for radio communication and, by extension, for both commercial and military aircraft operations.

Absorption happens when free electrons in the plasma, driven by the radio wave, lose energy through collisions with ions and neutral particles. The density of free electrons and the collision frequency are greatest in the D-region (approximately 2×10^6 s^{-1}), which is why radio waves below 10 MHz are primarily attenuated there [1]. In this region, the frequency of electron collisions is mainly influenced by the density and temperature of the neutral particles. The most intense external force causing further ionization and absorption in the ionosphere is related to solar flares, which release a substantial amount of electromagnetic energy across a wide range of wavelengths for a brief duration, typically lasting from about 30 min to 1 h [2]. According to the emitted X-Rays flux in the 0.1–0.8 nm band measured by the GOES spacecrafts, flares can be classified as large (X, >10×10^{-4} W/m^2), medium-size (M, ~10×10^{-5}–10×10^{-4} W/m^2) and small (C, ~10×10^{-5}–10×10^{-5} W/m^2). Furthermore, with numbers from 0 to 9 within the above categories applying a finer logarithmic scale. Total radio fade-outs lasting for tens of minutes or even hours can occur during flare events as a result of the increased ionization and attenuation of the radio waves caused by the extra radiation [2–7]. Enhancement of the D-region electron density and consequently increased radio wave absorption can also be caused by collision with energetic auroral electrons (Auroral Absorption [8]), or with solar energetic protons (Polar Cap Absorption—PCA).

Generally, ionospheric absorption is measured by riometers (Relative Ionospheric Opacity meters), which record the cosmic radio noise at certain frequencies (~20–50 MHz frequency range) [9]. Nevertheless, these instruments are generally located at higher latitudes, thus they give information about the changes of the ionospheric absorption that occurred there [10]. Using data from the Imaging Riometer for Ionospheric Studies (IRIS) at Kilpisjärvi (Finland) from solar cycle 23, Ogunmodimu et al. [11] identified that M4 class flares are the less intense ones that still can bear significant influence on cosmic noise absorption. They also presented the magnitude of absorption that the different classes of solar flares can produce. Data from imaging riometers were also used to investigate the frequency and spatial dependence of short-wave fadeouts (e.g., [12,13]).

The minimum frequency (fmin) recorded as the lowest frequency at which an o-mode echo is observed on the ionograms measured by the ionosonde is also a commonly used proxy for the "non-deviative" absorption of the ionosphere [3–7,14,15]. Another way to measure the ionospheric absorption variations by the ionosondes is the analysis of the signal-to-noise ratio of radio waves recorded on the ionograms (SNR method) during solar flares and its comparison with reference quiet days [16]. This method was further developed and used to study absorption changes caused by flare events by de Paula et al. [17]. On the other hand, a new method has recently been developed to assess changes in ionospheric absorption using amplitude data collected by European Digisondes (DPS-4D) [18]. The amplitude changes recorded at the three sites (Juliusruh, Průhonice and San Vito) ranged from 68% to 1430% (these variations can be primarily associated with increased absorption in the D-layer). The amplitude variations at 4 MHz—reflecting cumulative absorption variations in the E- and D-layers—ranged from 18% to 565% at the same stations, which is slightly smaller than the changes observed at 2.5 MHz. However, the amplitude method has two main limitations: (1) it is not applicable at higher frequencies (>3 MHz) during dawn and dusk periods, and during negative ionospheric storms caused by geomagnetic disturbances when the value of the foF2 drops below 4 MHz; (2) no amplitude data is detected in the lower frequency band (2.5–3.5 MHz) following more intense flare events (greater than M3 class) due to partial or complete radio fade-outs, resulting in a minimum frequency (fmin) increasing above 4 MHz.

There are multiple ways to model the ionospheric absorption. One of the better-known models is the so-called D-RAP (NOAA D-region Absorption Prediction) model, which uses the 0.1–0.8 nm solar soft X-Ray flux observed by the NOAA GOES satellites. In the papers by Fiori et al. [19,20], a semi-empirical, data-based method to further refine the absorption prediction is introduced. Based on riometer data and taking into account the solar zenith angle of the site of observation, Fiori et al. [19,20] was able to model the impact of a flare with a given X-Ray intensity on the ionosphere on different spatial and temporal scales for a higher frequency (30 MHz). Furthermore, there are physics-based absorption models, which solve dispersion equations to evaluate the enhanced ionization by solving radiative transport equations [21–23]. These models can be more accurate than the semi-empirical ones; however, their run time is much longer, which can be a problem in their use for operational purposes [19].

In this study, we compare the results of the amplitude method with absorption changes measured by the Finnish Riometer Network and assessed using the NOAA D-RAP model during 13 solar flare events that took place from 4 to 10 September 2017. The paper is structured as follows: next, we will describe the used measurement methods and data, and detail the applied model. The comparison of the ionosonde and riometer data with the D-RAP model for the same locations will be detailed in Section 3. Then, besides the discussion of the results, the main conclusions will be summarized in Section 4.

2. Methods and Data

We selected thirteen flare events that occurred between 4 and 10 September 2017 for the current analysis. The following aspects were considered during the selection: the solar zenith angle (SZA) had to be <90° at the investigated ionosonde stations at the time of the flare events (thus the Sun to be over the horizon); M class flares were selected for investigation, furthermore, one C7.7 flare from 4 September, because it occurred around 12 UT and it caused a detectable impact on the absorption. Figure 1 displays the X-Ray and proton flux recorded by the GOES satellites during the analyzed period. We will use X-Ray and proton flux names during the manuscript which refer to the solar ones. The time of the selected flares with their class are indicated by vertical dashed lines. Table 1 provides a detailed overview of the key properties of the selected events.

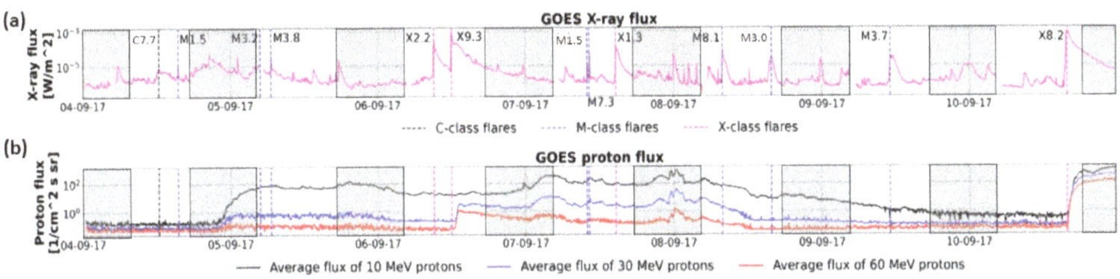

Figure 1. Solar X-Ray (panel (**a**)) and solar proton flux (panel (**b**)) during the investigated time period (4–10 September 2017). Panel (**a**) shows the X-Ray flux in the 0.1–0.8 nm wavelength band as measured by the GOES spacecrafts, whereas panel (**b**) shows the solar proton fluxes at 10, 30 and 60 MeV energies recorded by the same GOES spacecrafts. The dashed lines—magenta, blue and black—indicate the peak times of the X-, M- and the C7.7 class flare, respectively. The shaded areas indicate the time periods out of the scope of this study.

Table 1. The key parameters of the selected solar flares (based on Buzás et al. [18]).

Date	Start Time [UT]	Peak Time [UT]	End Time [UT]	Duration (min)	Class
4 September 2017	11:54	12:22	14:00	126	C7.7
4 September 2017	15:11	15:30	15:33	22	M1.5
5 September 2017	4:33	4:53	5:07	34	M3.2
5 September 2017	6:33	6:40	6:43	10	M3.8
6 September 2017	8:57	9:10	9:17	20	X2.2
6 September 2017	11:53	12:02	12:10	17	X9.3
7 September 2017	9:35	9:54	11:28	113	M1.5
7 September 2017	10:11	10:15	10:18	7	M7.3
7 September 2017	14:20	14:36	14:55	35	X1.3
8 September 2017	7:40	7:49	7:58	18	M8.1
8 September 2017	15:09	15:47	16:04	55	M3.0
9 September 2017	10:50	11:04	11:42	52	M3.7
10 September 2017	15:35	16:06	16:31	56	X8.2

2.1. Ionosonde Measurements—Amplitude Method

Amplitude data from three European ionosonde stations operating with DPS-4D Digisondes (a novel type of ionosonde) were used to calculate the ionospheric absorption. The three stations are (more-or-less) located along a meridional line: Juliusruh (JR, Germany, 54.63°N, 13.37°E), Průhonice (PQ, Czechia, 49.98°N, 14.55°E) and San Vito (VT, Italy, 40.6°N, 17.8°E) (marked by green stars on Figure 2). The amplitudes (in dB) of the reflected echoes emitted at different sounding frequencies (typically between 1.2 and 12 MHz) and reflected at different heights are received and stored in the Digisonde systems. An ionogram was made every 5 and 15 min in the case of the JR and VT stations, respectively. The sampling rate was 2 and 15 min in the case of the PQ station during the investigated period. The denser, 2 min campaign took place on 08–09 September.

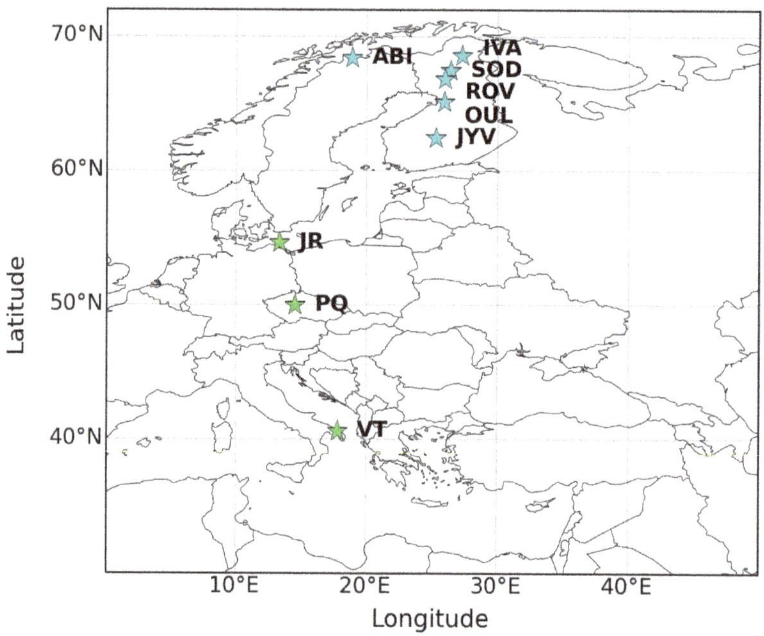

Figure 2. The geographical locations of the used riometer (blue stars) and ionosonde (green stars) stations, with the abbreviated names (codes) of the observation sites.

The amplitude data recorded by the Digisondes at the three investigated stations (JR, PQ, VT) were used to calculate the relative ionospheric absorption of the emitted radio waves following the method proposed by Sales [24]. This method is called the Sales-absorption method or amplitude method and it is based on the Friis-transmission equation [25]. From the Friis-equation, one can derive the loss of a given signal transmitted from an antenna at a certain wavelength which travels along a certain propagation path to arrive at the receiver station, as:

$$10log\,(L) = 10log\,(P_tG_tG_r) + 20log\left(\frac{\lambda}{4\pi 2h}\right) - 10log\,(P_r) \quad (1)$$

where L is the loss term, P_t is the power of the transmitted signal, G_t and G_r are the gain of the transmitter and receiver, respectively, λ is the wavelength of the transmitted signal, h is the height of the reflection, and P_r is the power of the transmitter.

However, in order to obtain the loss term, one must determine the $P_tG_tG_r$ product first as it is not known a priori. To do this, each of the Digisonde systems was calibrated based on quiet (i.e., solar X-Ray flux < 10 × 10^6 Wm^{-2}), nighttime (18–3 UT, which is 19–4 and 20–5 in the local time of the Digisonde measurements during winter and summer, respectively) data. In total, 173 quiet days from the years 2017 to 2018 were selected to this end. Assuming that the loss term was (relatively to the investigated disturbed period) zero during these quiet periods, one can calculate the $P_tG_tG_r$ product from Equation (1). (please, refer to Figure 3 in [18]). More details about the calibration are described in chapter 2.1 in Buzás et al. [18].

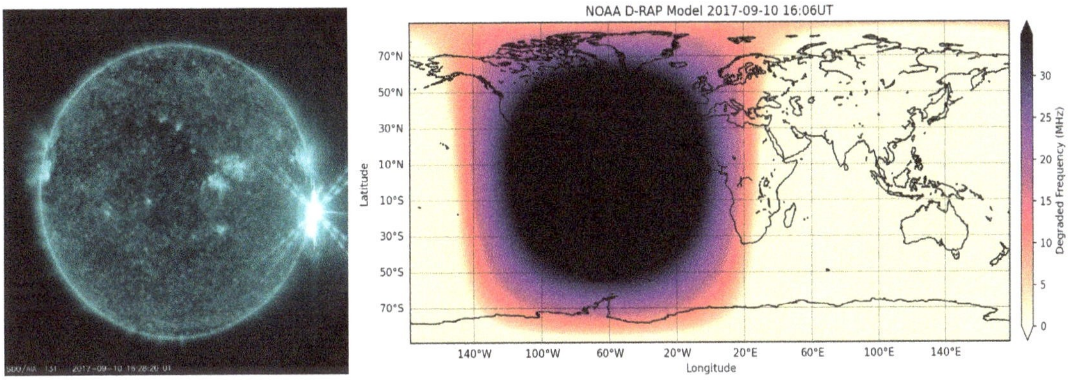

Figure 3. The left panel shows an X8.2 flare event detected by the Solar Dynamic Observatory (SDO) and the right panel shows its impact on the ionospheric absorption as modeled by the D-RAP at 16:06 UT on 10 September.

Ultimately, after the $P_tG_tG_r$ product was quantified in the case of every Digisonde system and at different frequency ranges (namely at 2.5, 3.0, 3.5, 4.0, 4.5, 5.0, and 5.5 MHz ± 200 kHz, respectively), the loss term in Equation (1), which relates to the (relative) absorption, could be calculated in dB for the investigated time period.

2.2. Riometer Measurements

The Sodankylä Geophysical Observatory (https://www.sgo.fi/ [26], accessed on 30 August 2024) operates several traditional La Jolla riometers and novel spectral riometers in Scandinavia (mostly in Finland) forming the Finnish Riometer Network (blue stars in Figure 2). In the present study, data from the La Jolla riometers are used from the 6 stations listed in Table 2. These narrow band, widebeam (60°) systems are equipped with a half-wavelength dual dipole antenna. Most of the stations are located in the 25–28°E longitude range (except for ABI). Four stations (ABI, IVA, SOD, OUL) operate at a frequency of 30 MHz, while 2 stations (ROV, JYV) operate at a frequency of 32.4 MHz. Absorption values are given in dB relative to the quiet day level in 1 min time resolution.

Table 2. Detailed information about the La Jolla riometer stations used in this study. Note that the coordinates (Latitude, Longitude) are geographic.

Station	Abbreviation	Latitude	Longitude	Frequency
Abisko	ABI	68.40°N	18.90°E	30.0 MHz
Ivalo	IVA	68.55°N	27.28°E	30.0 MHz
Sodankylä	SOD	67.42°N	26.39°E	30.0 MHz
Rovaniemi	ROV	66.78°N	25.94°E	32.4 MHz
Oulu	OUL	65.08°N	25.90°E	30.0 MHz
Jyväskylä	JYV	62.42°N	25.28°E	32.4 MHz

2.3. D-RAP Model

The Global D-Region Absorption Prediction (D-RAP) model operated by NOAA's Space Weather Prediction Center (SWPC) predicts the absorption values occurring mainly in the D-region. The model determines the combined absorption caused by X-Ray changes and proton events. The D-RAP is a complex prediction model, nevertheless, it still has several weaknesses according to its validation report [27], as follows: at high-latitude stations, the model occasionally overestimates the peak ionospheric absorption by more than a factor of 2; in the European sector the model appears to systematically underestimate the absorption; absorption at midlatitude appears to be also systematically underestimated; the model does not consider additional absorption due to auroral particle precipitation. Figure 3 shows the picture of an X8.2 flare event observed by the Solar Dynamic Observatory (SDO) and the ionospheric absorption caused by the enhanced X-Ray flux as determined by the D-RAP model at 16:06 UT on 10 September.

The following program was used to download the D-RAP data: https://github.com/hayesla/NOAA-DRAP-Analysis/tree/master [28] (accessed on 30 August 2024). This Python code was modified in order to collect the data from the coordinates closest to the ionosonde and riometer stations for comparison between the modeled and measured absorption values. D-RAP originally shows all the frequencies where the absorption is at least 1 dB. However, for the comparison, we needed the absorption that occurred at certain frequencies (2.5 and 4 MHz for the ionosonde, and 30 or 32 MHz for the riometer measurements). To obtain the absorption for a fixed frequency the following equation was used [4]:

$$f_1 = f_C \cdot \left(\frac{A_C}{A_1}\right)^{\frac{2}{3}} \tag{2}$$

where f_1 is the original frequency for the 1 dB absorption, while A_1 = 1 dB, since the absorption value is equal to 1 dB at that frequency. A_C and f_C are the input values. Based on Equation (2), we can arrive to the following formula for A_C:

$$A_C = \left(\frac{f_1}{f_C}\right)^{\frac{3}{2}} \cdot A_1 \tag{3}$$

where the input data are f_1 and A_1 = 1 dB. Therefore, at a certain frequency (f_c), A_C is going to be the absorption value for that given frequency. Based on this approach, the time series of the absorption values at the given frequencies (2.5 and 4 MHz, and 30 or 32 MHz) and locations (close to the ionosonde and riometer stations) were calculated and analyzed (compared with the measured values there).

3. Results

The main aim of the current study is to compare the ionospheric absorption changes occurring during solar flares and determined by the recently developed amplitude method with the absorption values predicted by the NOAA D-RAP model. The magnitude of the observed differences between the measured and modeled values can give us feedback about the applicability of the developed method. We also analyze the modeled absorption values and the measured changes recorded by the Finnish Riometer Network According to the validation report [27] of D-RAP, the model was evaluated by observations of riometer stations in the American and European sectors. Finally, we compare the magnitude of the differences between the riometer and the D-RAP values with the differences between the amplitude method and the D-RAP values.

3.1. Comparing the Results of the Amplitude Method with D-RAP

The most intense flare events, consequently the largest variations in the ionospheric absorption occurred during the X2.2 and X9.3 flares on 06 September. However, we do not show the occurred changes recorded at 2.5 and 4 MHz by the European ionosondes at midlatitude, because total radio fade-out was detected almost at every investigated

Digisonde station in most part of the day (see Table 3). Therefore, no measured values are available, especially at 2.5 MHz, for comparison with the D-RAP model on that day. Nevertheless, the X-class flares caused increased fmin values (Table 3) and increased amplitude data (Table 4) even after the long blackouts. Because of the quite long duration of the fade-outs at 2.5 and 4 MHz, we compared the measured values with the modeled ones at the time of the first record after the blackout (Table 4). In the case of the measured values at 4 MHz, it is clearly seen that one of the largest absorptions was detected after the X-class flares on 6 September (with 17–23 dB at the different stations).

Table 3. Duration of the total radio fade-outs and the fmin value's changes compared to reference quiet days at the Digisonde stations: JR, PQ and VT, respectively.

Peak Time and Class of Flares	Duration of Total Fade-Out at JR [min]	Change in fmin at JR [%]	Duration of Total Fade-Out at PQ [min]	Change in fmin at PQ [%]	Duration of Total Fade-Out at VT [min]	Change in fmin at VT [%]
09.04.17 12:22 C7.7	-	+73%	-	+68%	-	+213%
09.04.17 15:30 M1.5	-	+31%	-	+18%	-	+62%
09.05.17 4:53 M3.2	-	+0.4%	-	+9%	-	+33%
09.05.17 6:40 M3.8	-	+65%	-	+53%	-	+100%
09.06.17 9:10 X2.2	30 min	+258%	45 min	+132%	45 min	+227%
09.06.17 12:02 X9.3	90 min	+275%	105 min	+203%	120 min	+249%
09.07.17 9:54 M1.5	-	+128%	-	+80%	-	+95%
09.07.17 10:15 M7.3	-	+283%	30 min	+12%	-	+318%
09.07.17 14:36 X1.3	10 min	+252%	-	+179%	-	+280%
09.08.17 7:49 M8.1	15 min	+119%	30 min	+59%	-	+208%
09.08.17 15:47 M3.0	-	+55%	-	+27%	-	+20%
09.09.17 11:04 M3.7	30 min	+135%	48 min	+59%	30 min	+176%
09.10.17 16:06 X8.2	25 min	+185%	45 min	+124%	30 min	+243%

The absorption variation measured by the three ionosonde stations and determined by the D-RAP model on 08 September is seen in Figure 4. Two M-class flares occurred on that day (M8.1 and M3.0 with peak times of 7:49 and 15:47, respectively), as can be recognized in the X-Ray flux changes (Figure 4a). Total and partial radio blackouts occurred at every investigated station after the M8.1 flare (Table 3). and examples of ionograms are shown in Supplementary Figure S1). The time of the fade-out at 2.5 or 4 MHz was determined by a manual check of the ionograms. However, the ionograms from Juliusruh are quite noisy, thus, we have data points from the amplitude method during that period as well when the traces disappeared from the ionograms (some examples for the ionograms are presented in the Supplementary Materials, Supplementary Figure S3). The observed values varied between ~10 and 25 dB in the periods before and after the investigated M8.1 and M3.0 class flares. The increased values of the absorption as a result of the flares are more remarkable at 2.5 MHz (Figure 4c–e), while it is difficult to recognise it at 4 MHz (Figure 4f–h). It is also worth mentioning here that the time resolution of the recorded data is also an important factor, e.g., PQ with 2 min resolution nicely follows the changes of the X-Ray curve during the M3.0 flare at 2.5 MHz, while the changes of the observed values at VT, with 15 min resolution, are not that remarkable, but still can be recognized. It is also important to note here, that there was a negative ionospheric storm in the F-region daytime on 08 September caused by the geomagnetic storm taking place then. The foF2 were below 4 MHz at some ionograms at JR and PQ stations in the afternoon hours leading to the increased values and large changes in the amplitude data there during those periods (Figure 4f,g, more details in [18]). Comparing the measured absorption values with the values determined by D-RAP, we can conclude that the recorded values exceed the estimated ones almost always during the day. Therefore, the model underestimates the ionospheric absorption, e.g., at the peak time of the M3.0 flare (which did not cause total radio blackouts) the D-RAP predicted 9–10 dB absorption at 2.5 MHz at the location of the three European Digisonde stations, while the observed values varied between ~17 and 23 dB at 2.5 MHz (Figure 4, and Table 4).

Therefore, there are 2–4 times differences between the modeled and observed values during that event.

Table 4. The absorption values measured by the Digisondes and determined by D-RAP at 2.5 MHz at the peak time of the flares, or—in the case of fade-out—at the first measurement time after them. The X-Ray flux values are from 0.1 to 0.8 nm wavelength band measured by the GOES spacecrafts.

Measured Values [dB] at 2.5 MHz						
Date	Class	Time (UT) of Peak or of First Record After Fade-Out	X-Ray Flux at That Time [Wm^{-2}]	JR [dB]	PQ [dB]	VT [dB]
2017-9-4	C7.7	12:22	7.79×10^{-6}	17.120	23.828	21.122
2017-9-4	M1.5	15:30	1.53×10^{-5}	8.127	20.256	12.7977
2017-9-5	M3.2	4:53	3.24×10^{-5}	3.99734	6.773	8.239
2017-9-5	M3.8	6:40	3.82×10^{-5}	13.6911	18.288	16.611
2017-9-6	X2.2	15:15	2.26×10^{-5}	15.41	20.3	19.4613
2017-9-6	X9.3	15:15	2.26×10^{-5}	15.41	20.3	19.4613
2017-9-7	M1.5	10:30	1.70×10^{-6}	20.44	22.124	22.4674
2017-9-7	M7.3	10:30	1.70×10^{-6}	20.44	22.124	22.4674
2017-9-7	X1.3	15:15	2.75×10^{-5}	15.96	23.6961	15.4766
2017-9-8	M8.1	8:45	4.65×10^{-5}	15.67	19.46	22.0385
2017-9-8	M3.0	15:47	3.00×10^{-5}	17.4162	23.627	18.15
2017-9-9	M3.7	12:15	6.86×10^{-6}	15.06	24.01	21.6762
2017-9-10	X8.2	16:45	2.49×10^{-4}	19.04	22.3796	12.7332
D-RAP Determined Values [dB] at 2.5 MHz						
Date	Class	Time (UT) of Peak or of First Record after Fade-Out	X-Ray Flux at That Time [Wm^{-2}]	JR [dB]	PQ [dB]	VT [dB]
2017-9-4	C7.7	12:22	7.79×10^{-6}	15.14	17.58	19.16
2017-9-4	M1.5	15:30	1.53×10^{-5}	7.88	8.85	7.88
2017-9-5	M3.2	4:53	3.24×10^{-5}	2.10	1.66	3.00
2017-9-5	M3.8	6:40	3.82×10^{-5}	11.99	13.54	16.80
2017-9-6	X2.2	15:15	2.26×10^{-5}	12.83	13.54	12.83
2017-9-6	X9.3	15:15	2.26×10^{-5}	12.83	13.54	12.83
2017-9-7	M1.5	10:30	1.70×10^{-6}	6.27	7.17	8.00
2017-9-7	M7.3	10:30	1.70×10^{-6}	6.27	7.17	8.00
2017-9-7	X1.3	15:15	2.75×10^{-5}	12.83	14.40	13.54
2017-9-8	M8.1	8:45	4.65×10^{-5}	-	-	-
2017-9-8	M3.0	15:47	3.00×10^{-5}	9.61	10.39	9.61
2017-9-9	M3.7	12:15	6.86×10^{-6}	14.40	16.80	18.36
2017-9-10	X8.2	16:45	2.49×10^{-4}	6.60	6.83	3.35

Figure 5 shows the X-Ray (a), proton flux (b) and absorption changes at 2.5 (c,d,e) and 4 MHz (f,g,h) based on the amplitude method during an X8.2 flare on 10 September (examples of ionograms recorded during the flare event are seen in Supplementary Figure S2). Increased amplitude values compared to the before the flare period were detected at both frequencies at every station. The detected absorption values increased from ~10 dB to ~20 dB at both frequencies at Juliusruh. It means that the flare caused ~10 dB changes at both frequencies. Thus, the more intense flare compared to the previous cases has a clear impact on the measured data. The impact of the energetic particles can be recognized also at Juliusruh sub-auroral station at 2.5 MHz, since the amplitude values do not decrease, but remain enhanced for a longer time period than at the other two stations. The model underestimated the measured absorption values most of the time according to the careful comparison of the measured (green dots) and the modeled values (red curve). It predicts the absorption close to zero before the flare, while the detected values are ~10 dB at most of the stations even before the flare. It is worth noting that the recorded values have the largest variation (within the period before the flare) at the VT station, where the cadence of the measurements is the longest (15 min) among the three stations (Figure 5e). Furthermore, the predicted values at the peak time of the flare were ~16 and ~8 dB at 2.5 and 4 MHz, respectively, while the observed ones were a bit larger (~19 dB) at 2.5 MHz and much larger (~23 dB) at 4 MHz. Moreover, D-RAP does not take into account the impact of the particles at Juliusruh; however, it can be seen in the observation at 2.5 MHz (Figure 5c).

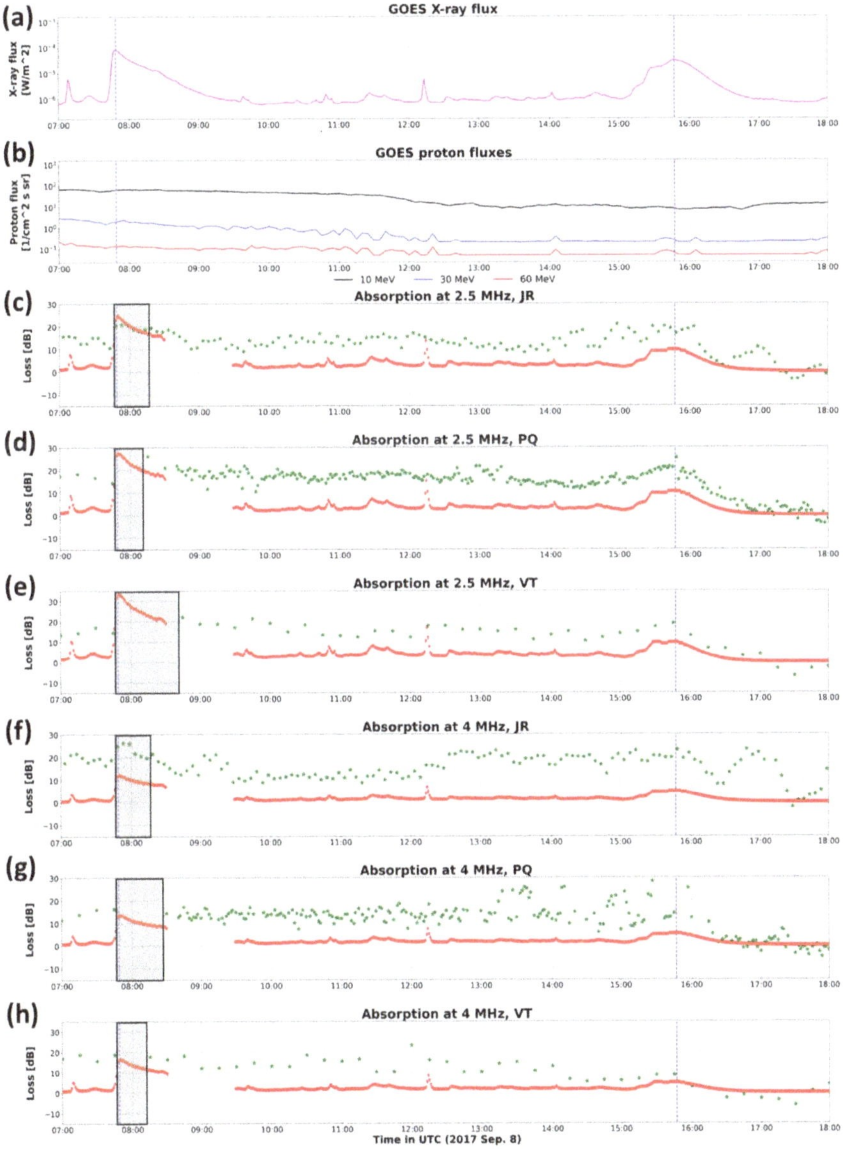

Figure 4. X-Ray (panel (**a**)) and proton flux (panel (**b**)) on 08 September, and the changes measured by the European Digisondes (green stars) and determined by the D-RAP model (red diamonds) at 2.5 MHz (panels (**c**–**e**) and at 4 MHz (panels (**f**–**h**)). The radio fade-outs that occurred at certain frequencies are indicated by gray rectangles (left). The X-Ray data plotted on panel (**a**) are the 0.1–0.8 nm X-Ray flux measured by the GOES spacecrafts. The blue dashed lines indicate the peak time of the M-class flares.

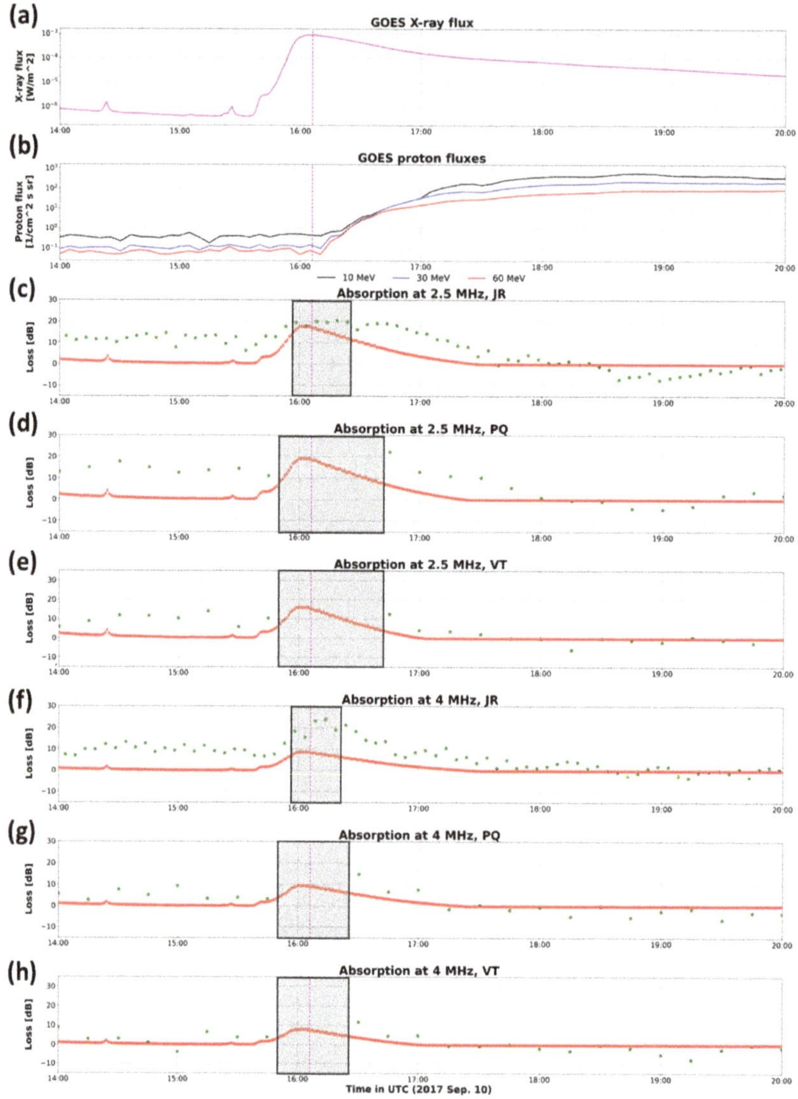

Figure 5. X-Ray (panel (**a**)) and proton flux (panel (**b**)) on 10 September, and the changes measured by the European Digisondes (green stars) and determined by the D-RAP model (red diamonds) at 2.5 MHz (panels (**c**–**e**)) and at 4 MHz (panels (**f**–**h**). The radio fade-outs that occurred at certain frequencies are indicated by gray rectangles. The X-Ray data plotted on panel (**a**) are the 0.1–0.8 nm X-Ray flux measured by the GOES spacecrafts. The magenta dashed line indicates the peak time of the X-class flare.

The increased ionospheric absorption caused total radio fade-outs and increased values of the fmin in the ionograms measured by the ionosondes. Table 3 presents the duration of the total radio fade-outs and the percentage changes of the fmin parameter, recorded at the three stations at the peak time of the 13 flares or after the blackout, compared to its values during the quiet reference period (X-Ray < 1×10^{-6}). Please see the details in Buzas et al. (2023) [18].

Table 4 presents the measured (upper part) and the modeled (lower part) values at 2.5 MHz at the three ionosonde stations. The date and the class of the selected flares are seen, respectively, at the first two columns. In some cases, long total radio fade-out occurred during and after the flares (see Table 3), thus, we decided to do the comparison between the measured and modeled values at the time of the first record (ionogram) after the blackout (measured and modeled values at the peak time of the investigated flares are shown in Supplementary Table S2). The time of the peak, or the first record after the fade-outs are shown in the 3rd column, while the X-Ray flux at that time is in the 4th column. The absorption values calculated by the amplitude method at JR, PQ and VT are shown, respectively, in the next columns. Table 5 shows the same but for 4 MHz. Looking through the values, the picture seems to be quite complicated. According to the D-RAP values (lower part of Tables 4 and 5), the more intense flares caused larger changes in the ionospheric absorption. However, the time of the investigated impact also plays an important role, e.g., the effect of the C7.7 flare which occurred around noon (peak time is 12:22 UT) is larger than the response to the X-class flares in the afternoon hours. This is true also for the measured values, which are larger at the time of a C7.7 flare than, e.g., at 15:15 after an X1.3 flare on 07. September. This is not that evident for the values recorded at 4 MHz (upper part of Table 5).

Table 5. The absorption values measured by the Digisondes and determined by D-RAP at 4 MHz at the peak time of the flares, or—in the case of fade-out—at the first measurement time after them. The X-Ray flux values are from 0.1 to 0.8 nm wavelength band measured by the GOES spacecrafts.

Measured Values [dB] at 4 MHz						
Date	**Class**	**Time (UT) of Peak or of First Record After Fade-Out**	**X-Ray Flux at That Time [Wm^{-2}]**	**JR [dB]**	**PQ [dB]**	**VT [dB]**
2017-9-4	C7.7	12:45	4.73×10^{-6}	13.46	11.86	16.06
2017-9-4	M1.5	15:30	1.53×10^{-5}	6.24	9.29	9.08
2017-9-5	M3.2	4:53	3.24×10^{-5}	3.91	2.89	5.63
2017-9-5	M3.8	6:40	3.82×10^{-5}	9.85	8.45	6.24
2017-9-6	X2.2	10:15	2.02×10^{-5}	21.23	21.69	22.73
2017-9-6	X9.3	14:15	4.56×10^{-5}	19.76	17.08	17.30
2017-9-7	M1.5	9:54	1.56×10^{-5}	15.33	13.42	14.82
2017-9-7	M7.3	10:30	1.70×10^{-6}	15.38	12.33	17.67
2017-9-7	X1.3	15:15	2.76×10^{-5}	15.76	10.47	10.25
2017-9-8	M8.1	8:30	4.45×10^{-6}	17.03	14.45	16.35
2017-9-8	M3.0	15:57	2.27×10^{-5}	21.60	25.34	2.74
2017-9-9	M3.7	12:00	1.14×10^{-5}	17.84	16.16	16.15
2017-9-10	X8.2	16:30	4.19×10^{-4}	18.34	14.94	11.81
D-RAP Determined Values [dB] at 4 MHz						
Date	**Class**	**Time (UT) of Peak or of First Record after Fade-Out**	**X-Ray Flux at That Time [Wm^{-2}]**	**JR [dB]**	**PQ [dB]**	**VT [dB]**
2017-9-4	C7.7	12:45	4.73×10^{-6}	5.93	6.69	7.12
2017-9-4	M1.5	15:30	1.53×10^{-5}	3.89	4.38	3.89
2017-9-5	M3.2	4:53	3.24×10^{-5}	1.04	0.82	1.48
2017-9-5	M3.8	6:40	3.82×10^{-5}	5.93	6.69	8.30
2017-9-6	X2.2	10:15	2.02×10^{-5}	11.86	13.43	15.44
2017-9-6	X9.3	14:15	4.56×10^{-5}	10.68	11.86	12.64
2017-9-7	M1.5	9:54	1.56×10^{-5}	8.30	9.47	11.10
2017-9-7	M7.3	10:30	1.70×10^{-6}	3.10	3.55	3.95
2017-9-7	X1.3	15:15	2.76×10^{-5}	6.34	7.12	6.69
2017-9-8	M8.1	8:30	4.45×10^{-6}	7.12	7.93	9.47
2017-9-8	M3.0	15:57	2.27×10^{-5}	3.95	4.38	3.83
2017-9-9	M3.7	12:00	1.14×10^{-5}	9.07	10.27	11.43
2017-9-10	X8.2	16:30	4.19×10^{-4}	5.13	5.52	3.66

In order to show the results of the comparison in a clearer way, we show diagrams about the flare intensity dependence of the modeled and measured values (Figure 6 for the results at 2.5 MHz, while Figure 7 for the results at 4 MHz). Here we considered also the solar zenith angle (SZA) of the station at the observation time (peak time of the flare or after the fade-out) because according to previous results, the SZA plays an important role in the caused changes in the parameters derived from the ionograms (e.g., [5,6]). Figures 6 and 7 present that the observed values (red symbols) are almost always larger than the modeled ones (except in two cases at 4 MHz, Figure 7). Consequently, the model underestimates the absorption that occurred during and after increased X-Ray flux events. Furthermore, a slightly increasing trend with the X-Ray*cos (SZA) can be recognized, thus during the more intense flares the caused changes seem to be larger.

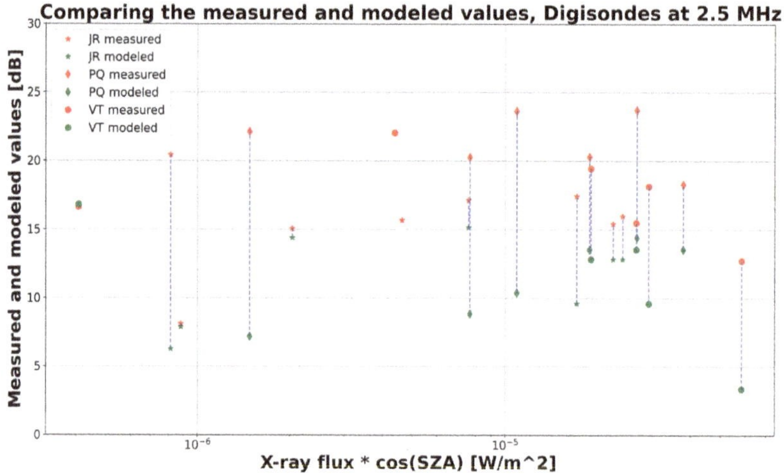

Figure 6. Comparison of the modeled and measured values at 2.5 MHz (y-axis) with the X-Ray flux taking into account also the solar zenith angle of the Digisonde stations at the peak time of the flare or after the fade-out (x-axis).

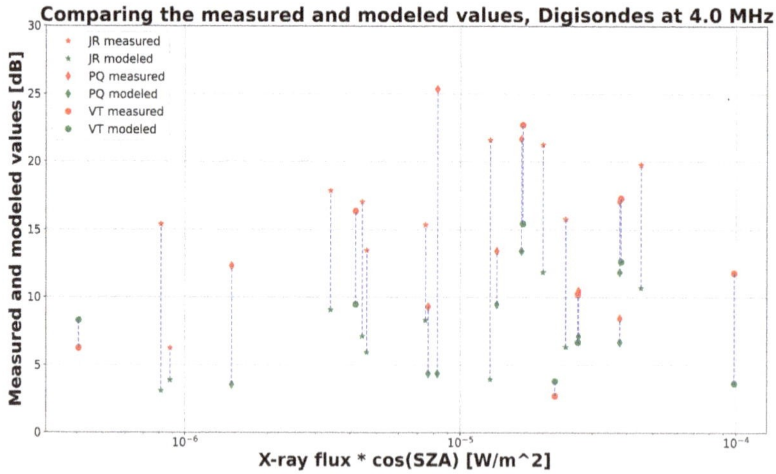

Figure 7. Comparison of the modeled and measured values at 4 MHz (y-axis) with the X-Ray flux taking into account also the solar zenith angle of the Digisonde stations at the peak time of the flare or after the fade-out (x-axis).

3.2. Comparing the Riometer Data with D-RAP

On 06 September, the X9.3 flare caused ~1.6–2.6 dB changes in the ionospheric absorption according to the riometer data. The largest changes (2.6 dB) occurred at the ABI station, while the smallest one (1.6 dB) was recorded at the JYV station. Nevertheless, no clear latitudinal (or solar zenith angle) dependence can be recognized based on the results (Figure 8, Table 6). The flare was accompanied by an energetic particle event as it can be seen in the 30 and 60 MeV proton flux measured by the GOES satellite (Figure 8b). The energetic particles caused ~0.3–1.2 dB variation in the absorption data. Clear latitude dependence can be recognized in the ionospheric response in this case, the larger the latitude, the greater the detected absorption changes were.

The D-RAP model predicted ~1–2.4 dB for the ionospheric absorption during and after the X9.3 flare, with clear latitudinal dependence, the larger the latitude, the greater the predicted values. However, it determined smaller absorption almost in all cases than was detected by the riometers at the same frequency and the same location (Table 6). The absorption varied between ~0.7 and 2.2 dB with clear latitudinal dependence (from North to South) after the increased proton flux based on the D-RAP model. However, it is very striking from Figure 8 that the model overestimates the absorption generated by the energetic particles at every latitude compared with the values recorded at the riometer stations. The difference is larger at higher latitudes than at lower latitudes.

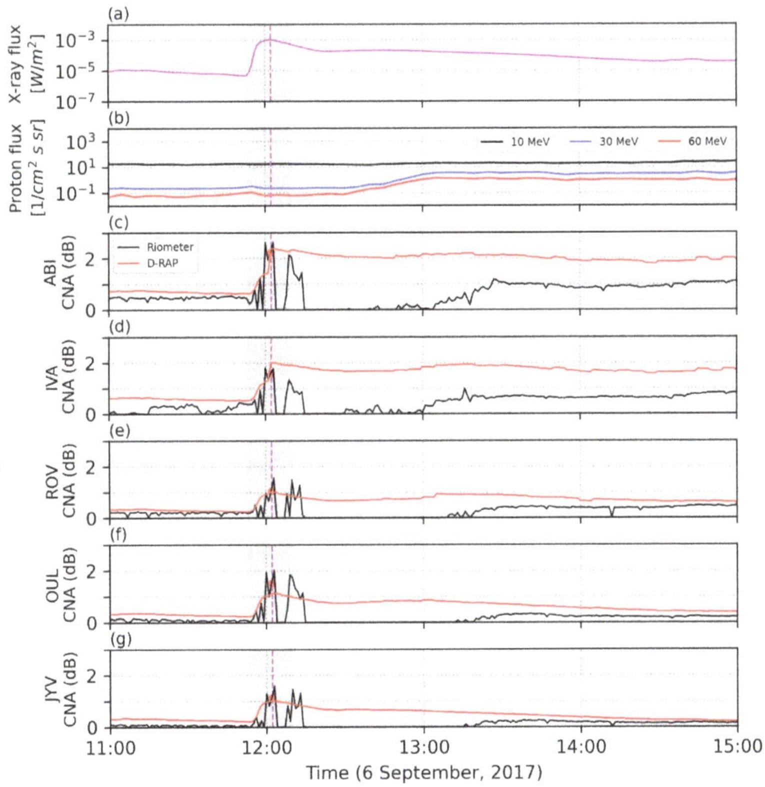

Figure 8. X-Ray (panel (**a**)) and proton flux (panel (**b**)) on 06 September, and the changes measured by the Finnish Riometer Chain (black curve) and determined by the D-RAP model at the same frequencies (30 and 32 MHz) and at the same locations (red curves) (panels (**c**–**g**)). The X-Ray flux values are from 0.1 to 0.8 nm wavelength band measured by the GOES spacecrafts. The magenta dashed line indicates the peak time of the X-class flare.

Table 6. The absorption values measured by the riometers and determined by D-RAP at 30/32.4 MHz at the peak time of the investigated flares and two energetic particle events that followed the X-class flares on 06 and 10 September. The X-Ray flux values are from 0.1 to 0.8 nm wavelength band measured by the GOES spacecrafts.

Measured Values [dB] by the Riometers (30–32 MHz)									
Date	**Class**	**Peak Time (UT)**	**X-Ray Flux [Wm^{-2}]**	**ABI [dB]**	**IVA [dB]**	**SOD [dB]**	**ROV [dB]**	**OUL [dB]**	**JYV [dB]**
2017-9-4	C7.7	12:22	7.79×10^{-6}	0.14	0.13	-	0.24	0.14	0.1
2017-9-4	M1.5	15:30	1.53×10^{-5}	0.59	0.11	-	0.57	0.1	0.08
2017-9-5	M3.2	4:53	3.24×10^{-5}	1.22	1.06	-	0.36	0.19	0.09
2017-9-5	M3.8	6:40	3.82×10^{-5}	1.0	0.64	-	0.36	0.23	0.17
2017-9-6	X2.2	9:10	2.26×10^{-4}	1.53	1.12	-	1.36	1.47	1.45
2017-9-6	X9.3	12:02	9.33×10^{-4}	2.62	1.79	-	1.56	2.01	1.58
2017-9-6	Proton	13:05	1.55×10^{-4}	1.16	0.97	-	0.45	0.33	0.26
2017-9-7	M1.5	9:54	1.56×10^{-5}	-	-	-	-	-	-
2017-9-7	M7.3	10:15	7.36×10^{-5}	-	-	-	-	-	-
2017-9-7	X1.3	14:36	1.39×10^{-4}	2.11	1.46	1.86	1.210	0.65	0.48
2017-9-8	M8.1	7:49	8.12×10^{-5}	1.3	1.04	1.8	1.18	0.85	0.38
2017-9-8	M3.0	15:47	3.00×10^{-5}	4.13	3.08	3.79	1.81	1.18	-
2017-9-9	M3.7	11:04	3.79×10^{-5}	0.65	0.46	0.78	0.5	0.7	0.54
2017-9-10	X8.2	16:06	8.28×10^{-4}	1.06	0.37	1.04	0.72	0.57	0.6
2017-9-10	Proton	17:20–17:30	1.01×10^{-4}	2.29	1.1	1.48	0.71	0.2	0.22
Values [dB] Determined by D-RAP (30–32 MHz)									
Date	**Class**	**Peak Time (UT)**	**X-Ray Flux [Wm^{-2}]**	**ABI [dB]**	**IVA [dB]**	**SOD [dB]**	**ROV [dB]**	**OUL [dB]**	**JYV [dB]**
2017-9-4	C7.7	12:22	7.79×10^{-6}	0.249	0.249	0.249	0.225	0.268	0.257
2017-9-4	M1.5	15:30	1.53×10^{-5}	0.159	0.122	0.127	0.113	0.13	0.118
2017-9-5	M3.2	4:53	3.24×10^{-5}	2.598	2.074	0.336	0.299	0.219	0.146
2017-9-5	M3.8	6:40	3.82×10^{-5}	2.963	2.308	0.46	0.41	0.426	0.324
2017-9-6	X2.2	9:10	2.26×10^{-4}	-	-	-	-	-	-
2017-9-6	X9.3	12:02	9.33×10^{-4}	2.368	1.998	1.138	1.014	1.122	1.014
2017-9-6	Proton	13:05	1.55×10^{-4}	2.177	1.923	1.04	0.927	0.849	0.669
2017-9-7	M1.5	9:54	1.56×10^{-5}	-	-	-	-	-	-
2017-9-7	M7.3	10:15	7.36×10^{-5}	-	-	-	-	-	-
2017-9-7	X1.3	14:36	1.39×10^{-4}	3.602	2.963	0.636	0.567	0.4	0.343
2017-9-8	M8.1	7:49	8.12×10^{-5}	5.387	5.422	1.597	1.423	0.675	0.533
2017-9-8	M3.0	15:47	3.00×10^{-5}	2.145	2.138	2.138	1.905	2.138	1.911
2017-9-9	M3.7	11:04	3.79×10^{-5}	0.675	0.636	0.565	0.503	0.52	0.495
2017-9-10	X8.2	16:06	8.28×10^{-4}	0.268	0.164	0.167	0.148	0.169	0.153
2017-9-10	Proton	17:20–17:30	1.01×10^{-4}	2.32	1.656	0.636	0.567	0	0

Similar results were obtained during the X8.2 solar flare that occurred at ~16 UT on 10 September. The X-class flare caused ~0.4–1 dB of changes in the ionospheric absorption according to the riometer data (Figure 9, Table 6). Larger detected changes seem to be occurred at the higher latitudes (ABI, SOD ~1 dB; IVA is an exception with ~0.4 dB) than at the lower latitudes (OUL, JYV, ~0.6–0.7 dB). This flare was also accompanied by an energetic particle event. Proton flux (at 10, 30 and 60 MeV as well) enhanced ~20 min after the peak time of the flare. The induced absorption values recorded at the riometer stations as a result of the energetic particles varied between ~0.2 and 2.3 dB. The observed changes showed a clear latitude dependence, at higher latitudes (ABI) reached 2.3 dB and it decreased with latitude until it became negligible at the OUL and JYV stations. The D-RAP model underestimated the effect caused by the increased X-Ray flux at every station, since it determined only ~0.15–0.26 dB values (Figure 9 and Table 6). On the contrary, the model overestimated the impact of the energetic particles, especially at higher latitudes (ABI, IVA) where the predicted values for 18 UT reached ~2.6–3.4 dB, which is

~2 times higher than the observed values at the same time. In summary, the D-RAP model underestimates the values obtained from the riometers during the solar flares (X-Ray flux changes) while it overestimates the caused impact during the particle events based on the investigated events.

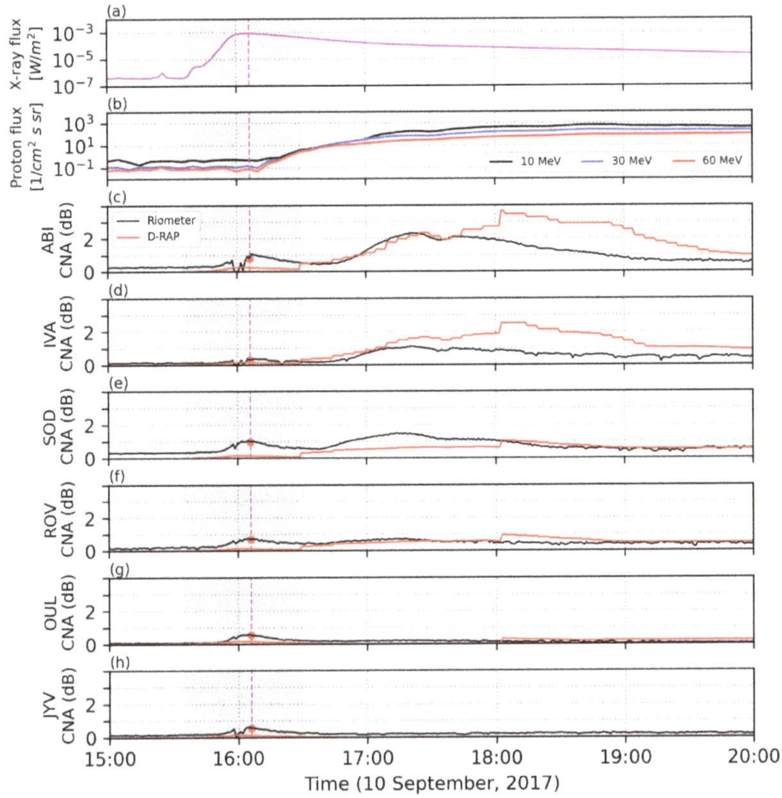

Figure 9. X-Ray (panel (**a**)) and proton flux (panel (**b**)) on 10 September, and the changes measured by the Finnish Riometer Network (black curve) and determined by the D-RAP model at the same frequencies (30 and 32 MHz) and at the same locations (red curves) (panels (**c**–**h**)). The X-Ray flux values are from 0.1 to 0.8 nm wavelength band measured by the GOES spacecrafts. The magenta dashed line indicates the peak time of the X-class flare.

Looking through the absorption values recorded by the riometers during the selected events (Table 6), we can conclude that, generally, the X-class solar flares had a larger impact (~1–2.6 dB) than the M-class flares (~0.1–0.7 dB). (Detailed information for the flare and energetic particle events is seen in Supplementary Table S1). However, the largest detected values (~1.2–4.1 dB) were observed during an M3.0 class flare (peak time: 15:47 UT) on 08 September (this conclusion is also true for the ionosonde data). Nevertheless, the duration of the increased X-Ray flux took almost two hours (Figure 4a) during this event, which could explain why the integrated effect on the ionospheric absorption was the highest in this case. Thus, the duration of the flare event seems to be an important parameter.

In the case of the riometers, we present the comparison of the modeled and measured values separately for the different stations (Figure 10) because the stations operated at different frequencies (ROV and JYV riometers operated at 32.4 MHz, while the others operated at 30.0 MHz). Furthermore, the maximum absorption levels should be compared to the level of the Quiet Day Curve (QDC) before onset, and the QDC level also depends on the station. Looking through the plots, a slightly increasing trend can be recognized at almost every station, the larger the X-Ray flux of the flare, the larger the absorption changes measured by the riometers (the impact of the above detailed M3.0 flare is not presented in Figure 10). Another important thing to note is that the difference between the modeled and observed values seems to be larger at the higher latitude stations (ABI (a), IVA (b), SOD (c)) than at the lower latitude station, JYV (f). In the case of ABI (a) and IVA (b), the model values are much larger than the observed ones in some cases, which agrees with the conclusion of the validation report [27] for the high-latitude stations. The comparison of the model and the measurements during the two investigated energetic particle events were detailed above.

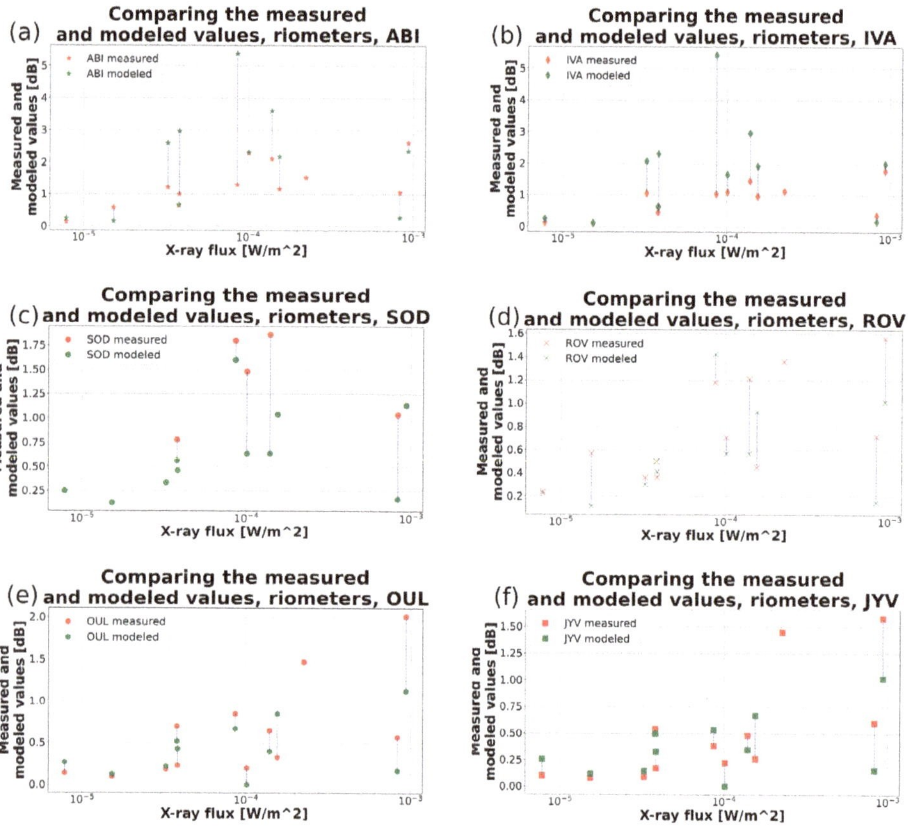

Figure 10. Comparison of the modeled and measured values at the different riometer stations (y-axis, ABI (**a**), IVA (**b**), SOD (**c**), ROV (**d**), OUL (**e**), JYV (**f**)) with the X-Ray flux (x-axis). ROV (**d**) and JYV (**f**) riometers operated at 32.4 MHz, while other riometers operated at 30.0 MHz.

4. Discussion

The ionospheric response to 13 solar flares that occurred in the period of 4–10 September 2017 was investigated in the current study with a special focus on the ionospheric absorption changes. A new observation method to determine the absorption variation was developed by Buzás et al. (2023) [18] based on the amplitude data detected by the Digisondes. The absorption changes that occurred during the 13 above-mentioned flares were carefully analyzed in the previous study. However, the method has not been compared with a model for the ionospheric absorption. The present study aims to fill this gap.

The most commonly used model for ionospheric absorption determination globally is the D-RAP (D Region Absorption Predictions) developed by the Space Weather Prediction Center of NOAA. Previous studies found that the D-RAP model underestimates the ionospheric absorption changes during solar flares [22,29]. Our results generally agree with this, the D-RAP model underestimates the caused ionospheric absorption changes as a result of the enhanced X-Ray flux in the 2.5–4 MHz range (ionosonde data, Figures 4 and 5). Moreover, the observed values are larger most of the time than the predicted ones at 2.5 and at 4 MHz not only during the periods under the impact of the solar flares (Figures 4 and 5), which confirms the conclusion of the validation report [27] for the European region. The main reason for the underestimation can be that the D-RAP model was originally based on the results of Stonehocker [30] who used a limited data set from one station from New Mexico (low-latitude) detected at 5 MHz (instead of the 30 MHz generally used by riometers) during noon in summertime [19]. As the validation report of D-RAP described [27], the model has been evaluated using data from several riometer stations during several historic representative events. However, riometers operate only at high latitudes and at high frequencies (the data investigated in the report were measured at stations with a latitude higher than 50° and at $f \geq 30$ MHz), thus, their data are not necessarily representative for the whole globe (especially for lower latitudes). A further explanation is that the absorption changes related to auroral particle (electron) precipitation have not been considered in the model [27].

The period between 4 and 10 September 2017 was the most active period of the Sun to produce solar flares in Solar Cycle 24. The X9.3 class flare occurred at noon (peak time is 12:02 UT) on 06 September, and it was the most intense detected solar flare of the decade. According to our results, it caused ~1.5–2 h of total radio fade-out at every investigated ionosonde station. While the duration of the blackout during the X2.2 flare (peak time 09:10 UT on 06 September) varied between 30 and 45 min (shortest at Juliusruh) depending on the latitude (solar zenith angle, Table 3), our results agree well with previous findings. The ionospheric impact of these two events was carefully discussed in the literature because of their severity. Kolarski et al. [31] showed that the electron density in the lower ionosphere (70–80 km height) increased by ~2–3 orders of magnitude during the X-class solar flares on 06 September. Fagundes et al. [32] reported that the total radio blackouts lasted for about 45 min during the X2.2 flare, while the duration of the blackout caused by the X9.3 flare was ~1.5 h. The detected duration of the fade-outs in our study agrees very well with this. According to their results, the fade-out was seen to a global extent, which demonstrates the severity of the caused ionospheric absorption changes. Curto et al. [16] demonstrated that the HF absorption took more than 3 h at Ebro station during the X9.3 class event. Based on their results, although the time interval when the X-Ray flux agreed with a category of that of an X-class lasted only for a few minutes, the radiation remained in the category of class M, consequently, the HF absorption is sensitive also to the fainter (M-class) solar flares. Our results also confirm this statement, since increased values of the fmin parameter (Table 3) and even 10–14 dB enhancement of the loss values determined by the amplitude data were found during the M-class flares of the investigated period. The riometers detected 1.1–1.5 dB of cosmic noise absorption during the X2.2 flare, which agrees well with the values reported by Sinha et al. [33], being 1–2 dB during X2 and X2.2 solar flares. The measured values varied between 1.6 and 2.6 dB in the case of the X9.3 flare event (Table 6).

It has been extensively discussed in the literature that ionospheric absorption changes depend on the class of the flare and, thus, on the enhancement of the X-Ray flux that causes the increased ionization and attenuation of the electromagnetic waves in the ionospheric plasma. This was demonstrated by using a riometer (e.g., [33,34]) and different ionosonde data (e.g., [5,6]) as well. In the case of the ionosonde data, a slightly increasing trend with the X-Ray*cos (SZA) can be recognized in Figures 6 and 7, thus during the more intense flares the caused changes seem to be larger. Nevertheless, this trend is not that evident for these results from the amplitude method, like in the other cases reported in the literature [5,6]. Our results regarding the absorption values recorded by the riometers also seem to agree with this, generally, the X-class solar flares had a larger impact (~1–2.6 dB) at most of the stations than the M1–M4-class flares (~0.1–0.7 dB, Table 6, Figure 10).

Besides the X-Ray flux enhancement itself, the solar zenith angle of the observation site at the time of the flare events is also an important factor in determining the caused changes. The impact of 43 solar flares (37 M-class and 6 X-class) on the ionospheric absorption was investigated using riometer data from Abisko in the northern hemisphere [33]. According to their results, the solar zenith angle (SZA) plays a crucial role in the ionospheric response. The detected absorption was smaller in the case of higher SZA (e.g., 1 dB when SZA ~82.3° for X2 class) than under smaller SZA but with similar intensity (2 dB when SZA ~45.6° for X2.2 class). This agrees well with previous findings based on ionosonde data (fmin), that the larger the SZA, the larger the caused impact in the ionospheric absorption [5,6]. Our present results in the case of the ionosonde data also confirm this SZA dependence (Figures 6 and 7). However, the role of the SZA for the caused changes in the ionospheric absorption as a result of the flares is not that clear based on the riometer data in the present case. For example, the impact of the X8.2 flare on 10 September seems to be higher at the higher latitude stations (e.g., 1 dB at ABI and SOD) than at the lower latitude stations (~0.6 dB at OUL and JYV, Figure 9 and Table 6). Nevertheless, Sinha et al. [33] discussed that other factors, like the latitudinal position of the station and background conditions, can also play an important role in cosmic radio noise absorption at high latitudes besides the flare intensity and the solar zenith angle.

Some of the investigated flare events were accompanied by increased flux of energetic protons, the X9.3 flare (peak time at 12:02 UT) on 06 September and the X8.2 flare (peak time at 16:06 UT) on 10 September as is shown in Figures 5b, 8b and 9b. The absorption changes caused by the energetic protons on 10 September were even detectable at the sub-auroral region at 2.5 MHz at the Juliusruh station (Figure 5). This extra absorption related to the particle events was not determined by the D-RAP model. The impact of the increased proton flux was recorded at the riometer data in both cases with a clear latitude dependence, the larger the latitude the greater the observed values. The latitude-dependent enhancement of the ionospheric absorption at higher latitudes as a result of the increased proton flux (solar proton events) has already been discussed in the literature (see, e.g., [35–37]). In the case of the 06 September event, the D-RAP model overestimated the effect of the energetic particles at every station (Figure 8) while in the other case, it over- and underestimated the caused impact depending on the stations. Furthermore, the model predicted a second peak for the absorption around 18:00 UT on 10 September which is not visible in the observed data at all (Figure 9). At this time, the model overestimated the increased absorption at almost every station (except JYV). The determined values were almost two times higher at the higher latitude stations (ABI, IVA) than the observed ones. The D-RAP global model uses a linear relationship between absorption and the square root of the integral proton flux for the proton events, based on the observation of a single high-latitude station. Akmaev et al. [27] discussed that D-RAP often misrepresents the occurred absorption changes and it can only be considered as a qualitative indicator of absorption. Recently, Rogers and Honary [38] suggested an optimized D-RAP modeling approach based on data observed during 94 solar proton events and reached a better agreement between the measured and modeled values. Fiori and Danskin [36] also highlighted that a single constant of proportionality derived from data of one high latitude station [39] can no longer be used to determine the ionospheric absorption. They analyzed data from lower latitude stations (up to 58° MLAT) to define a factor that varies as a function of magnetic latitude leading to a much-improved fit compared to the D-RAP model.

Despite the large discrepancy between the riometer and other observations and the modeled values, D-RAP is the most accepted and widely used model to determine the ionospheric absorption, thus, the degradation of HF waves used for radio communication and aviation (see, e.g., [40]). Our results agree with previous papers suggesting that the D-RAP model should be significantly improved to carefully determine the ionospheric absorption changes as a result of solar flares [20–23]. Nevertheless, based on the present results, the amplitude method seems to be applicable to detect the ionospheric absorption changes during flare events taking into consideration its limitations (that it can produce high values during twilight and during negative ionospheric storms when the foF2 is low, more details in [18]).

5. Conclusions

The most important results of the study can be summarized as follows:

1. The X-class flares resulted in 1–2.6 dB of attenuation at 30–32.5 MHz according to the riometer data (Figures 8 and 9), whereas absorption changes derived from the Digisondes ranged from 10 to 15 dB in the 2.5–4.5 MHz frequency range (indicating a tenfold increase compared to previous cases, Figures 4 and 5).
2. The ionospheric response to X-Ray flux changes (flares) at midlatitude measured by the ionosondes shows a solar zenith angle dependence (Figures 4–7). This dependence is not evident for riometer measurements (Figures 8 and 9). A clear SZA dependence can not be determined in the modeled values at either frequency (Figures 6–9).
3. The effects of energetic particles following solar flares are evident in the riometer data (Figures 8 and 9), although they can only be detected at Juliusruh (~55°N) after some events (Figure 5) among the investigated Digisonde stations. Consequently, the absorption changes due to solar proton events are pronounced at high latitudes but diminish quickly as latitude decreases, becoming undetectable below the sub-auroral region.
4. The absorption changes measured by the riometers and determined by the D-RAP model as a result of the increased energetic particle flux seem to show a clear latitude dependence, the higher the latitude the larger the caused impact (Figures 8 and 9).
5. By comparing the results of the amplitude method with the D-RAP model, it is clear that the observed values almost always exceed the absorption values given by the model both at 2.5 MHz and at 4 MHz (Figures 4–7).
6. By comparing the riometer data with the modeled values, the model overestimates the impact of the solar flares at the highest latitude stations (IBA, AVI), while generally, it underestimates the caused effect at lower latitudes (Figure 10). However, it overestimates the caused impact during the particle events (Figures 8 and 9).
7. According to the current results, the amplitude method seems to be applicable for detecting the ionospheric absorption changes during flare events, taking into consideration its limitations.

Supplementary Materials: The following supporting information can be downloaded at: https://www.mdpi.com/article/10.3390/rs16213975/s1, Figure S1: Consecutive ionograms recorded at Pruhonice ionosonde station during and after an M8.1 class flare (between 07:45 and 08:40) on 08 September. The total radio fade-out is clearly seen on the ionograms. Figure S2: Consecutive ionograms recorded at VT ionosonde station during and after an X8.2 class flare (between 15:45 and 16:30 UT) on 10 September. The total radio fade-out is clearly seen on the ionograms. Figure S3: Consecutive ionograms recorded at Juliusruh ionosonde station during and after an M8.1 class flare (between 07:45 and 08:40) on 08 September. The total radio fade-out, thus the disappearance of the traces from the Ionograms can be seen, despite the fact that the Ionograms are very noisy. Table S1: Detailed data of the investigated flare and energetic particle events. Date, maximum X-Ray or proton flux (measured at different channels) and the exact time of the maximum. Table S2: The values measured by the Digisondes and determined by D-RAP at 2.5 and at 4 MHz at the peak time of the investigated flares. Table S3: The solar zenith angle (SZA) calculated at the peak time of the flare or, in the case of a fade-out, at the time of the first measurement after the fade-out (2.5 MHz measurements), and at the location of the Digisonde stations. The X-Ray flux values are from 0.1 to 0.8 nm wavelength band measured by the GOES spacecrafts. Table S4: The solar zenith angle (SZA) calculated at the peak time of the flare or, in the case of a fade-out, at the time of the first measurement after the fade-out (4 MHz measurements), and at the location of the Digisonde stations. The X-Ray flux values are from 0.1 to 0.8 nm wavelength band measured by the GOES spacecrafts. Table S5: The solar zenith angle (SZA) calculated at the peak time of the flare and at the location of the riometer stations.

Author Contributions: V.B. was the conceptual author of the study, performed the scientific analysis and comparison of the different data and wrote the main part of the text. A.B. performed the amplitude method analysis of the Digisonde data. T.B. analyzed the data of the Finnish Riometer Network. D.P.S. downloaded the data of the D-RAP model and produced the time series of them for the investigated events. V.B., A.B., T.B. and D.P.S. produced the Figures and the Tables of the study. J.M. was responsible for the data from Juliusruh. D.K. was responsible for the data from Pruhonice. T.R. was responsible for the data of the Finnish Riometer Network. All authors contributed to the article and approved the submitted version. All authors have read and agreed to the published version of the manuscript.

Funding: The present work was funded by OTKA, Hungarian Scientific Research Fund (Grant No. PD 141967) of the National Research, Development and Innovation Office. The contribution of VB was also supported by the Bolyai Fellowship (GD, No. BO/00461/21). The authors express their gratitude for the support of the bilateral project between the Czech Academy of Sciences and the Hungarian Academy of Sciences, title: Multi-instrumental investigation of the midlatitude ionospheric variability (n. MTA-19-03 and NKM 2018-28) in facilitating scientific communication. TB's contribution was supported by the ÚNKP-23-4 New National Excellence Program of the Ministry for Culture and Innovation from the source of the National Research, Development and Innovation Fund.

Data Availability Statement: The data sets used in this work are available in online repositories. The names of the repository/repositories and access number(s) are the following: Global Ionospheric Radio Observatory http://giro.uml.edu (accessed on 30 August 2024); NASA/GSFC's OMNI data set through OMNIWeb https://omniweb.gsfc.nasa.gov/ (accessed on 30 August 2024); GOES data, National Oceanic and Atmospheric Administration https://soho.nascom.nasa.gov/sdb/goes/xray/ (accessed on 30 August 2024); Hinode flare catalog https://hinode.isee.nagoya-u.ac.jp/flare_catalogue/ (accessed on 30 August 2024); SAO-X software https://ulcar.uml.edu/SAO-X/SAO-X.html (accessed on 30 August 2024), Finnish Riometer Network https://www.sgo.fi/ (accessed on 30 August 2024), D Region Absorption Prediction (D-RAP) model, operated by the Space Weather Prediction Center (SWPC) of the National Oceanic and Atmospheric Administration (NOAA) https://www.swpc.noaa.gov/products/d-region-absorption-predictions-d-rap, https://www.swpc.noaa.gov/content/global-d-region-absorption-prediction-documentation (accessed on 30 August 2024).

Acknowledgments: The authors wish to express their gratitude to the late Gary Sales who made a substantial contribution in developing the absorption method used in this study and to Bodo Reinisch who attracted our attention to this method. The authors wish to thank the OMNIWeb data center for providing web access to the solar data of the Geostationary Operational Environmental Satellites (GOES) satellites.

Conflicts of Interest: The authors declare no conflicts of interest.

References

1. Zolesi, B.; Cander, L. *Ionospheric Prediction and Forecasting*; Springer Geophysics: Berlin/Heidelberg, Germany, 2014. [CrossRef]
2. Tsurutani, B.T.; Verkhoglyadova, O.P.; Mannucci, A.J.; Lakhina, G.S.; Li, G.; Zank, G.P. A brief review of "solar flare effects" on the ionosphere. *Radio Sci.* **2009**, *44*, 1–14. [CrossRef]
3. Rishbeth, H.; Garriot, O.K. Introduction to Ionospheric Physics. In *International Geophysics Series*; Academic Press: Cambridge, MA, USA, 1969; Volume 14.
4. Davies, K. *Ionospheric Radio*; IET: London, UK, 1990; No. 31.
5. Barta, V.; Sátori, G.; Berényi, K.A.; Kis, Á.; Williams, E. Effects of solar flares on the ionosphere as shown by the dynamics of ionograms recorded in Europe and South Africa. *Ann. Geophys.* **2019**, *37*, 747–761. [CrossRef]
6. Tao, C.; Nishioka, M.; Saito, S.; Shiota, D.; Watanabe, K.; Nishizuka, N.; Tsugawa, T. Statistical analysis of short-wave fadeout for extreme space weather event estimation. *Earth Planets Space* **2020**, *72*, 173. [CrossRef]
7. Barta, V.; Natras, R.; Srećković, V.; Koronczay, D.; Schmidt, M.; Šulic, D. Multi-instrumental investigation of the solar flares impact on the ionosphere on 05–06 December 2006. *Front. Environ. Sci.* **2022**, *10*, 904335. [CrossRef]
8. Fiori, R.A.D.; Trichtchenko, L.; Balch, C.; Spanswick, E.; Groleau, S. Characterizing auroral-zone absorption based on global Kp and regional geomagnetic hourly range indices. *Space Weather* **2020**, *18*, e2020SW002572. [CrossRef]
9. Little, C.G. High latitude ionospheric observations using extra-terrestrial radio waves. *Proc. IRE* **1954**, *42*, 1700.
10. Stauning, P. Investigations of ionospheric radio wave absorption processes using imaging riometer techniques. *J. Atmos. Sol. Terr. Phys.* **1996**, *58*, 753–764. [CrossRef]

11. Ogunmodimu, O.; Honary, F.; Rogers, N.; Falayi, E.O.; Bolaji, O.S. Solar flare induced cosmic noise absorption. *NRIAG J. Astron. Geophys.* **2018**, *7*, 31–39. [CrossRef]
12. Honary, B.; Hargreaves, J.; Browne, S. An imaging riometer for ionospheric studies. *Electron. Com. Eng. J.* **1995**, *7*, 209–217. [CrossRef]
13. Kero, A.; Enell, C.F.; Ulich, T.; Turunen, E.; Rietveld, M.T.; Honary, F.H. Statistical signature of active. D-region HF heating in IRIS riometer data from 1994–2004. *Ann. Geophys.* **2007**, *25*, 407–415. [CrossRef]
14. Sripathi, S.; Balachandran, N.; Veenadhari, B.; Singh, R.; Emperumal, K. Response of the equatorial and low-latitude ionosphere to an intense X-class solar flare (X7/2B) as observed on 09 August 2011. *J. Geophys. Res. Space Phys.* **2013**, *118*, 2648–2659. [CrossRef]
15. Nogueira, P.A.B.; Souza, J.R.; Abdu, M.A.; Paes, R.R.; Sousasantos, J.; Marques, M.S.; Bailey, G.J.; Denardini, C.M.; Batista, I.S.; Takahashi, H.; et al. Modeling the equatorial and low-latitude ionospheric response to an intense X-class solar flare. *J. Geophys. Res. Space Phys.* **2015**, *120*, 3021–3032. [CrossRef]
16. Curto, J.J.; Marsal, S.; Blanch, E.; Altadill, D. Analysis of the solar flare effects of 6 September 2017 in the ionosphere and in the Earth's magnetic field using Spherical Elementary Current Systems. *Space Weather* **2018**, *16*, 1709–1720. [CrossRef]
17. de Paula, V.; Segarra, A.; Altadill, D.; Curto, J.J.; Blanch, E. Detection of Solar Flares from the Analysis of Signal-to-Noise Ratio Recorded by Digisonde at Mid-Latitudes. *Remote Sens.* **2022**, *14*, 1898. [CrossRef]
18. Buzás, A.; Kouba, D.; Mielich, J.; Burešová, D.; Mošna, Z.; Koucká Knížová, P.; Barta, V. Investigating the effect of large solar flares on the ionosphere based on novel Digisonde data comparing three different methods. *Front. Astron. Space Sci.* **2023**, *10*, 1201625. [CrossRef]
19. Fiori, R.A.D.; Chakraborty, S.; Nikitina, L. Data-based optimization of a simple shortwave fadeout absorption model. *J. Atmos. Sol. Terr. Phys.* **2022**, *230*, 105843. [CrossRef]
20. Fiori, R.A.D.; Rogers, N.C.; Nikitina, L.; Lobzin, V.; Rock, E. Duration and extent of solar X-Ray flares and shortwave fadeouts likely to impact high frequency radio wave propagation based on an evaluation of absorption at 30 MHz. *J. Atmos. Sol. Terr. Phys.* **2023**, *252*, 106148. [CrossRef]
21. Eccles, J.V.; Hunsucker, R.D.; Rice, D.; Sojka, J.J. Space weather effects on midlatitude HF propagation paths: Observations and a data-driven D-region model. *Space Weather* **2005**, *3*. [CrossRef]
22. Levine, E.V.; Sultan, P.J.; Teig, L.J. A parameterized model of X-Ray solar flare effects on the lower ionosphere and HF propagation. *Radio Sci.* **2019**, *54*, 168–180. [CrossRef]
23. Chakraborty, S.; Baker, J.B.H.; Fiori, R.A.D.; Ruohoniemi, J.M.; Zawdie, K.A. A model framework to estimate ionospheric HF absorption produced by solar flares. *Radio Sci.* **2021**, *56*, 1–15. [CrossRef]
24. Sales, G.S. *HF Absorption Measurements Using Routine Digisonde Data, Conference Material, XII*; International Digisonde Forum, University of Massachusetts: Lowell, MA, USA, 2011.
25. Friis, H.T. A note on a simple transmission formula. *Proc. IRE* **1946**, *34*, 254–256. [CrossRef]
26. Available online: https://www.sgo.fi/ (accessed on 30 August 2024).
27. Akmaev, R.A.; Newman, A.; Codrescu, M.; Schulz, C.; Nerney, E. D-RAP Model Validation: I. Scientific Report. 2010. Available online: https://www.ngdc.noaa.gov/stp/drap/DRAP-V-Report1.pdf (accessed on 30 August 2024).
28. Available online: https://github.com/hayesla/NOAA-DRAP-Analysis/tree/master (accessed on 30 August 2024).
29. Schumer, E.A. Improved Modeling of Midlatitude D-Region Ionospheric Absorption of High Frequency Radio Signals during Solar X-Ray Flares. Ph.D. Thesis, U.S. Air Force, Wright-Patterson Air Force Base, Dayton, OH, USA, 2009.
30. Stonehocker, G.H. Advanced telecommunication forecasting technique in AGY, 5th., Ionospheric forecasting. *AGARD (Advis. Group Aerosp. Res. Dev) Conf. Proc.* **1970**, *29*, 27–31.
31. Kolarski, A.; Veselinović, N.; Srećković, V.A.; Mijić, Z.; Savić, M.; Dragić, A. Impacts of extreme space weather events on september 6th, 2017 on ionosphere and primary cosmic rays. *Remote Sens.* **2023**, *15*, 1403. [CrossRef]
32. Fagundes, P.R.; Pezzopane, M.; Habarulema, J.B.; Venkatesh, K.; Dias MA, L.; Tardelli, A.; Seemala, G.K. Ionospheric disturbances in a large area of the terrestrial globe by two strong solar flares of September 6, 2017, the strongest space weather events in the last decade. *Adv. Space Res.* **2020**, *66*, 1775–1791. [CrossRef]
33. Sinha, S.; Vichare, G.; Sinha, A.K. Hemispheric comparison of solar flare associated cosmic noise absorption (SCNA) from high latitude stations: Maitri (70.75°S, 11.75°E) and Abisko (68.4°N, 18.9°E). *Front. Astron. Space Sci.* **2024**, *11*, 1347874. [CrossRef]
34. He, L.; Nishino, M.; Zhang, B.; Liu, R.; Liu, Y.; Liu, S.; Yamagishi, H. Absorption events associated with solar flares. *Chin. Sci. Bull.* **2001**, *46*, 369–372. [CrossRef]
35. Romano, V.; Cerrone, M.; Perrone, L.; Pietrella, M. Magnetic and solar effects on ionospheric absorption at high latitude. *Ann. Geophys.* **2002**, *45*. [CrossRef]
36. Fiori RA, D.; Danskin, D.W. Examination of the relationship between riometer-derived absorption and the integral proton flux in the context of modeling polar cap absorption. *Space Weather* **2016**, *14*, 1032–1052. [CrossRef]
37. Heino, E.; Verronen, P.T.; Kero, A.; Kalakoski, N.; Partamies, N. Cosmic noise absorption during solar proton events in, WACCM-D. and riometer observations. *J. Geophys. Res. Space Phys.* **2019**, *124*, 1361–1376. [CrossRef]
38. Rogers, N.C.; Honary, F. Assimilation of real-time riometer measurements into models of 30 MHz polar cap absorption. *J. Space Weather Space Clim.* **2015**, *5*, A8. [CrossRef]

39. Sellers, B.; Hanser, F.A.; Stroscio, M.A.; Yates, G.K. The night and day relationships between polar cap riometer absorption and solar protons. *Radio Sci.* **1977**, *12*, 779–789. [CrossRef]
40. National Research Council. *Severe Space Weather Events—Understanding Societal and Economic Impacts: A Workshop Report*; National Academies Press: Washington, DC, USA, 2008; p. 144. [CrossRef]

remote sensing

MDPI

Article

Subauroral and Auroral Conditions in the Mid- and Low-Midlatitude Ionosphere over Europe During the May 2024 Mother's Day Superstorm

Kitti Alexandra Berényi [1,2,*], Veronika Barta [2], Csilla Szárnya [2], Attila Buzás [1,2,3] and Balázs Heilig [2,3,4]

1 HUN-REN-ELTE Space Research Group, H-1117 Budapest, Hungary; buzas.attila@epss.hun-ren.hu
2 HUN-REN Institute of Earth Physics and Space Science, H-9400 Sopron, Hungary; barta.veronika@epss.hun-ren.hu (V.B.); szarnya.csilla@epss.hun-ren.hu (C.S.); heilig.balazs@epss.hun-ren.hu (B.H.)
3 Space Research Group, Department of Geophysics and Space Science, Institute of Geography and Earth Sciences, Eötvös Loránd University (ELTE), H-1117 Budapest, Hungary
4 Institute of Electrophysics, Kandó Kálmán Faculty of Electrical Engineering, Óbuda University, H-1034 Budapest, Hungary
* Correspondence: berenyi.kitti@epss.hun-ren.hu

Abstract: This study focuses on the mid- and low-midlatitude ionospheric response to the 2024 Mother's Day superstorm, utilizing ground-based and Swarm satellite observations. The ground-based ionosonde measured F1, F2-layer, B0 and B1 parameters, as well as isodensity data, were used. The ionospheric absorption was investigated with the so-called amplitude method, which is based on ionosonde data. Auroral sporadic E-layer was the first time ever recorded at Sopron. Moreover, the auroral F-layer appeared at exceptionally low latitude (35° mlat, over San Vito) during the storm main phase. These unprecedented detections were confirmed by optical all-sky cameras. The observations revealed that these events were linked to the extreme equatorward shift of the auroral oval along with the midlatitude trough. As a result, the midlatitude ionosphere became confined to the trough itself. Three stages of F2-layer uplift were identified during the night of 10/11 May, each caused by different mechanisms: most probably by the effect of prompt penetration electric fields (PPEFs) (1), the travelling ionospheric disturbances (TIDs) (2) and the combination of electrodynamic processes and decreased O/N_2 ratio (3). After a short interval of G-condition, an unprecedented extended disappearance of the layers was observed during daytime hours on 11 May, which was further confirmed by Swarm data. This phenomenon appeared to be associated with a reduced O/N_2 along with the influence of disturbance dynamo electric fields (DDEFs) and it cannot be explained only by the increased ionospheric absorption according to the results of the amplitude method.

Keywords: Mother's Day superstorm; geomagnetic storm; auroral ionosphere; ionospheric storm; swarm

1. Introduction

Space weather events such as the May 2024 Mother's Day Superstorm, caused by a series of interplanetary coronal mass ejections (ICMEs) and high-speed streams, induce hazardous perturbations in the coupled thermosphere–ionosphere–magnetosphere system. Analyzing such unique events offers valuable insight into the evolution of extreme geospace dynamics.

The Mother's Day storm event originated from the solar active region classified as AR13664 [1], which produced at least four M-class flares and one X1.0 flare on 8 May 2024, followed by further strong solar flares (including six X-class flares) between 9 and 11 May, launching a series of fast ICMEs. For a detailed analysis of this storm, including a list of the ICMEs and their associated flares, see Spogli et al. [1]. Another study by Das et al. [2] examines the role of storm-time electric fields in the formation of the equatorial ionization anomaly (EIA) during this storm. Additionally, Aa et al. [3] investigate the expansion and merging of the EIA in the American and Atlantic longitude sectors through multi-instrumental data analysis. Further aspects of this event have been studied by [4–15].

When an ICME collides with Earth's magnetic field, it triggers a geomagnetic storm. The resulting disturbance in the ionosphere is referred to as an ionospheric storm, which is traditionally classified into two types: positive and negative ionospheric storms, depending on whether the storm causes an increase or decrease in the electron density in the ionospheric layers, respectively [16,17]. Their effects can last from 1 to 10 days. There are several comprehensive reviews and studies on the potential physical and chemical mechanisms that may arise during ionospheric storms (see, e.g., [16–23] and references therein).

The ionosphere interacts with the plasmasphere through plasma diffusion and transport processes. Magnetospheric phenomena, such as field-aligned currents (FACs), particle precipitation, plasma drifts, and enhanced electric fields near the plasmapause (PP), are closely linked to ionosphere dynamics. For example, the ionospheric footprint of the nightside PP was found to be aligned with the midlatitude ionospheric trough (MIT) minimum [24] and the equatorward boundary of small-scale FACs ([25,26]). These boundaries are highly sensitive to geomagnetic activity.

This article examines the ionospheric F-layer, which splits into F1 and F2 sublayers during daylight. The F1-layer's electron density is primarily induced by solar radiation, while the F2-layer is more affected by transport processes driven by neutral winds or electric fields [27]. The F2-layer's peak density (NmF2) usually surpasses that of the F1-layer (NmF1). However, under special conditions (called G-conditions), such as intense geomagnetic storms, the F1-layer can blanket the F2-layer due to the latter's electron density dropping below the former's [28,29], with an occurrence rate of ~0.34% [30]. Auroral reflections, another unique feature, appear on ionograms at low- and mid-latitudes only during the strongest geomagnetic storms [29,31].

By employing ground-based Digisonde as well as Swarm satellite data, this study aims to identify the primary mechanisms responsible for the observed mid-latitude ionospheric perturbations arising due to the Mother's Day superstorm. Extreme geomagnetic storms, like the Mother's Day Superstorm, occur on average once in a solar cycle, providing a unique opportunity for studying the ionospheric response to such extraordinary conditions. Good quality Digisonde data in Europe and Swarm data were not available when the last same magnitude superstorm occurred in October/November 2003.

The rest of the paper is organized as follows: Section 2 introduces the data utilized for this study, Section 3 presents the observations, and Section 4 provides the discussion. Finally, Section 5 summarizes the results and concludes the paper.

2. Data and Methods

Manually scaled ionograms from Sopron, Hungary (SO148, quasi-dipole geomagnetic coordinates: 43.95°N, 90.69°E) and San Vito, Italy (VT139, geomagnetic coordinates: 35.94°N, 90.72°E) Digisonde stations [32–34] were analyzed. Key ionospheric parameters, including foF1, foF2, h'F1, and h'F2, which represent peak electron density and maximum layer altitude, were examined. Monthly median values of foF1, foF2, h'F2 and h'F2 were calculated to use them as references for each station. The time resolution is 5 min for Sopron (SO) and 7.5 min for San Vito (VT). In addition, Digisonde-derived B0 and B1 parameters were used, where B0 shows the height difference between hmF2 and the height where the electron density has dropped down to 0.24·NmF2 (maximum of electron density in the F2 layer) [29]. B1 characterizes the shape of the electron density profile: the larger B1 is the larger are the electron densities right below the F2 peak from hmF2 (F2 layer peak height) down to (hmF2 − B0) [29,35]. See more about the calculation and usage of B0, B1 parameters in the article of [36,37]. Figure A10 shows all the stations used in this study.

In addition, the manually scaled data of the ionosondes, the received amplitudes at SO and VT stations, were also taken into account to determine the ionospheric absorption changes. At every sounding, the amplitude values were taken in the following frequency bins: 2, 2.5, 3, 3.5, 4, 4.5, 5, and 5.5 MHz +/−200 kHz, respectively. From these amplitudes, the relative loss of the electromagnetic signal was estimated according to the following equation based on the Friis Transmission Formula (for a more detailed explanation see [38]):

$$10\lg(\mathrm{L}) = 10\lg(\mathrm{P_t G_t G_r}) + 20\lg(\frac{\lambda}{4\pi 2h}) - 10\lg(\mathrm{P_r}) \quad (1)$$

In Equation (1), the known terms are λ (the wavelength of the received signal), h (height of the reflection), and $\mathrm{P_r}$ which is the power of the receiver. If we would like to calculate L, which is the loss in dB units, we have to determine the $\mathrm{P_t G_t G_r}$ (power of the transmitter, gain of the transmitter, and gain of the receiver, respectively) product first. To do so, one can estimate that during nighttime, under quiet solar conditions, the loss is relatively "zero" (that is the reason that this method can only be used for the 2–5.5 MHz frequency range, where there is reflected signal from the F-layer during nighttime, too, the method is described in more detail in Buzás et al. [38]). With this assumption, we calculated the $\mathrm{P_t G_t G_r}$ product based on Equation (1) for SO and VT stations using nighttime data from 173 quiet days in 2019 and 2020 (the same days as described in the Supplementary Material of Buzás et al. [38], just for the years of 2019 and 2020). With the $\mathrm{P_t G_t G_r}$ product now established for each station, we can calculate the loss during the investigated period, which is proportional to the relative ionospheric absorption [38]. Subsequently, the absorption values are compared to the X-ray and integrated proton fluxes as measured by the GOES-16 satellite. The X-ray and integrated proton fluxes and the absorption values are determined as a median diurnal variation of three quiet days around the storm (24–26 May 2024) for the sake of better comparison.

For the analysis of the auroral region's storm-time condition, we used the IE, IU and IL indices (source: https://space.fmi.fi/image/www/il_index_panel.php (accessed on 8 July 2025). These auroral electrojet indicators are simple estimates of the total eastward and westward currents crossing the magnetometer network [39,40], managed by the International Monitor for Auroral Geomagnetic Effects (IMAGE), described by [41]. The definition for IE, IU and IL is like the standard AL, AU, and AE indices [42]. For our study, the data of the so-called PPN–SOR meridional station chain, namely the SOR, MAS, MUO, PEL, RAN, OUJ, HAN, NUR, TAR, BRZ, SUW, PPN stations, were used (Figure A10b).

Swarm satellite observations [43,44] provided a global context for the local ground-based data. During the analyzed period, Swarm C orbited at ~476 km altitude, and Swarm B at ~516 km, both with a 95-min orbital period. The study examined in-situ plasma electron density and temperature from Langmuir probes, vertical total electron content (vTEC) integrated between Global Navigation Satellite System (GNSS) and Swarm satellite altitudes, small-scale FAC (ssFAC) density from magnetic field data [26], and the MIT location, which aligns with the nightside plasmapause footprint [24]. Detailed information on these data products is available in the Swarm Product Data Handbook at https://swarmhandbook.earth.esa.int (accessed on 8 July 2025).

Global Ultraviolet Imager (GUVI) measurements from the Thermosphere, Ionosphere, Mesosphere, Energetics and Dynamics (TIMED) satellite were used to investigate the changes in the thermospheric O/N_2 ratio. GUVI measurements provide dayside O/N_2

composition (note that this is a column integrated value for an altitude range) and temperature profile of the Mesosphere and Lower Thermosphere/Ionosphere (MLTI) region, as well as the auroral energy inputs (see more in [20]).

3. Observations

The storm under investigation, also referred to as the Gannon Superstorm [45], commenced at 17:05 UT (sudden storm commencement–SSC time) on 10 May 2024 [46]. Following a brief initial phase with positive Dst, the main phase began at 18:04 UT and lasted until Dst reached its minimum (−412 nT) at 02:00 UT on 11 May. During the subsequent recovery phase, Dst gradually returned to pre-storm values. Figure 1 displays the Bz and By component of the Interplanetary magnetic field (IMF), the IU, IL indices and the Dst, Kp indices during the evolution of the storm. This study considers the 10–12 May interval, including a pre-storm period, the initial, main, and the early recovery phases of the event.

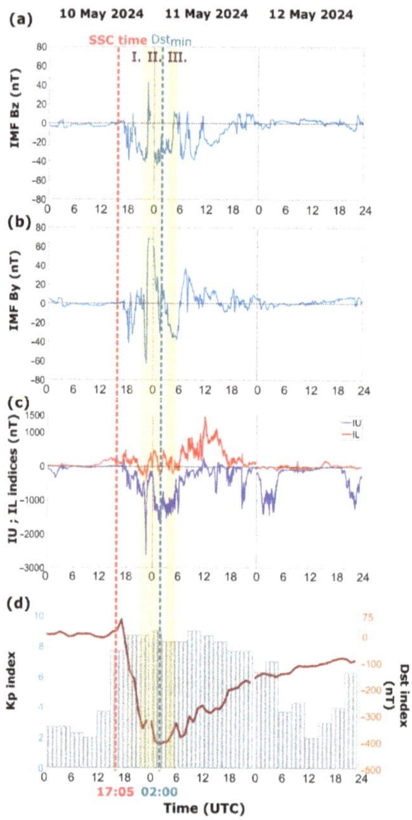

Figure 1. The Bz (panel (**a**)) and By (panel (**b**)) component of the interplanetary magnetic field, the IU and IL indices (panel (**c**)) for the characterization of auroral electrojet (over Europe) and the geomagnetic indices (Kp and Dst; panel (**d**)) are portrayed during the Mother's Day storm. Red and blue vertical dashed lines represent the SSC time and the time of the Dst minimum, respectively. Light yellow bars (I, II, III) are indicating the three analyzed F-layer uplift phases (see the detailed description of these phases below in the Section 3.1).

The temporal evolution of the Kp index is shown in Figure 1d. Following the SSC, the Kp value jumped close to 8 from the pre-storm 4^-, and then oscillated between 8–9 until the evening of 11 May.

3.1. Digisonde Measurements

3.1.1. Overall Storm-Time Condition in the Ionosphere

Figure 2 illustrates the Digisonde observations of the F1- and F2-layers with their monthly median values, highlighting G-condition, the "fade-out" and auroral F- and Es-layer intervals. Following the SSC, the F2-layer's electron density drastically decreased at both stations until 04:50 UT at SO and 04:07 UT at VT on 11 May, when the G-condition emerged. At SO, G-condition intervals were observed at 04:50–07:05 UT, 14:30–14:50 UT, and 16:10 UT (marked by magenta bars on Figure 2), while at VT, they occurred at 04:07:30–08:00 UT, 14:15–15:00 UT, and 16:07:30 UT.

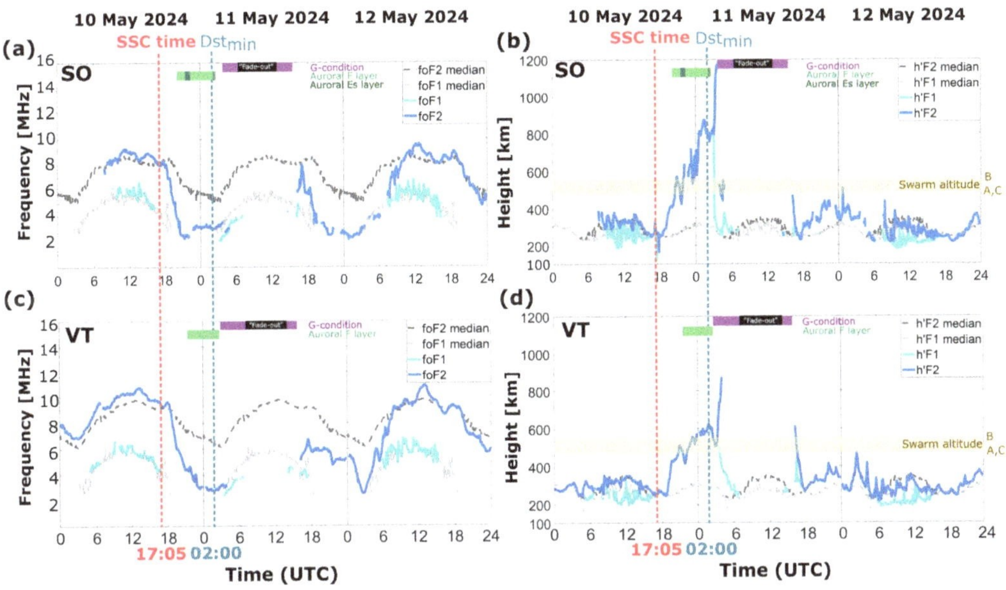

Figure 2. Digisonde F1- and F2-layer measurements during the Mother's Day storm at Sopron (panel (**a**,**b**)) and San Vito (panel (**c**,**d**)). F1-layer data: turquoise line for foF1 and h'F1; F2-layer data: dark blue line for foF2 and h'F2. Gray dashed lines represent median values. Auroral layers (green bars), G-condition (magenta bars) and fade-outs (black bars) are shown, with the altitude range of the three Swarm satellites (A, B and C) marked by a yellow horizontal bar.

These intervals were followed by peculiar total "fade-out" periods, where all ionogram traces disappeared. At SO, the exceptional "fade-outs" occurred from 07:10–14:10 UT and 15:05–16:05 UT, while at VT, they spanned 08:07:30–14:07:30 UT and 15:07:30–16:00 UT. Notably, the "fade-out" began one hour later at VT than at SO, though their end times coincided. Before the "fade-out," the F2-layer rose to an extreme altitude (h'F2 > 800 km at VT and >1000 km at SO), while the F1-layer rapidly descended (below 250 km), see Figure 2b,d. An unusually fast recovery in ionospheric electron density was observed on 12 May.

To quantify the storm-time changes in respect to the quiet day level in percentage value, the so-called deltafoF1, deltafoF2, and deltahF1 and deltahF2 parameters were calculated using the well-known equation from previous studies, e.g., [47,48]:

$$\text{deltafoF2} = \left(\frac{\text{foF2}_{\text{storm}} - \text{foF2}_{\text{quiet}}}{\text{foF2}_{\text{quiet}}} \right) * 100\% \quad (2)$$

On Figure A1 the calculated deltafoF1, deltafoF2, deltahF1, and deltahF2 parameters are shown. In Appendix A section the detailed description of the results can be found. In conclusion, the height values in percentage in Figure A1b,d show quite the same pattern as in Figure 2b,d in the virtual height data.

Section 3.1.3 below describes in detail the nighttime uplift phases on 10/11 May.

3.1.2. Auroral Conditions

The appearance of auroral layers on mid- and low-midlatitude ionograms makes this event exceptional. The equatorward shift of the auroral oval (AO) is evident in ionograms (Figure 3), including the first-ever recorded auroral sporadic E at SO. Auroral F-layer intervals, typically seen at high latitudes, were observed from 20:30 to 04:45 UT at SO and 22:07:30–03:52:30 UT at exceptionally low latitude, at VT on the night of 10/11 May (green bars in Figure 2a–d).

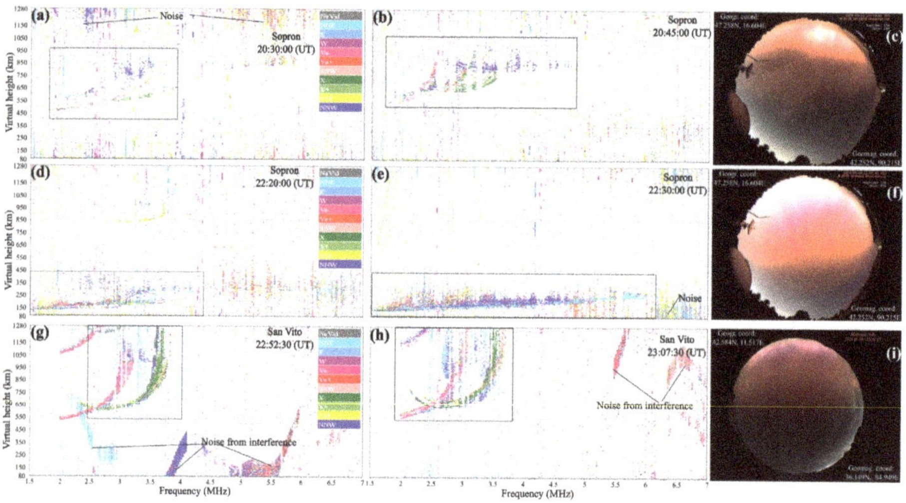

Figure 3. Compilation of optical all-sky camera images (right column, (**c**,**f**,**i**)) and ionograms (first two columns (**a**,**b**,**d**,**e**,**g**,**h**)). Geographic and geomagnetic coordinates of the optical observation sites are shown in the subplots. Auroral reflections are marked with black frames in the ionograms. Rows 1 and 2 are for Sopron (**a**–**f**), and row 3 is for San Vito (**g**–**i**). Light/dark blue reflection points indicate NNE/NNW direction, respectively, while red points represent vertical reflections in the ionograms.

Using the four cross-loop receiver antennas of the Digisonde systems, the direction (vertical and azimuthal) and polarization of the echoes were analyzed [29,49,50]. At 20:30 UT, SO ionograms indicated auroral activity in the F-layer, with oblique spread reflections from north–northeast (NNE, faint blue) and north–northwest (NNW, dark blue) directions (Figure 3a,b). At this time, only a reddish aurora, with an elevation angle of ~46°, was visible with optical all-sky cameras from Szombathely, Hungary (Figure 3c), suggesting the southern edge of the oval may have crossed 49.45°N mlat, the Prague–Krakow line (the detailed calculation steps are in Appendix B section and see also [51,52]). By 22:15 UT, a green aurora was also observed from Doba, Hungary (see Figure A9b), with an elevation angle of ~40° (corresponding to the latitude of Brno, 48.27°N mlat, map of Figure A10a). The red aurora reached the zenith at ~22:30 UT, extending to ~40° mlat (Figure 3f).

Auroral sporadic E appeared on ionograms by this time (Figure 3d,e) as a diffuse, rising trace [53]. Reflections from auroral ionization, including spread signals and oblique echoes from NNE/NNW, confirmed that the northern lights reached SO's latitude (43.95°N mlat).

The auroral intensity was remarkable, it was optically detected from Pesaro, Italy (22:52 UT, 39.48°N mlat, Figures A9c and A10) and after the horizontal distance calculation, the aurora line was above Manciano, Italy (mlat 36.14°N, see Appendix B section and Figure A10). San Vito's ionograms also recorded auroral ionization as spread-F and oblique reflections from NNW and NNE directions (Figure 3g,h).

3.1.3. The Three Identified F-Layer Uplift Phases

On Figure 4 we are focusing on the ionospheric layer uplift phases during the night of 10/11 May using the B0, B1, the isodensity and the h'F, h'F2 parameter data. At first, the uplift phases were identified in virtual height data (Figure 4e,f). The Digisonde-derived B0 and B1 parameters can be used as indicators for rapid ionospheric changes, which appears as a modification of the shape of the electron density profile. In Figure 4a,b, it seems to be clear that the B0 (thickness parameter) is strongly modified by the geomagnetic storm in the main phase. Our results also strengthen the conclusion of [29,37], that the B0 parameter is the most affected by space weather effects. The B1 parameter does not show any significant variability during storm conditions. Interestingly the three uplift phases are also nicely seen in the B0 parameter variability (Figure 4a,b).

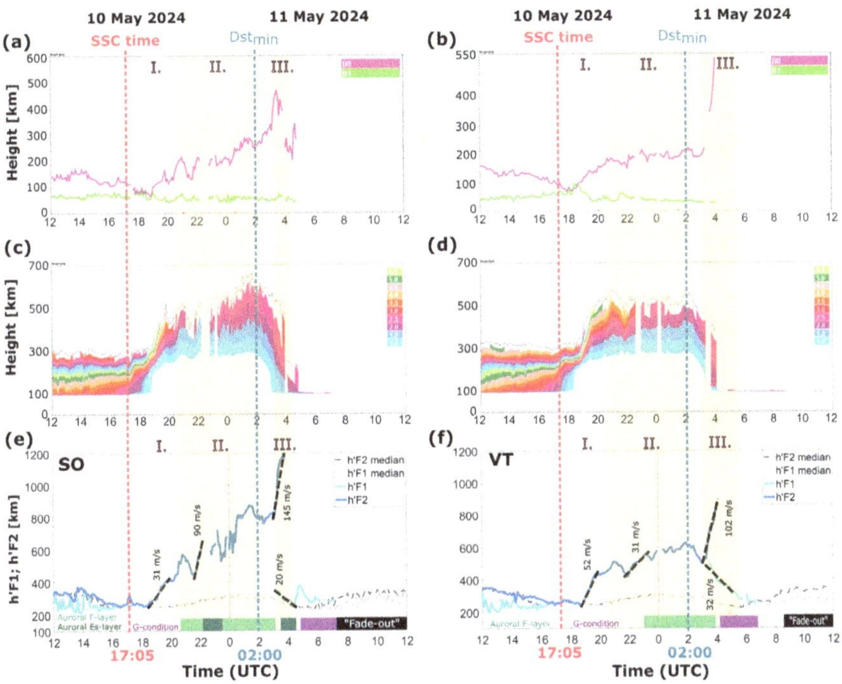

Figure 4. The Digisonde-derived B0, B1 (**a**,**b**), isodensity data (**c**,**d**) and the h'F1 and h'F2 parameter (**e**,**f**) are displayed for SO and VT stations during a 24 h time interval (from 12:00 UT 10 May 2024 to 12:00 UT 11 May 2024). Light yellow bars (I, II, III) are indicating the three analyzed F-layer uplift phases.

The isodensity contours in Figure 4c,d are commonly used to investigate wave-like perturbations in the ionosphere, like large-scale Travelling Ionospheric Disturbances (LSTIDs), see e.g., the paper of [54,55]. LSTIDs mostly occur during geomagnetic storms and appear in the ionosonde data as height variations of the ionospheric layers in time. The second uplift phase identified in Figure 4e,f is such a wave-like variability of the layers, which most probably connected to LSTIDs. Isodensity and B0 data strengthen this possibility at both stations. The third identified uplift phase is the clearest in virtual height data (Figure 4e,f), therefore, those parameters are the most suitable for the investigation of that type of uplift variation during storms.

3.1.4. Ionospheric Absorption Investigation

During the above listed periods of the "fade-out", not only the F1 and F2 layers disappeared from the ionograms at the investigated stations but the E layer as well. A known reason for the fade-out phenomena, especially in the lower frequency range (1–5 MHz), is the increased absorption occurring in the lower ionosphere caused by enhanced X-ray or proton flux (see e.g., [56,57]). During this Mother's Day storm period, the background X-ray flux was quite high. Furthermore, several strong solar flares were registered, consequently the ionospheric absorption is probably also increased (see the article of [58] for the list of the incident flares). To investigate the reason for the fade-out in the current case, the "amplitude method" [38] has been applied for the ionosonde data recorded at SO and VT during the studied period (Figures 5 and A9). Figure 5 shows the X-ray and proton flux, furthermore, the absorption detected at 3 MHz at SO during 3 consecutive reference days (upper 3 subplots) and during the time of the geomagnetic storm (lower 3 subplots). The daily variation of the absorption can be clearly recognized in the data measured during the reference days (Figure 5c). The time of the "fade-out" is indicated by dark grey, while the periods when the maximum frequency of the F layer (foF2 or foF1) was below 3 MHz are indicated by faint grey shadow. Increased values of the absorption can be seen in the early morning hours and before the "fade-out" (02–06 UT) on 11 May (Figure 5f). The period of enhanced absorption agrees well with the time when the integrated proton flux increased according to the GOES data (Figure 5e). It is also important to note that the X-ray flux was increased as well throughout the day on 11 May (Figure 5d, time of terminators are indicated by black dashed lines). In summary, according to the results of the current analysis, we cannot exclude that the increased absorption caused by the enhanced proton and X-ray flux (during the early morning hours and the whole day, respectively) has an important role in the disappearance of the E-layer from the ionograms during the fade-out period.

3.2. *Swarm Measurements*

The electron density during the pre-storm and storm period (9–15 May 2024), observed by Swarm B at MLT = 23.5, is presented in Figure 6. Dotted red and blue lines mark the SSC and Dst minimum times, respectively. The quasi dipole (QD) mlat for SO and VT is indicated by grey horizontal lines. Black dots represent trough minima from the Swarm MIT product, with additional manual selections where the automated algorithm failed.

Swarm observations reveal that the trough moved rapidly equatorward during the storm's main phase, nearing VT's latitude by phase end. This is supported by all-sky records (Figure 3i), which show the Stable Auroral Red (SAR) arc approaching the zenith at Manciano (36.14°N). In the Southern Hemisphere, the MIT also reached low latitudes below 35°. During the recovery phase, the MIT gradually retreated poleward.

Enhanced density was observed equatorward of the MIT at night in both hemispheres, particularly early in recovery, while a density hole formed around the equator. This latter phenomenon was partially discussed by Das et al. [2] and is not the focus here. Poleward of the MIT, the ionosphere was severely depleted at the satellite's altitude.

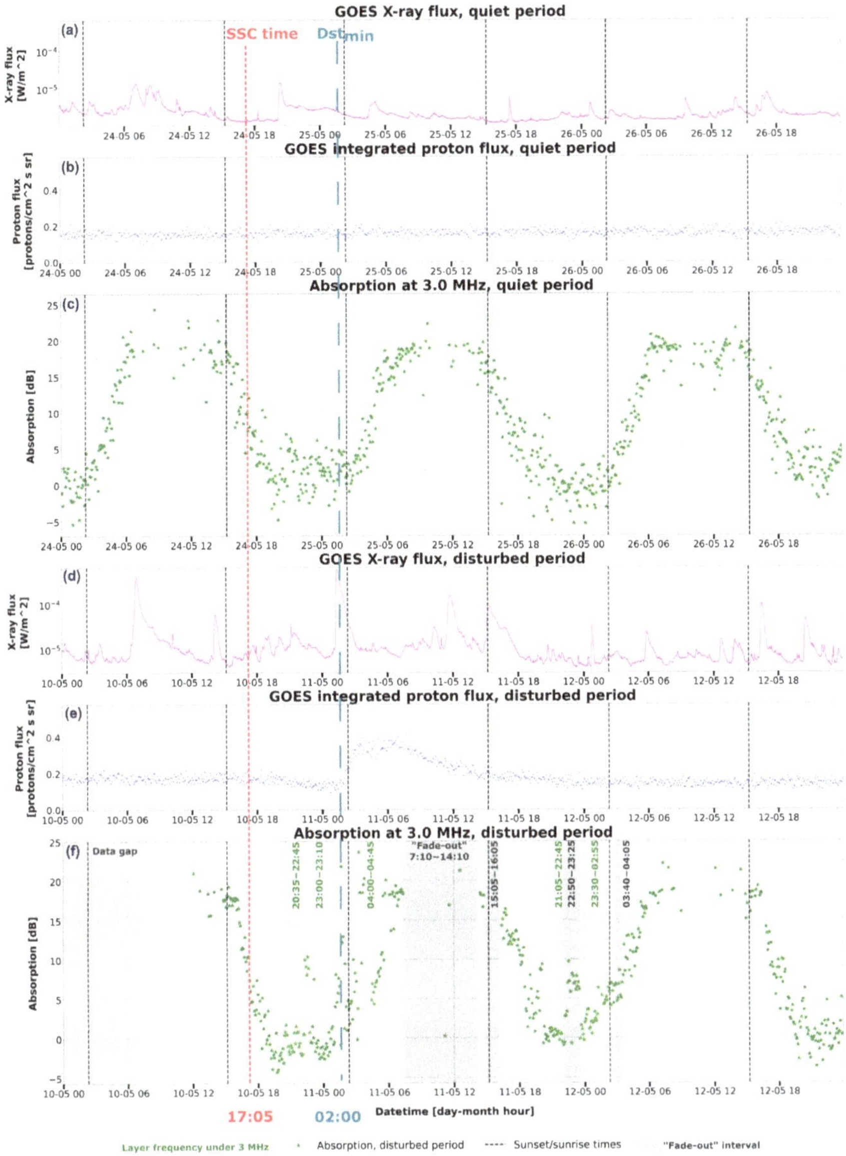

Figure 5. The ionospheric absorption investigation with the amplitude method for Sopron (**a**,**b**,**d**,**e**) and the portray of the GOES satellite X-ray flux and integrated proton flux data for the quiet and the disturbed period, respectively. Panel (**c**) shows the calculated ionospheric absorption data at 3 MHz for the quiet period and panel (**f**) the calculated ionospheric absorption for the disturbed period at 3 MHz. The green bars mark the intervals when the layer frequency went under 3 MHz; the grey bars mark the intervals of the "fade-out periods".

Swarm passed over VT and SO four times within a day following the SSC. The first pass occurred shortly after the SSC at 17:47 UT (SO) and 17:49 UT (VT) at MLT 19.1. Compared to the previous day (shown in grey in Figure 7a–c), EIA crests on 10 May (shown in black) were more pronounced in both Swarm C's plasma density and LEO vTEC observations (Figure 7a–c). Additionally, ssFACs intensified poleward of mlat $-60°$ (Figure 7a).

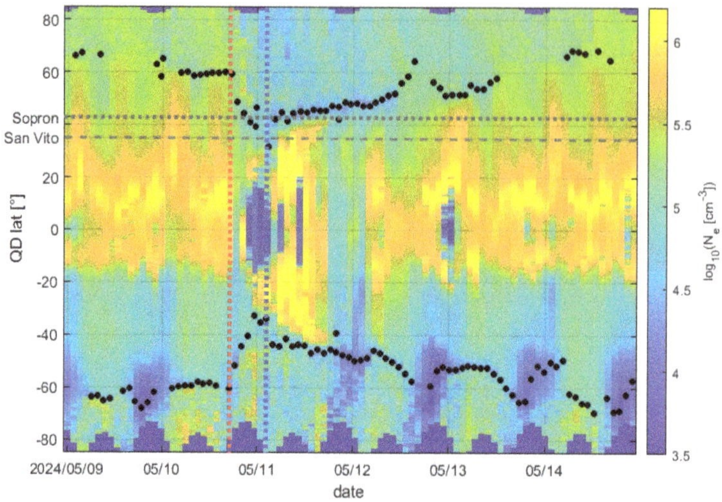

Figure 6. Ionospheric plasma density observed by Swarm B at MLT 23.5 between 9–15 May 2024.

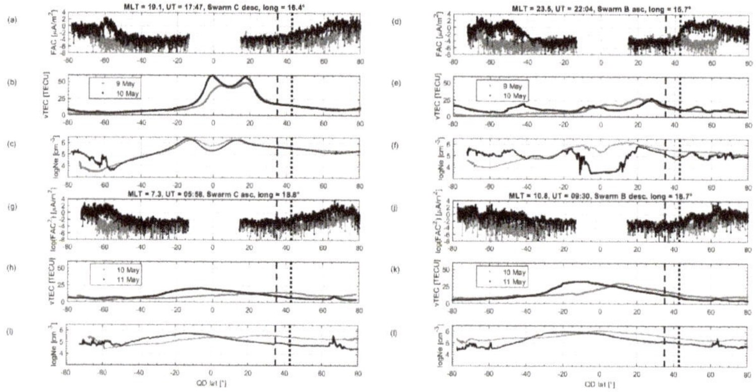

Figure 7. Swarm passes over Sopron and San Vito between 9–11 May 2024. Black curves show observations following the SSC, while grey curves represent observations a day earlier for comparison. For each of the four passes (**a–c**), (**d–f**), (**g–i**), (**j–l**), ssFAC density (log scale), vertical TEC, and plasma density (log scale) are shown. MLT, UT (at Sopron), and orbit longitude are provided above each block. Vertical dashed and dotted lines mark the positions of San Vito and Sopron, respectively.

More dramatic changes were observed a few hours later at 22:04 UT when Swam B passed over VT and SO (Figure 7d–f). Poleward of 40° mlat in both hemispheres, ssFAC intensity increased by orders of magnitude. This zone of intense ssFACs (and particle precipitation) corresponds to the auroral zone. Notably, its equatorward boundary shifted to as low as 40° mlat (Figure 7d), south of SO. Local density peaks in both vTEC and plasma density (see Figure 7e,f), occurring between 40–60° mlat, were likely generated by storm-enhanced particle precipitation. During this flyby, MIT was located at around 40° mlat, marking the equatorward boundary of the auroral zone.

On 11 May, Swarm C, followed by Swarm B, passed over the ground stations again. In both cases, the in-situ plasma density observed near the stations was three to four times lower (log density decreased by 0.44 at MLT 7.3 and by 0.63-.64 at MLT 10.8) than the previous day (Figure 7i,l). This decrease was also noted in the LEO TEC in the Northern Hemisphere, poleward of 30° at MLT 7.3 (5:58 UT) (Figure 7h) and 10° at MLT 10.8 (9:30 UT)

(Figure 7k). Meanwhile, excess plasma accumulated near the equator and in the Southern Hemisphere. The density distribution was asymmetric on both days, but in opposite senses.

4. Discussion

Multi-instrumental observations captured unique ionospheric conditions at mid- and low-midlatitudes during the storm's main phase. Initially, F2 density rapidly decreased, followed by an unusual G-condition at both SO and VT, where F2 density fell below F1 in the early recovery phase. This was followed by an exceptionally long fade-out of the layers during daytime in the early recovery phase.

By 22 UT, the F2-layer raised to the altitude of Swarm, and Swarm confirmed the severe nighttime depletion of this layer (Figure 7f). Meanwhile, a horizontal redistribution also took place. By this time, MIT reached SO's latitude. Afterwards, the equatorward shift of the MIT went to the extremely low 35° mlat (Figure 6), which was indicative of the simultaneous dramatic shrinkage of the plasmasphere [24]. The observed density depletion was limited to the extent of the MIT and to the F2-layer as vTEC remained nearly unchanged. The low F2 density observed by ionosondes in the main phase can also be primarily attributed to the arrival of MIT. Our findings are in good agreement with the result of [58].

Vichare and Bagiya [59] recorded IMF By turning at 22:30 UT on 10 May which also coincided with northward turning of IMF Bz, see also on Figure 1a,b.

Storm-enhanced plasma density (SED) developed at nighttime low latitudes and expanded toward mid-latitudes (Figure 6), approaching the counter-shifting MIT, while a deepening equatorial plasma depletion (EPD) evolved at the equator [2]. Das et al. [2] found that the daytime EIA extended up to +/−40° mlat due to a "Strong Super fountain Effect" which were linked to eastward PPEFs and DDEFs during the main and the early recovery phase of the present storm. EPDs generally develop due to the generalized Rayleigh-Taylor instability (e.g., [60–62]) and prefer the early morning [63–66] and post sunrise sectors [67] and can extend to +/−36° mlat [68]. By the end of the main phase, Swarm observed that the poleward-expanding SED and the equatorward-shifting ssFAC-dominated (precipitation) zone nearly merged, with the MIT sandwiched between them (Figures 6 and 7). At this point, the nightside midlatitude ionosphere was confined to the MIT.

The movement of the MIT was closely followed by the AO. The poleward wall of MIT is believed to be shaped by auroral precipitation in the evening MLT sector [69,70]. The observations confirm the rapid equatorward shift of the AO (optical images, auroral Es- and F-layers and intense Swarm ssFAC). The unique observations included the first-ever Digisonde recording of auroral Es at SO, and auroral F observations at unprecedented low latitudes (down to 35° mlat), confirmed by optical all-sky observations (see Figure 3c,f,i). The source of these recorded auroral Es was most probably due to particle impact to the region of auroral oval [15].

Using multi-instrumental observations, Aa et al. [3] examined the American–Atlantic longitude sector during this event and found that the EIA expanded and merged with the AO. Karan et al. [14] reported repeated overlaps between the poleward-expanding EIA and the AO during the storm's main phase, based on NASA's Global-scale Observations of the Limb and Disk (GOLD) mission data. The northern and southern EIA crests reached approximately 38°N and 35°S mlat around UT midnight (20 MLT), while the AO extended to 40°S and 38°N (the latter estimated from a GOLD image taken at 22:10 UT, presented in Figure 5 of Karan et al. [14]). These observations confirm a remarkable absence of a mid-latitude ionosphere also on the dayside. Previous studies during extremely intense storms have also registered similar auroral expansions up to ~40–50° mlat [71–73].

Regarding the vertical plasma redistribution, three uplift phases can be distinguished (Figure 4):

(1) 18:30–19:30 UT (SO) and 18:37–19:37 UT (VT): Simultaneous parallel variations with similar speeds (around 31 m/s at SO and 52 m/s at VT), likely driven by the same global electric fields, initially PPEF, later with contribution from DDEF. The study of Astafyeva et al. [74] strengthens this scenario, while they observed signatures of strong unshielded PPEF (between 17:05 UT, and until ~19:40 UT). This caused an increase of the equatorial ExB drifts up to 95 m/s, which led to the occurrence of a strong ionospheric super-fountain effect around the equator. Most probably, the upward drift effect of PPEF manifested in our F-layer height data of VT and SO stations with the listed velocities.

Plasma uplifts typically enhance the density, while in this case a decrease was observed. The depletion can be attributed to the horizontal plasma redistribution, the arrival of MIT to the latitude of SO and VT.

This phase can be clearly observed in isodensity and virtual height data too (Figure 4c–f). The B0 parameter is also sensitive for this type of change, but with not that significant increase in its value.

(2) 21:30–02:15 UT (SO) and between 21:52–02:10 UT (VT): Uplift with strong fluctuations with significantly greater F2 heights (890 km vs. 630 km) and speeds (around 90 m/s vs. 31 m/s) at SO and VT, suggesting spatially inhomogene sources like traveling ionospheric disturbances (TIDs) or storm-time meridional wind surges [75]. During this phase, the stations were near the AO (as indicated by the presence of auroral layers), where storm-time winds and TIDs are initiated. The examination of Paul et al. [58] strengthens the presence of MS- and LSTIDs during the main and early recovery phase of this storm. Horvath et al. [76] during the analysis of the Halloween 2003 superstorm also identified TIDs which created large plasma depletions.

As for the analysis of the different ionosonde parameters, our conclusion is that for the identification of the TID effect, the isodensity data are found to be the best indicator (Figure 4c,d).

(3) 03:05–03:50 UT (F2), 03:10–04:25 UT (F1) (SO) and 03:00–04:00 UT (F2), 03:22–05:30 UT (F1) (VT): During this period, the F2-layer uplifted with 145 m/s velocity at SO and 102 m/s velocity at VT. An important difference between the two stations is that at SO the increase in the F2-layer height is not as steady as at VT, but much faster and has three uplift subsections (showing wave-like pattern between 261–546 km, see Figure 4). The steady uplift at VT is also peculiar, since the altitude of the F2-layer rose from 509 to 877 km reaching its highest altitude, thus, the layer went out of the range of the ionograms. In the meantime, the F1-layer at VT and SO descended with velocities of 20 m/s at SO and 32 m/s at VT as it is portrayed in Figure 4e,f. This height decrease was most probably due to the effect of westward DDEF as its presence was registered by Aa et al. [3]. The overall decrease, seen in the F2 density, the Swarm-observed density and Swarm vTEC is the result of the downward drift of the dominant part of the plasma. The anomalous rapid (around 145 m/s at SO and 102 m/s at VT) and steady uplift of the F2 layer at SO and VT may be linked to local heat sources which affect just the F2-layer heights, like frictional heating by subauroral polarization jets streaming along the MIT, or heat downward conducted from ring current (RC) [24]. Karan et al. [68] and Paul et al. [58] proposed that the rapid poleward shift of the EIA (with 450 m/s) together with the equatorward placement of the aurora caused the dramatic ionospheric uplift, which resulted in the prolonged disappearance of the layers at midlatitude during daytime. Note that during this phase, the Dst index was close to its minimum, and the RC to its maximum intensity. This peculiar feature requires further investigation.

The third uplift phase can be identified in all three types of ionosonde parameters portrayed on Figure 4, however, it is important to highlight that the virtual height (h'F, hF2) data is the most suitable for the identification of the uplifts and F1-layer height decreases.

Vichare and Bagiya [59] analyzed the equatorial ionosphere at Asian longitudes (54–76° W) with ionosonde and Swarm data during the main phase of this storm and they identified three IMF phases. They also analyzed the height variation of the F2-layer, but they came up with different conclusions as we were during our study. This is most probably connected to the fact that, during this superstorm interval, the equatorial and the low- and midlatitude ionosphere reacted differently to the incident interplanetary effects. Between 21.4 to 21.8 UT, IMF Bz was mostly constant at ~-40 nT and IMF By fluctuated around zero, the study of Vichare and Bagiya [59] named this interval as Condition I. They observed almost steady height in ionosonde data taken from near the magnetic equator in contrast to our observations at European longitudes. Between 22.2 and 22.5 UT the IMF Bz turned northward without much variation in IMF By (strongly negative with ~-60 nT, Vichare and Bagiya [59] named this as Condition II). In their ionosonde data, they observed a variable response. During northward IMF Bz turnings, around the equator due to overshielding, electron density decrease should be observed. Between 22.5 and 23 UT, IMF Bz turned northward to zero, and IMF By turned sharply from negative to positive (Condition III after Vichare and Bagiya [59]). They observed in ionospheric layer height data a very large increase at all three analyzed stations [59]. Vichare and Bagiya [59] concluded that this enhancement probably occurred due to the post-sunset pre-reversal enhancement (PRE) in that local time sector. Our observations reveal similar ionospheric height increasement during the period of Condition III (Figure 4), at SO station the height increased from 597.7 km to 642.2 km during this half hour interval, in the meantime VT height data show enhancement from 505 to 549.9 km indicating that the European midlatitude ionosphere behave similarly as the Asian longitude one. Ionosonde data of BVJ04 (Geographic: 2.8°S, 60.6° W; Geomagnetic: 11.98°N) (see Figure 2 of Vichare and Bagiya [59]) show the most similar variation to our investigated height parameters. It is important to note that our data show height fluctuations around the interval of Condition III, which can be attributed to TID activity as it discussed above. Therefore, this increasement cannot be clearly connected to the effect of positive IMF By turning in our case. In the European sector the effect of the strong eastward PPEF (caused by the strong positive IMF By turning, reported by Vichare and Bagiya [59]) between 22.5 and 23 UT manifested as electron density decrease at the latitude of VT (4 to 3.12 MHz), on the contrary no plasma density variation was observed at the latitude of Sopron.

In the early recovery phase, the G-condition was followed by an exceptionally prolonged disappearance of the layers from the ionograms ("fade-out"), lasting for 6–7 h at SO (07:10–14:10 UT) and VT (08:07–14:07 UT), which uniquely occurred during daytime. Meantime, the Swarm C and B observed a density decrease of a factor of 3 to 4 around 6:00 and 9:30 UT (7 and 11.5 MLT) compared to the previous day's quiet values (Figure 7i,l). Swarm's vTEC and electron density measurements also showed extreme (30–50% decrease in TEC) and latitudinally extended plasma depletion in the Northern Hemisphere (Figure 7h,k) during this interval. The "fade-out" from the ionograms likely resulted also from the fact that electron density dropped below the daytime detectability level of the Digisondes, approximately 2.3 MHz ($6.58 \cdot 10^4$ 1/cm^3) [20,77]. This prolonged depletion was mainly attributed to the extremely decreased O/N_2 ratio [45] and the resulting high recombination rate as it can be seen in GUVI O/N_2 data as well (see Figure A10b). Additionally, the low density may have been reinforced by the push of the layers exerted by DDEF, to lower altitudes, where the recombination rate is higher, as reflected in the h'F data in Figure 2b,d. However, it is important to note that increased background X-ray flux and solar flares during the investigated period could also contribute to the disappearance of the lower layers through the resulting increased absorption [38,56]. In the recent study of Pierrard et al. [15], different longitudinal sectors were investigated including one in Europe. In their observation they also registered "fade-outs" during quite the same interval during the May superstorm. Our results fit nicely into their ionosonde station line and complementing them. At high latitudes (53.95 mlat), the disappearance of the layers registered shortly after the SSC and the G-condition phase lasted much longer than at SO (46.67 mlat) and VT (35.94 mlat), which is in line with expectations. G-condition phase starts earlier and lasts longer at higher latitudes and appears during intense geomagnetic storms which are causing negative ionospheric storms in the main and early recovery phase [29]. Besides, the appearance of G-condition is dependent on the local time of the storm onset, as was concluded by Pierrard et al. [15] during the investigation of the October 2024 storm, where no G-condition was observed. A similar "fade-out" scenario was reported, e.g., during the nighttime of St. Patrick's Days 2015 and the November 2012 storm. It was found to be related to both the equatorward movement of the MIT and the O/N_2 ratio decrease as they caused significant negative ionospheric storms [20]. The Swarm-detected asymmetric electron density distribution can also be connected to the summer-to-winter solar radiation-driven circulation which causes the seasonal difference in the ionospheric plasma density, however, this needs further investigation [78].

However, it is important to note that increased background X-ray flux and solar flares during the investigated period could also contribute to the disappearance of the lower layers through the resulting increased absorption [38,56]. Increased absorption values were detected during the early morning hours (02 –06 UT) on 11 May, and the reason for the enhanced values was not the negative ionospheric storm in this case (the critical frequency of the F-layer was not below 3 MHz during this period). Thus, the reason could be the enhanced proton (early morning hours, Figure 5e) or X-ray flux (after sunrise, indicated by the black dashed line, Figure 5d). However, no increased values compared to the reference days were observed directly before the fade-out (please compare Figure 5c and f). Therefore, the reason for the disappearance of the E- and F1-layers from the ionograms cannot be clearly connected to the increased ionospheric absorption based on the present results, more investigation is needed.

5. Conclusions

During this current study, we aimed to investigate the mid- and low-midlatitude ionospheric response to the 2024 Mother's Day superstorm. For this purpose, ground-based ionosonde measured F1, F2-layer, B0 and B1 parameters and isodensity data along with satellite measured electron density and O/N_2 ratio data were used. The main conclusions of our study are as follows:

- A series of unique ionospheric conditions were determined by multi-instrumental observations and discussed in the current study.
- A sudden and extreme decrease in the F2 layer electron density and a simultaneous uplift were detected with the Digisondes above SO and VT right after the SSC. The initial density decrease was mostly caused by the arrival of MIT to these latitudes.
- Right after the end of the main phase, a peculiar G-condition developed in the ionosphere. This appears only during intense geomagnetic storms and is connected to extremely depleted plasma in the F2-layer.
- Auroral F- and auroral Sporadic E-layers were observed at unprecedented low latitudes (42°, 35° mlat, respectively) during the main phase, while the analyzed stations were below or near the AO. The first ever observation of auroral Es at SO occurred between 22:15–23:10 UT, while auroral F layers were recorded at SO (22:30–03:45 UT)

and as extremely low latitude as VT station (22:37–03:52 UT) on 10/11 May 2024. Besides the optical observations and the Swarm measurements confirmed that the MIT, with the AO at its equatorward boundary, moved down as low as 35.9°N mlat.

- The auroral and low-latitude nightside ionospheres converged at the MIT, creating an extraordinary situation where the midlatitude ionosphere was confined to the MIT itself.
- Exceptionally prolonged disappearance of the ionospheric layers was observed through Digisonde measurements uniquely during daytime hours in the early recovery phase of the superstorm. The joint effect of the extremely decreased O/N_2 ratio and density, as well as along with the influence of DDEFs was the most probable cause of the disappearance and not the increased ionospheric absorption caused by the incident solar flares, which was confirmed by the amplitude method.
- Three uplift phases were distinguished during the main and early recovery phases. The first phase was driven by global electric fields. The second phase was most likely caused by the spatially varying effect of storm-time winds and TIDs. While the decrease of the F1-layer height during the third uplift phase must have been driven by the westward DDEF. The underlying process responsible for the peculiar third-phase F2-layer uplift remains unclear and requires further investigation, involving additional observations from the plasmasphere.
- The isodensity and virtual height (h′F, h′F2) data are most suitable for the investigation of the ionospheric layer height variations.
- Effects of TIDs during geomagnetic storms are best observed with isodensity data.
- The B0 parameter is sensitive to all the three types of ionospheric layer uplift (and overall, to geomagnetic storm effects), however, it is the most suitable for the indication of TID related ionospheric height variations.
- In a future study, we plan to investigate the prolonged daytime disappearance of the F-layer and its anomalous 3rd-phase uplift in detail, incorporating further experimental data and models like the thermosphere–ionosphere electrodynamic general circulation model (TIEGCM, see the article of [79,80], and the Horizontal Wind Model 2007 (HWM07 [81,82]).

Author Contributions: K.A.B. was the main writer of the article, which contained the preparation of the event, manual scaling of ionosonde data, figure making and discussion of the results. V.B. helped in the manual scaling of VT data, in the preparation of the manuscript and reviewed the paper. B.H. gave ideas, made the figures from the Swarm data, wrote the paragraphs about the Swarm observations and reviewed the manuscript. A.B. made the calculation for the ionospheric absorption data. C.S. calculated the aurora distances and wrote that connected section. Therefore, every author of this manuscript contributed substantially to this study. All authors have read and agreed to the published version of the manuscript.

Funding: The contribution of VB was supported by the National Research Excellence Programme (grant no. STARTING 150984) and Bolyai Fellowship (GD, no. BO/00461/21). BH was supported through ESA Swarm DISC (Contract SW-CO-DTU-GS-33).

Data Availability Statement: All data used in this study are publicly available. Solar and geomagnetic indices (Kp, Dst and IMF) were taken from http://wdc.kugi.kyoto-u.ac.jp/wdc/Sec3.html (accessed on 8 July 2025), https://space.fmi.fi/image/www/il_index_panel.php (accessed on 8 July 2025) and https://omniweb.gsfc.nasa.gov/form/dx1.html (accessed on 8 July 2025). The GOES-16 X-ray (XRS-B channel) and integrated proton fluxes plotted on Figures 6 and 7 are available from NOAA's data center: https://data.ngdc.noaa.gov/platforms/solar-space-observing-satellites/goes/goes16/l2/data/xrsf-l2-flx1s_science/ (accessed on 8 July 2025) and https://data.ngdc.noaa.gov/platforms/solar-space-observing-satellites/goes/goes16/l2/data/sgps-l2-avg1m/ (accessed on 8 July 2025). The absorption data calculated by the amplitude method supporting the conclusions of this article may be requested from the contact author via e-mail (Attila Buzás, buzas.attila@epss.hun-ren.hu). All

Swarm data are freely accessible on the ESA Swarm server https://swarm-diss.eo.esa.int (accessed on 8 July 2025). AACGM coordinates of ground stations were computed at https://sdnet.thayer.dartmouth.edu/aacgm/aacgm_calc.php#AACGM (accessed on 8 July 2025). Ionosonde data are available via the Global Ionospheric Radio Observatory https://giro.uml.edu/didbase/scaled.php (accessed on 8 July 2025). Aurora borealis pictures are from https://www.idokep.hu/hirek/a-pentek-esti-buli-sztarja-a-sarki-feny-volt-iden-otodszorre (accessed on 8 July 2025), https://www.idokep.hu/hirek/zold-sarki-feny-eszakon (accessed on 8 July 2025), https://www.youtube.com/watch?v=yTQO1dr1IVk (accessed on 8 July 2025), https://www.youtube.com/watch?v=B9diWHKlQXM (all accessed on 8 July 2025). The GUVI data used here are provided through support from the NASA MO&DA program (https://guvitimed.jhuapl.edu/data_fetch_l3_on2_gif?y=2015&d=076) (accessed on 8 July 2025). The GUVI instrument was designed and built by The Aerospace Corporation and The Johns Hopkins University. The Principal Investigator is Andrew B. Christensen and the Chief Scientist and co-PI is Larry J. Paxton. The manuscript's Figure 1, Figures 2 and 4–7 were produced using Matlab R2019a.

Acknowledgments: We thank the data centers (OMNIWEB, WDC for Geomagnetism, Kyoto, and GIRO) for the high-quality data. SSC data are partly from Observatori de l'Ebre (OE, Spain).

Conflicts of Interest: The authors declare no conflicts of interest.

Abbreviations

The following abbreviations are used in this manuscript:

AO	auroral oval
DDEF	disturbance dynamo electric field
EIA	equatorial ionization anomaly
EPD	equatorial plasma depletion
ICME	interplanetary coronal mass ejection
GNSS	global navigation satellite system
HWM07	Horizontal Wind Model 2007
IMAGE	International Monitor for Auroral Geomagnetic Effects
IMF	interplanetary magnetic field
FAC	field-aligned current
GOLD	global-scale observations of the limb and disk
LSTID	large-scale TID
NNE	north–northeast
NNW	north–northwest
MIT	midlatitude ionospheric trough
MLT	magnetic local time
MSTID	medium-scale TID
PPEF	prompt penetration electric field
QD	quasi dipole
PRE	pre-reversal enhancement
RC	ring current
SAR	stable auroral red
SED	storm-enhanced plasma density
SSC	sudden storm commencement
ssFAC	small-scale FAC
SO	Sopron
TID	travelling ionospheric disturbance
TIEGCM	thermosphere-ionosphere- electrodynamic general circulation model
UT	universal time
VT	San Vito
vTEC	vertical total electron content

Appendix A

In Figure A1, the calculated deltafoF2, deltafoF2, deltahF1, and deltahF2 parameters are shown. The sudden decrease in F2-layer electron density after the SSC went up to −63% (at 20:40 UT) and 58% (at 23:22:30) at SO and VT, respectively (Figure A1a,c). A slight increase can be tracked in the data until G-condition mainly at SO. In F1-layer electron density it varies between 13% and −36% during the day of 11 May. At SO, a second nighttime electron density minimum with −60% can be observed from similarly around 20 UT until 02 UT during the night of 11/12 May. At VT, the second minimum in F2-layer electron density with −61% (at 00:50 UT) is more expressed and sharper after midnight on 12 May, which continues with a significant increase up to the quiet day values (0%) around 9 UT.

The height values in percentage in Figure A1b,d show quite the same pattern as in Figure 2b,d the virtual height data. After the SSC both station data show a sudden increase from 18:30 and 18:52 UT up to 50% and 90% at SO and VT, respectively. This phase continues with a wave-like, more oscillatory scenario until 03:05 UT and 03:22:30 UT on 11 May. Right after that an extreme increase follows, the deltahF2 values go up to 358% and 240% at SO and VT until the start time of the G-condition (Figure A1b,d). During this F2 layer height increase, the F1-layer decreased from 113% to 3% at VT, so the storm-time increased layer altitude went back to its quiet day value right before the disappearance/"fade-out" of the layers started. At SO in the deltahF1 data, we do not see such a significant decrease. After the fade-out scenario during the afternoon of 11 May, deltahF2 values quickly went back to their quiet day level (from 18 UT). Section 3.1.3 above describes in detail the nighttime uplift phases on 10/11 May.

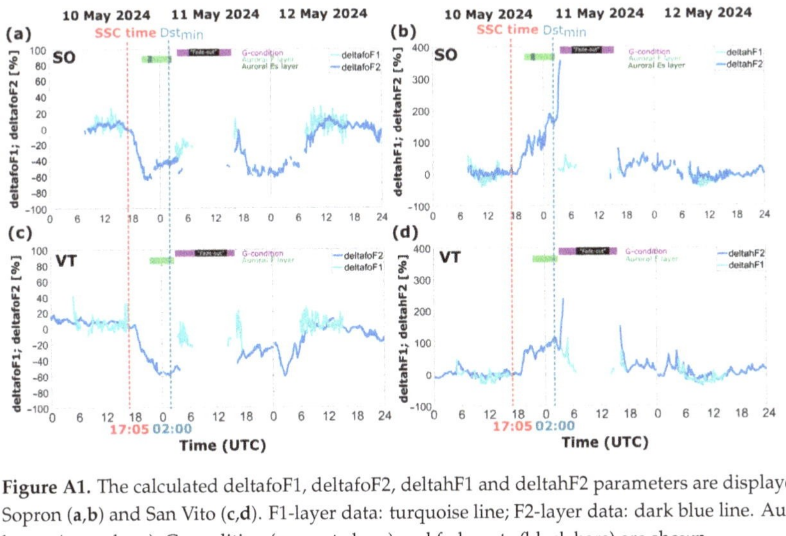

Figure A1. The calculated deltafoF1, deltafoF2, deltahF1 and deltahF2 parameters are displayed at Sopron (**a**,**b**) and San Vito (**c**,**d**). F1-layer data: turquoise line; F2-layer data: dark blue line. Auroral layers (green bars), G-condition (magenta bars) and fade-outs (black bars) are shown.

Appendix B

This is the additional explanatory text for the exact calculation steps, how the geographic and geomagnetic latitudes corresponding to the green and red auroras during the main phase of the superstorm were calculated.

Most studies regarding the aurora concentrate on the determination of the peak emission height of the aurora borealis and not the height boundaries of each type (color) of aurora. Since the accurate height of a given type (color) of aurora for this superstorm is currently not known and since the behavior of the aurora can differ storm by storm (and there can be differences at mid-latitude compared to high-latitude cases), here the upper boundary of the auroras is based on statistical observations (with this, the height is supposed to be overestimated). Regarding the green aurora, after the study of Whiter et al. [51] the correlation coefficient of the peak emission height of the green aurora is around 0 at 180 km height (see Figure 1 in the article of [51]), that is why 180 km was chosen as an overestimation for the upper height boundary of the green aurora.

For the height determination of the red aurora, we used the study of Jackel et al. [52]. The correlation coefficient of the peak emission height was close to 0 at 300 km (see Figure 2 in the article of [52]), therefore, the 300 km height was chosen as upper boundary height for the red aurora position calculations in our study.

To estimate the geographical location of the auroral borealis, the images of all-sky cameras were used (e.g., Figure A2). Based on the configuration of the all-sky images (Figure A3a) the distances (b, c) and the viewing angles (β, γ) are directly proportional and can be determined. Therefore, the equation is

$$\frac{b}{r} \cong \frac{\beta}{90^{\circ}}, \frac{c}{r} \cong \frac{\gamma}{90^{\circ}} \tag{A1}$$

In the next step x (the average of three different measurements taken from the picture) was used instead of r, since the radius of the all-sky image is unknown, because the pictures are from outside source, see Acknowledgement. On Figure A4a the x and y (the proportion of the aurora on the captured image) can be measured and quantified from the images. From x and y, the angle α (angle between the location of the all-sky camera and the top of the aurora, Figure A4b) can be calculated:

$$\frac{y}{x} = \frac{\alpha}{90^{\circ}} \tag{A2}$$

The h is the maximum height of the different type (color) of observed aurora, and were chosen as described above (180 km for the green aurora and 300 km for the red aurora), therefore the horizontal distance between the camera and the exact location of the aurora (d) can be estimated with this equation:

$$\frac{h}{d} = \tan(a) \tag{A3}$$

The exact calculation for 20:30 UT is as follows:

The subauroral arc (SAR) was at zenith at this time, and directly above Szombathely, Hungary (47.25°N lat, 42.25°N mlat, see Figure A8 for the location). For the red aurora with Equation (A2), where y = 4.8, x = 9.3 and after the calculation

$$\frac{4.8}{9.3} \cdot 90^{\circ} = 46.45^{\circ},$$

the top of the red aurora was visible around 46.45° degrees. Using the above-mentioned h = 300 km upper height for the red aurora (after [52]), the d value is determined with Equation (A3) as follows:

$$\frac{300}{\tan 46.45^{\circ}} = 285.7$$

As a result, the estimated horizontal distance between the camera (Szombathely) and the red aurora is around 285.7 km, so the aurora was above Prague (50.06°N lat, 49.45°N mlat, see Figure A8).

The ionograms from Pruhonice station (next to Prague, Figure A8) seems to confirm this, see Figure A5 (source: https://giro.uml.edu/ionoweb/# (accessed on 8 July 2025). An example for regular optical observation of the aurora recorded at Fonyód (Hungary, Figure A7) at 20:43 UT can be seen on Figure A7a.

The exact calculation for 22:15 UT is as follows:

At this time, the red aurora was at zenith, directly above Szombathely (see Figure A2).

For the green aurora with Equation (A2), where y = 4.2, x = 9.3 and after the calculation

$$\frac{4.2}{9.3} \cdot 90^{\circ} = 40.65^{\circ},$$

the top of the green aurora was visible around 40.65° degrees. Using the above-mentioned h = 180 km upper height for the green aurora (after [51]), the d value is determined with Equation (A3) as follows:

$$\frac{180}{\tan 40.65^{\circ}} = 214.51 \text{ km}$$

As a result, the estimated horizontal distance between the camera (Szombathely) and the green aurora is around 214.51 km, so the aurora was above Brno (49.2°N lat, 49.39°N mlat, see Figure A8 for the location). The ionogram recorded at Pruhonice at this time shows an intense auroral Es, see Figure A6. An optical observation of the aurora recorded at Doba (Hungary, Figure A7b) at 22:15 UT also shows the presence of green aurora (Figure A9b).

The exact calculation for 22:50 UT is as follows:

The SAR was almost at zenith at this time over Manciano, Italy (42.6°N lat, see Figure A8).

For the red aurora with Equation (A2), where y = 2.7, x = 8.3 and after the calculation

$$\frac{2.7}{8.3} \cdot 90^{\circ} = 29.28^{\circ},$$

the top of the red aurora was visible around 29.28° degrees. Using the above-mentioned h = 300 km upper height for the red aurora (after [52]), the d value is determined with Equation (A3) as follows:

$$\frac{300}{\tan 29.28^{\circ}} = 535.7 \text{ km}$$

As a result, the red aurora's estimated horizontal distance from Manciano is ~536 km (see Figure A8 for the location), which is in the line of Szombathely, Hungary. The all-sky camera records from Szombathely at this time strengthened this computation.

The closeness of the red aurora is confirmed by the image taken at Pesaro, Italy (Figure A7c, for the source, see Acknowledgement).

Figure A8 is a summary map, illustrating the locations of the used Digisondes, the all-sky cameras and optical observations for easier interpretation.

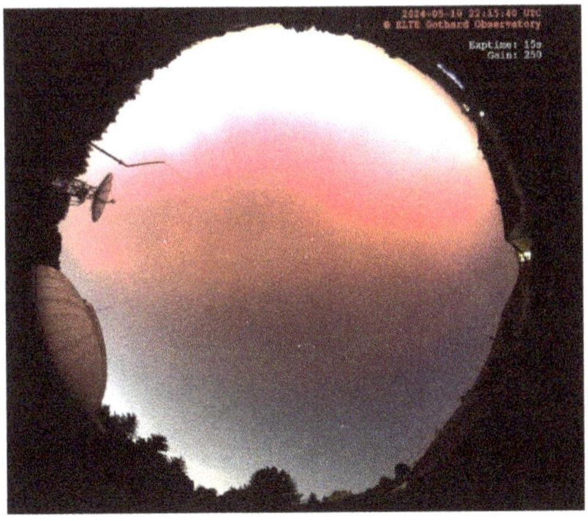

Figure A2. The all-sky camera image used for the estimations at 22:15 UT (10 May 2024), Szombathely, Hungary.

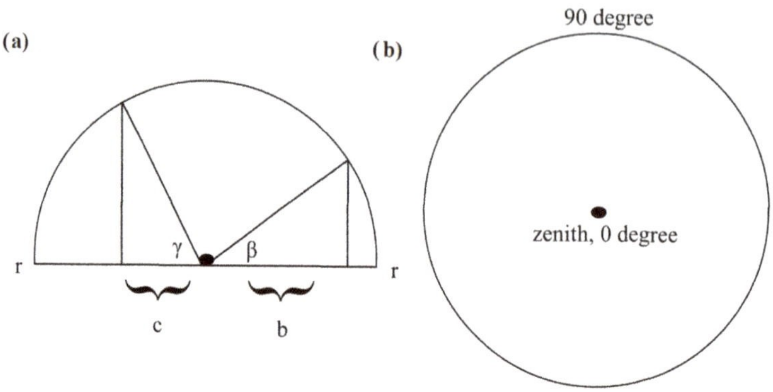

Figure A3. Configuration and interpretation of the all-sky camera. Panel (**a**) shows the configuration of the all-sky images, how the c and b parameters can be determined from them, and panel (**b**) shows the zenith and the 90 degree on such an image.

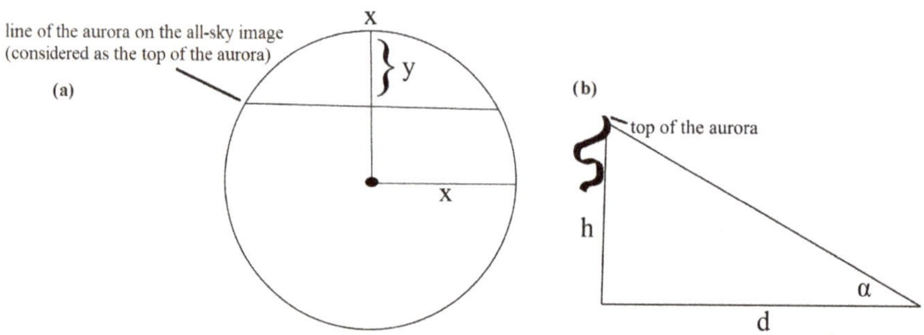

Figure A4. Determination of the parameters for geographic location determination from all-sky camera images. From panel (**a**) the x and y (the proportion of the aurora on the captured image) can be measured and quantified. Panel (**b**) shows the angle α (angle between the location of the all-sky camera and the top of the aurora) which is used for the calculation, see the details in the text.

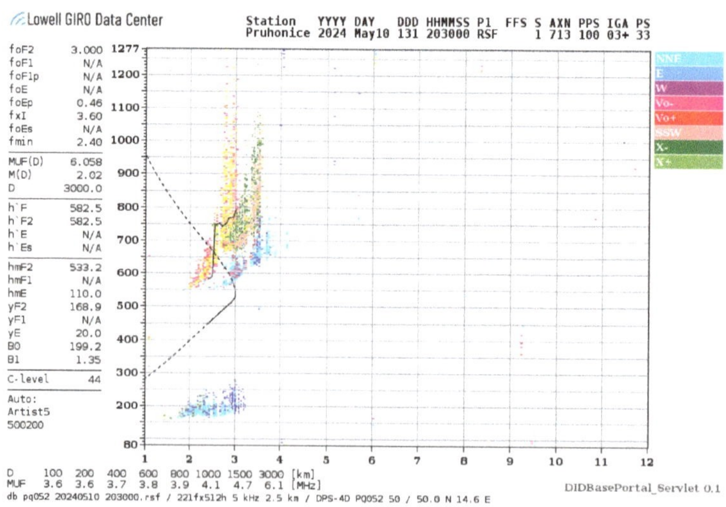

Figure A5. Ionogram from Pruhonice at 20:30 UT (10 May 2024).

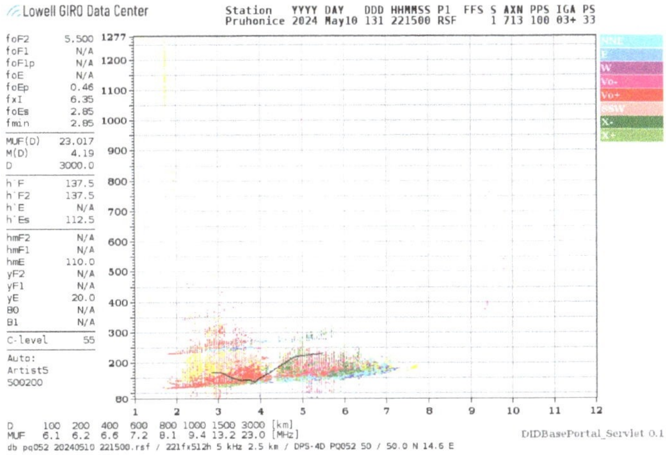

Figure A6. Ionogram from Pruhonice at 22:15 UT (10 May 2024).

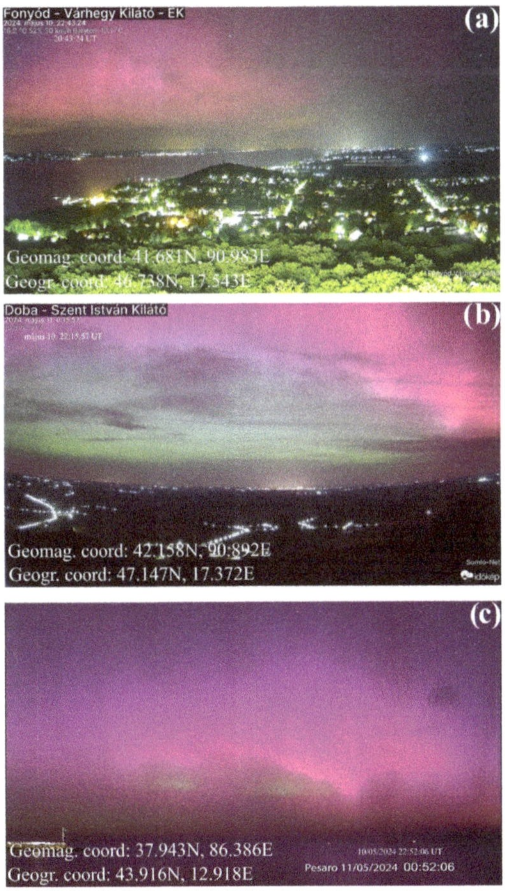

Figure A7. Optical aurora borealis observations, (**a**) from Fonyód, Hungary at 20:43 UT (10 May 2024), (**b**) from Doba, Hungary at 22:15 UT (10 May 2024), (**c**) from Pesaro, Italy at 22:52 UT (10 May 2024).

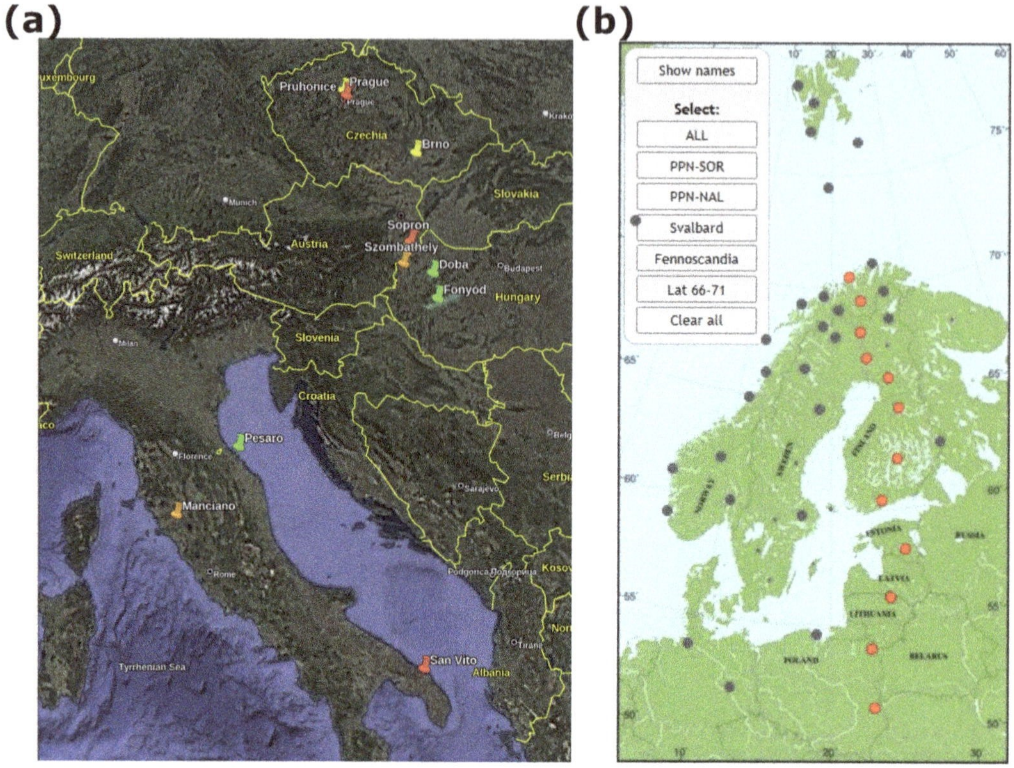

Figure A8. (**a**) Locations of the Digisondes, all-sky cameras and optical aurora observations. Red for the Digisondes, orange for the all-sky cameras, green for regular optical images and yellow for the reference points of the calculated position of the auroras. (**b**) The selected PPN-SOR meridional station chain used for the IU, IL indices for the studied interval.

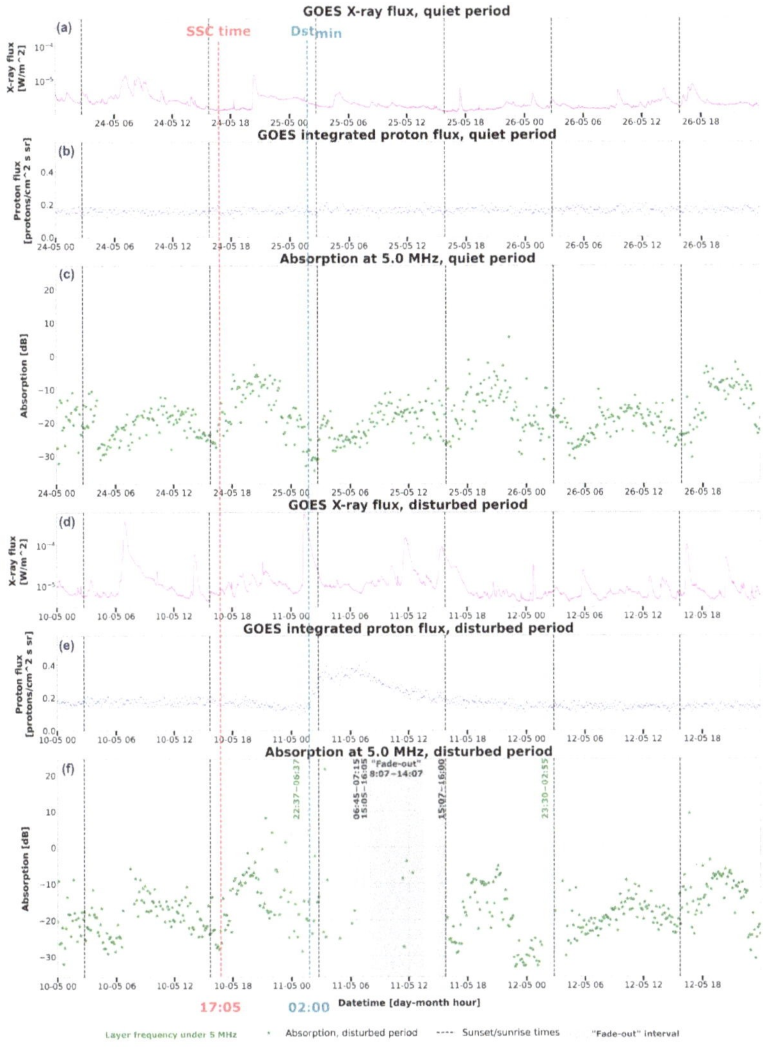

Figure A9. The amplitude method for San Vito at 5 MHz. The ionospheric absorption investigation with the amplitude method for Sopron. (**a**,**b**,**d**,**e**) portray the GOES satellite X-ray flux and integrated proton flux data for the quiet and the disturbed period, respectively. (**c**) shows the calculated ionospheric absorption data at 5 MHz for the quiet period and (**f**) the calculated ionospheric absorption for the disturbed period at 5 MHz. The green bars mark the intervals when the layer frequency went under 5 MHz; the grey bars mark the intervals of the "fade-out periods".

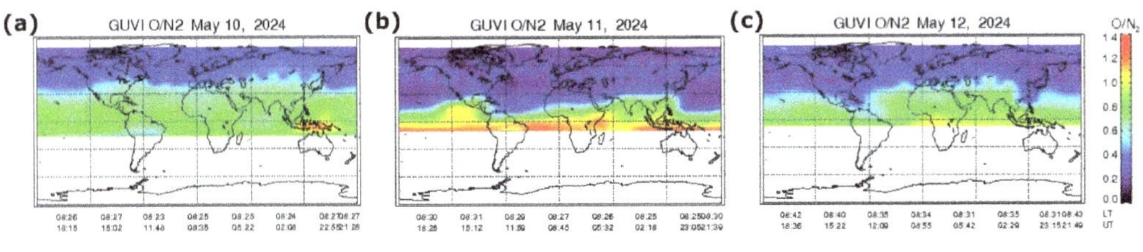

Figure A10. The GUVI measured O/N_2 ratio for the 2024 May storm period.

References

1. Spogli, L.; Alberti, T.; Bagiacchi, P.; Cafarella, L.; Cesaroni, C.; Cianchini, G.; Coco, I.; Di Mauro, D.; Ghidoni, R.; Giannattasio, F.; et al. The effects of the May 2024 Mother's Day superstorm over the Mediterranean sector: From data to public communication. *ESS Open Arch.* **2024**. [CrossRef]
2. Das, S.K.; Stolle, C.; Yamazaki, Y.; Rodríguez-Zuluaga, J.; Wan, X.; Kervalishvili, G.; Rauberg, J.; Zhong, J.; Perwitasari, S. On the F-region ionospheric plasma density distribution and irregularities response during the May-2024 geomagnetic storm observed by LEO satellites. *Geophys. Res. Lett.* **2025**, *52*. [CrossRef]
3. Aa, E.; Chen, Y.; Luo, B. Dynamic Expansion and Merging of the Equatorial Ionization Anomaly During the 10–11 May 2024 Super Geomagnetic Storm. *Remote. Sens.* **2024**, *16*, 4290. [CrossRef]
4. Aa, E.; Zhang, S.; Lei, J.; Huang, F.; Erickson, P.J.; Coster, A.J.; Luo, B.; Kerr, R.B. Significant Midlatitude Plasma Density Peaks and Dual-Hemisphere SED During the 10–11 May 2024 Super Geomagnetic Storm. *J. Geophys. Res. Space Phys.* **2024**, *129*, 3360. [CrossRef]
5. Bojilova, R.; Mukhtarov, P.; Pancheva, D. Global Ionospheric Response During Extreme Geomagnetic Storm in May 2024. *Remote. Sens.* **2024**, *16*, 4046. [CrossRef]
6. Pierrard, V.; Winant, A.; Botek, E.; de Bonhome, M.P. The Mother's Day Solar Storm of 11 May 2024 and Its Effect on Earth's Radiation Belts. *Universe* **2024**, *10*, 391. [CrossRef]
7. Carmo, C.S.; Dai, L.; Wrasse, C.M.; Barros, D.; Takahashi, H.; Figueiredo, C.A.O.B.; Wang, C.; Li, H.; Liu, Z. Ionospheric Response to the Extreme 2024 Mother's Day Geomagnetic Storm Over the Latin American Sector. *Space Weather.* **2024**, *22*, 4054. [CrossRef]
8. Gonzalez-Esparza, J.A.; Sanchez-Garcia, E.; Sergeeva, M.; Corona-Romero, P.; Gonzalez-Mendez, L.X.; Valdes-Galicia, J.F.; Aguilar-Rodriguez, E.; Rodriguez-Martinez, M.; Ramirez-Pacheco, C.; Castellanos, C.I.; et al. The Mother's Day Geomagnetic Storm on 10 May 2024: Aurora Observations and Low Latitude Space Weather Effects in Mexico. *Space Weather.* **2024**, *22*, 4111. [CrossRef]
9. Lazzús, J.; Salfate, I. Report on the effects of the May 2024 Mother's day geomagnetic storm observed from Chile. *J. Atmos. Sol.-Terr. Phys.* **2024**, *261*, 106304. [CrossRef]
10. Paul, K.S.; Moses, M.; Haralambous, H.; Oikonomou, C. Effects of the Mother's Day Superstorm (10–11 May 2024) over the Global Ionosphere. *Remote. Sens.* **2025**, *17*, 859. [CrossRef]
11. Liu, X.; Xu, J.; Yue, J.; Wang, W.; Moro, J. Mesosphere and Lower Thermosphere Temperature Responses to the May 2024 Mother's Day Storm. *Geophys. Res. Lett.* **2025**, *52*, 2179. [CrossRef]
12. Wang, X.; Aa, E.; Chen, Y.; Zhang, J.; Zhu, Y.; Cai, L.; Lu, X.; Luo, B.; Liu, S.; Li, M.; et al. Midlatitude Neutral Wind Response During the Mother's Day Super-Intense Geomagnetic Storm in 2024 Using Observations from the Chinese Meridian Project. *J. Geophys. Res. Space Phys.* **2025**, *130*, 3574. [CrossRef]
13. Singh, R.; Scipión, D.E.; Kuyeng, K.; Condor, P.; De La Jara, C.; Velasquez, J.P.; Flores, R.; Ivan, E.; Souza, J.R.; Migliozzi, M. Ionospheric Disturbances Observed Over the Peruvian Sector During the Mother's Day Storm (G5-Level) on 10–12 May 2024. *J. Geophys. Res. Space Phys.* **2024**, *129*, e2024JA033003. [CrossRef]
14. Karan, D.K.; Martinis, C.R.; Daniell, R.E.; Eastes, R.W.; Wang, W.; McClintock, W.E.; Michell, R.G.; England, S. GOLD Observations of the Merging of the Southern Crest of the Equatorial Ionization Anomaly and Aurora During the 10 and 11 May 2024 Mother's Day Super Geomagnetic Storm. *Geophys. Res. Lett.* **2024**, *51*, 632. [CrossRef]
15. Pierrard, V.; Verhulst, T.G.W.; Chevalier, J.-M.; Bergeot, N.; Winant, A. Effects of the Geomagnetic Superstorms of 10–11 May 2024 and 7–11 October 2024 on the Ionosphere and Plasmasphere. *Atmosphere* **2025**, *16*, 299. [CrossRef]
16. Buonsanto, M. Ionospheric Storms—A Review. *Space Sci. Rev.* **1999**, *88*, 563–601. [CrossRef]
17. Prölss, G.W. *Ionospheric F-region storms. Handbook of Atmospheric Electrodynamics*; CRC Press: Boca Raton, FL, USA, 1995; Volume 2, Chapter 8; pp. 195–248. Available online: https://cir.nii.ac.jp/crid/1570291225295033472 (accessed on 2 November 2022).
18. Balan, N.; Shiokawa, K.; Otsuka, Y.; Kikuchi, T.; Lekshmi, D.V.; Kawamura, S.; Yamamoto, M.; Bailey, G.J. A physical mechanism of positive ionospheric storms at low latitudes and midlatitudes. *J. Geophys. Res.* **2010**, *115*, A02304. [CrossRef]
19. Balan, N.; Liu, L.; Le, H. A brief review of equatorial ionization anomaly and ionospheric irregularities. *Earth Planet. Phys.* **2018**, *2*, 257–275. [CrossRef]
20. Berényi, K.A.; Heilig, B.; Urbář, J.; Kouba, D.; Kis, Á.; Barta, V. Comprehensive analysis of the ionospheric response to the largest geomagnetic storms from solar cycle 24 over Europe. *Front. Astron. Space Sci.* **2023**, *10*, 1092850. [CrossRef]
21. Danilov, A. Ionospheric F-region response to geomagnetic disturbances. *Adv. Space Res.* **2013**, *52*, 343–366. [CrossRef]
22. Prölss, G.W. Ionospheric Storms at Mid-Latitude: A Short Review. *Geophys. Monogr. Ser.* **2008**, *181*, 9–24. [CrossRef]
23. Tsurutani, B.T.; Verkhoglyadova, O.P.; Mannucci, A.J.; Saito, A.; Araki, T.; Yumoto, K.; Tsuda, T.; Abdu, M.A.; Sobral, J.H.A.; Gonzalez, W.D.; et al. Prompt penetration electric fields (PPEFs) and their ionospheric effects during the great magnetic storm of 30–31 October 2003. *J. Geophys. Res. Space Phys.* **2008**, *113*, A05311. [CrossRef]

24. Heilig, B.; Stolle, C.; Kervalishvili, G.; Rauberg, J.; Miyoshi, Y.; Tsuchiya, F.; Kumamoto, A.; Kasahara, Y.; Shoji, M.; Nakamura, S.; et al. Relation of the Plasmapause to the Midlatitude Ionospheric Trough, the Sub-Auroral Temperature Enhancement and the Distribution of Small-Scale Field Aligned Currents as Observed in the Magnetosphere by THEMIS, RBSP, and Arase, and in the Topside Ionosphere by Swarm. *J. Geophys. Res. Space Phys.* **2022**, *127*, e2021JA029646. [CrossRef]
25. Heilig, B.; Lühr, H. New plasmapause model derived from CHAMP field-aligned current signatures. *Ann. Geophys.* **2013**, *31*, 529–539. [CrossRef]
26. Heilig, B.; Lühr, H. Quantifying the relationship between the plasmapause and the inner boundary of small-scale field-aligned currents, as deduced from Swarm observations. *Ann. Geophys.* **2018**, *36*, 595–607. [CrossRef]
27. Polekh, N.; Chernigovskaya, M.; Yakovleva, O. On the formation of the F1 layer during sudden stratospheric warming events. *Solnechno-Zemnaya Fiz.* **2019**, *5*, 117–127. [CrossRef]
28. Deminov, M.G.; Romanova, E.B.; Tashchilin, A.V. Origination of G conditions in the ionospheric F region depending on solar and geomagnetic activity. *Geomagn. Aeron.* **2011**, *51*, 669–675. [CrossRef]
29. Mošna, Z.; Barta, V.; Berényi, K.A.; Mielich, J.; Verhulst, T.; Kouba, D.; Urbář, J.; Chum, J.; Knížová, P.K.; Marew, H.; et al. The March and April 2023 ionospheric storms over Europe. *Front. Astron. Space Sci.* **2024**, *11*, 1462160. [CrossRef]
30. Lobzin, V.V.; Pavlov, A.V. G condition in the F2 region peak electron density: A statistical study. *Ann. Geophys.* **2002**, *20*, 523–537. [CrossRef]
31. Moro, J.; Xu, J.; Denardini, C.M.; Resende, L.C.A.; Da Silva, L.A.; Chen, S.S.; Carrasco, A.J.; Liu, Z.; Wang, C.; Schuch, N.J. Different Sporadic-E (Es) Layer Types Development During the August 2018 Geomagnetic Storm: Evidence of Auroral Type (Es_a) Over the SAMA Region. *J. Geophys. Res. Space Phys.* **2022**, *127*, e2021JA029701. [CrossRef]
32. Reinisch, B.W.; Galkin, I.A. Global ionospheric radio observatory (GIRO). *Earth Planets Space* **2011**, *63*, 377–381. [CrossRef]
33. Reinisch, B.; Huang, X.; Galkin, I.; Paznukhov, V.; Kozlov, A. Recent advances in real-time analysis of ionograms and ionospheric drift measurements with digisondes. *J. Atmos. Sol.-Terr. Phys.* **2005**, *67*, 1054–1062. [CrossRef]
34. Bór, J.; Sátori, G.; Barta, V.; Szabóné-André, K.; Szendrői, J.; Wesztergom, V.; Bozóki, T.; Buzás, A.; Koronczay, D. Measurements of atmospheric electricity in the Széchenyi István Geophysical Observatory, Hungary. *Hist. Geo- Space Sci.* **2020**, *11*, 53–70. [CrossRef]
35. Bilitza, D.; Pezzopane, M.; Truhlik, V.; Altadill, D.; Reinisch, B.W.; Pignalberi, A. The International Reference Ionosphere Model: A Review and Description of an Ionospheric Benchmark. *Rev. Geophys.* **2022**, *60*, e2022RG000792. [CrossRef]
36. Bilitza, D.; Rawer, K. New options for IRI electron density in the middle ionosphere. *Adv. Space Res.* **1990**, *10*, 7–16. [CrossRef]
37. Verhulst, T.G.W.; Stankov, S.M.; Verhulst, T.G.W.; Stankov, S.M. The Changing Shape of the Ionosphere During a Solar Eclipse. *URSI Radio Sci. Lett.* **2023**, *4*, 68. [CrossRef]
38. Buzás, A.; Kouba, D.; Mielich, J.; Burešová, D.; Mošna, Z.; Knížová, P.K.; Barta, V. Investigating the effect of large solar flares on the ionosphere based on novel Digisonde data comparing three different methods. *Front. Astron. Space Sci.* **2023**, *10*, 1201625. [CrossRef]
39. Kallio, E.I.; Pulkkinen, T.I.; Koskinen, H.E.J.; Viljanen, A.; Slavin, J.A.; Ogilvie, K. Loading-unloading processes in the nightside ionosphere. *Geophys. Res. Lett.* **2000**, *27*, 1627–1630. [CrossRef]
40. Kauristie, K.; Sergeev, V.A.; Pulkkinen, T.I.; Pellinen, R.J.; Angelopoulos, V.; Baumjohann, W. Study on the ionospheric signatures of the plasma sheet bubbles. In *International Conference on Substorms, Proceedings of the 3rd International Conference, Versailles, France, 12–17 May 1996*; ESA SP-389; Rolfe, E.J., Kaldeich, B., Eds.; European Space Agency: Paris, France, 1996; p. 93. Available online: https://adsabs.harvard.edu/full/1996ESASP.389...93K (accessed on 13 May 2025).
41. Viljanen, A.; Hakkinen, L. *IMAGE Magnetometer Network, in Satellite-Ground Based Coordination Sourcebook*; ESA Publications—European Space Agency: Noordwijk, The Netherlands, 1997.
42. Tanskanen, E.I.; Viljanen, A.; Pulkkinen, T.I.; Pirjola, R.; Häkkinen, L.; Pulkkinen, A.; Amm, O. At substorm onset, 40% of AL comes from underground. *J. Geophys. Res.* **2001**, *106*, 13119–13134. [CrossRef]
43. Friis-Christensen, E.; Lühr, H.; Hulot, G. Swarm: A constellation to study the Earth's magnetic field. *Earth Planets Space* **2006**, *58*, 351–358. [CrossRef]
44. Olsen, N.; Friis-Christensen, E.; Floberghagen, R.; Alken, P.; Beggan, C.D.; Chulliat, A.; Doornbos, E.; da Encarnação, J.T.; Hamilton, B.; Hulot, G.; et al. The Swarm Satellite Constellation Application and Research Facility (SCARF) and Swarm data products. *Earth Planets Space* **2013**, *65*, 1189–1200. [CrossRef]
45. Evans, J.S.; Correira, J.; Lumpe, J.D.; Eastes, R.W.; Gan, Q.; Laskar, F.I.; Aryal, S.; Wang, W.; Burns, A.G.; Beland, S.; et al. GOLD Observations of the Thermospheric Response to the 10–12 May 2024 Gannon Superstorm. *Geophys. Res. Lett.* **2024**, *51*, e2024GL110506. [CrossRef]
46. Hayakawa, H.; Ebihara, Y.; Mishev, A.; Koldobskiy, S.; Kusano, K.; Bechet, S.; Yashiro, S.; Iwai, K.; Shinbori, A.; Mursula, K.; et al. The Solar and Geomagnetic Storms in 2024 May: A Flash Data Report. *Astrophys. J.* **2025**, *979*, 49. [CrossRef]
47. Buresova, D.; Lastovicka, J.; Hejda, P.; Bochnicek, J. Ionospheric disturbances under low solar activity conditions. *Adv. Space Res.* **2014**, *54*, 185–196. [CrossRef]

48. Berényi, K.A.; Perrone, L.; Sabbagh, D.; Scotto, C.; Ippolito, A.; Kis, Á.; Barta, V. Ionosonde Measurement Comparison during an Interplanetary Coronal Mass Ejection (ICME)- and a Corotating Interaction Region (CIR)-Driven Geomagnetic Storm over Europe. *Universe* **2024**, *10*, 344. [CrossRef]
49. Szárnya, C.; Mošna, Z.; Igaz, A.; Kouba, D.; Verhulst, T.G.W.; Knížová, P.K.; Podolská, K.; Barta, V. Technical Possibilities and Limitations of the DPS-4D Type of Digisonde in Individual Meteor Detections. *Remote Sens.* **2024**, *16*, 2658. [CrossRef]
50. Reinisch, B.W. *Digisonde 4D Technical Manual (Version 1.0)*; Lowell Digisonde International: Lowell, MA, USA, 2009.
51. Whiter, D.K.; Partamies, N.; Gustavsson, B.; Kauristie, K. The altitude of green OI 557.7 nm and blue N_2^+ 427.8 nm aurora. *Ann. Geophys.* **2023**, *41*, 1–12. [CrossRef]
52. Jackel, B.J.; Creutzberg, F.; Donovan, E.F.; Cogger, L.L. *Triangulation of Auroral Red-Line Emission Heights*; Sodankylä Geophysical Observatory Publications: Sodankylä, Finland, 2003; Volume 92, pp. 97–100. Available online: https://aurora.phys.ucalgary.ca/public/for_emma/Gillies_agu/LitReview/jackel_triangulation_oulu_2003.pdf (accessed on 5 February 2025).
53. Piggott, W.R.; Rawer, K. U.R.S.I. Handbook of Ionogram Interpretation and Reduction. 1978. Available online: https://repository.library.noaa.gov/view/noaa/10404 (accessed on 25 October 2024).
54. Belehaki, A.; Tsagouri, I.; Altadill, D.; Blanch, E.; Borries, C.; Buresova, D.; Chum, J.; Galkin, I.; Juan, J.M.; Segarra, A.; et al. An overview of methodologies for real-time detection, characterisation and tracking of traveling ionospheric disturbances developed in the TechTIDE project. *J. Space Weather Space Clim.* **2020**, *10*, 42. [CrossRef]
55. Ventriglia, V.; Guerra, M.; Cesaroni, C.; Spogli, L.; Altadill, D.; Segarra, A.; Galkin, I.; Barta, V.; Verhulst, T.G.; de Paula, V.; et al. An explainable Machine Learning model for Large-Scale Travelling Ionospheric Disturbances forecasting. *J. Space Weather Space Clim.* **2025**, *15*. [CrossRef]
56. Barta, V.; Sátori, G.; Berényi, K.A.; Kis, Á.; Williams, E. Effects of solar flares on the ionosphere as shown by the dynamics of ionograms recorded in Europe and South Africa. *Ann. Geophys.* **2019**, *37*, 747–761. [CrossRef]
57. Tao, C.; Nishioka, M.; Saito, S.; Shiota, D.; Watanabe, K.; Nishizuka, N.; Tsugawa, T.; Ishii, M. Statistical analysis of short-wave fadeout for extreme space weather event estimation. *Earth Planets Space* **2020**, *72*, 173. [CrossRef]
58. Paul, K.S.; Haralambous, H.; Moses, M.; Oikonomou, C.; Potirakis, S.M.; Bergeot, N.; Chevalier, J.-M. Investigation of the Ionospheric Response on Mother's Day 2024 Geomagnetic Superstorm over the European Sector. *Atmosphere* **2025**, *16*, 180. [CrossRef]
59. Vichare, G.; Bagiya, M.S. Manifestations of Strong IMF-By on the Equatorial Ionospheric Electrodynamics During 10 May 2024 Geomagnetic Storm. *Geophys. Res. Lett.* **2024**, *51*, e2024GL112569. [CrossRef]
60. Kelley, M.C.; Makela, J.J.; de La Beaujardière, O.; Retterer, J. Convective Ionospheric Storms: A Review. *Rev. Geophys.* **2011**, *49*. [CrossRef]
61. Woodman, R.F. Spread F – an old equatorial aeronomy problem finally resolved? *Ann. Geophys.* **2009**, *27*, 1915–1934. [CrossRef]
62. Woodman, R.F.; La Hoz, C. Radar observations of *F* region equatorial irregularities. *J. Geophys. Res.* **1976**, *81*, 5447–5466. [CrossRef]
63. Huang, C.-S. Occurrence of Equatorial Plasma Bubbles during Intense Magnetic Storms. *Int. J. Geophys.* **2011**, *2011*, 401858. [CrossRef]
64. Sripathi, S.; Abdu, M.A.; Patra, A.K.; Ghodpage, R.N. Unusual Generation of Localized EPB in the Dawn Sector Triggered by a Moderate Geomagnetic Storm. *J. Geophys. Res. Space Phys.* **2018**, *123*, 9697–9710. [CrossRef]
65. Ram, S.T.; Ajith, K.K.; Yamamoto, M.; Otsuka, Y.; Yokoyama, T.; Niranjan, K.; Gurubaran, S. Fresh and evolutionary-type field-aligned irregularities generated near sunrise terminator due to overshielding electric fields. *J. Geophys. Res. Space Phys.* **2015**, *120*, 5922–5930. [CrossRef]
66. Wu, K.; Xu, J.; Yue, X.; Xiong, C.; Wang, W.; Yuan, W.; Wang, C.; Zhu, Y.; Luo, J. Equatorial plasma bubbles developing around sunrise observed by an all-sky imager and global navigation satellite system network during storm time. *Ann. Geophys.* **2020**, *38*, 163–177. [CrossRef]
67. Carmo, C.S.; Pi, X.; Denardini, C.M.; Figueiredo, C.A.O.B.; Verkhoglyadova, O.P.; Picanço, G.A.S. Equatorial Plasma Bubbles Observed at Dawn and After Sunrise Over South America During the 2015 St. Patrick's Day Storm. *J. Geophys. Res. Space Phys.* **2022**, *127*, 9934. [CrossRef]
68. Karan, D.K.; Martinis, C.R.; Eastes, R.W.; Daniell, R.E.; McClintock, W.E.; Huang, C. GOLD Observations of Equatorial Plasma Bubbles Reaching Mid-Latitudes During the 23 April 2023 Geomagnetic Storm. *Space Weather.* **2024**, *22*, e2023SW003847. [CrossRef]
69. Jones, D.G.; Walker, I.K.; Kersley, L. Structure of the poleward wall of the trough and the inclination of the geomagnetic field above the EISCAT radar. *Ann. Geophys.* **1997**, *15*, 740–746. [CrossRef]
70. Karpachev, A. Longitudinal variations in the position and magnitude of the poleward wall of the ionospheric trough in the southern hemisphere. *Adv. Space Res.* **2023**, *71*, 4374–4381. [CrossRef]
71. Mannucci, A.J.; Tsurutani, B.T.; Iijima, B.A.; Komjathy, A.; Saito, A.; Gonzalez, W.D.; Guarnieri, F.L.; Kozyra, J.U.; Skoug, R. Dayside global ionospheric response to the major interplanetary events of October 29–30, 2003 "Halloween Storms". *Geophys. Res. Lett.* **2005**, *32*, L12S02. [CrossRef]

72. Martinis, C.; Baumgardner, J.; Mendillo, M.; Wroten, J.; Coster, A.; Paxton, L. The night when the auroral and equatorial ionospheres converged. *J. Geophys. Res. Space Phys.* **2014**, *120*, 8085–8095. [CrossRef]
73. Akasofu, S.-I. Electrodynamics of the magnetosphere: Geomagnetic storms. *Space Sci. Rev.* **1966**, *6*, 21–143. [CrossRef]
74. Astafyeva, E.; Maletckii, B.; Förster, M.; Ouar, I.D.; Huba, J.D.; Hairston, M.R.; Coley, W.R. Electrodynamic and Ionospheric Puzzles of the 10–11 May 2024 Geomagnetic Superstorm. *J. Geophys. Res. Space Phys.* **2025**, *130*, e2024JA033284. [CrossRef]
75. Panasenko, S.V.; Aksonova, K.D.; Burešová, D.; Bogomaz, O.V.; Zhivolup, T.G.; Koloskov, O.V. Large-scale traveling ionospheric disturbances over central and eastern Europe during moderate magnetic storm period on 22–24 September 2020. *Adv. Space Res.* **2023**, *72*, 4364–4378. [CrossRef]
76. Horvath, I.; Lovell, B.C. Traveling ionospheric disturbances and their relations to storm-enhanced density features and plasma density irregularities in the local evening and nighttime hours of the Halloween superstorms of 29–31 October 2003. *J. Geophys. Res.* **2010**, *115*, A09327. [CrossRef]
77. Berényi, K.; Barta, V.; Kis, Á. Midlatitude ionospheric F2-layer response to eruptive solar events-caused geomagnetic disturbances over Hungary during the maximum of the solar cycle 24: A case study. *Adv. Space Res.* **2018**, *61*, 1230–1243. [CrossRef]
78. Forbes, J.M. Dynamics of the Thermosphere. *J. Meteorol. Soc. Jpn. Ser. II* **2007**, *85B*, 193–213. [CrossRef]
79. Lu, G.; Zakharenkova, I.; Cherniak, I.; Dang, T. Large-Scale Ionospheric Disturbances During the 17 March 2015 Storm: A Model-Data Comparative Study. *J. Geophys. Res. Space Phys.* **2020**, *125*, e2019JA027726. [CrossRef]
80. Qian, L.; Burns, A.G.; Emery, B.A.; Foster, B.; Lu, G.; Maute, A.; Richmond, A.D.; Roble, R.G.; Solomon, S.C.; Wang, W. The NCAR TIE-GCM. In *Modeling the Ionosphere–Thermosphere System*; Huba, J., Schunk, R., Khazanov, G., Eds.; AGU Publications: Malden, MA, USA, 2014. [CrossRef]
81. Huang, C.; Xu, J.-Y.; Zhang, X.-X.; Liu, D.-D.; Yuan, W.; Jiang, G.-Y. Mid-latitude thermospheric wind changes during the St. Patrick's Day storm of 2015 observed by two Fabry-Perot interferometers in China. *Adv. Space Res.* **2018**, *61*, 1873–1879. [CrossRef]
82. Drob, D.P.; Emmert, J.T.; Crowley, G.; Picone, J.M.; Shepherd, G.G.; Skinner, W.; Hays, P.; Niciejewski, R.J.; Larsen, M.; She, C.Y.; et al. An empirical model of the Earth's horizontal wind fields: HWM07. *J. Geophys. Res.* **2008**, *113*. [CrossRef]

remote sensing

MDPI

Article

Radiative Transfer Simulation in the Near-Space Region for a Point Source at High Temperature Based on a Monte Carlo Method

Mingyang Liu [1], Bingqiang Sun [1,2,*] and Rui Lu [1]

[1] Department of Atmospheric and Oceanic Sciences, Fudan University, Shanghai 200438, China; 22213020014@m.fudan.edu.cn (M.L.); 23113020021@m.fudan.edu.cn (R.L.)
[2] CMA-FDU Joint Laboratory of Marine Meteorology, Shanghai 200438, China
* Correspondence: bingqsun@fudan.edu.cn

Abstract: When vehicles fly at hypersonic speeds, a high-temperature flow field forms and emits intense radiation across some spectral bands. This strong flow field can be seen as a point source and is transmitted to high altitudes through the radiative transfer process. In this study, the corresponding atmospheric profiles were introduced, the databases in high-temperature conditions were established, and the corresponding gas absorption characteristics were further calculated using the line-by-line method. Correspondingly, the radiative transfer processes were calculated by a Monte Carlo model with surface and point sources. The Monte Carlo model is rigorously validated against the VLIDORT model through detailed comparisons. For radiative simulations in the near-space environment, both high-temperature conditions and 3D scenarios are considered. The simulations reveal that the use of high-temperature databases could introduce an error of approximately 3%, and the radiation field exhibits stronger inhomogeneity in the infrared band.

Keywords: Monte Carlo radiative transfer method; upper-atmospheric radiative transfer process; high-temperature gas absorption database

1. Introduction

The near-space region, at altitudes of 20–100 km, is a significant region and the focus of many studies. Hypersonic vehicles have special capabilities in satellite transportation [1]. A vehicle flying at hypersonic speeds generates intense friction with the surrounding atmosphere, with a high-temperature flow field forming around it. Gases in this field undergo chemical reactions and emit intense radiation across the ultraviolet (UV) and infrared (IR) spectral regions [2,3]. The numerical simulation of bow-shock-layer emission is difficult, as multi-component gases within the high-temperature flow field undergo complex physical and chemical processes including internal energy-level excitation, dissociation, ionization, recombination, and photochemical reactions, accompanied by various radiative transition processes such as vibration, rotation, and electron energy-level transitions. As the relaxation time of the various processes differs, gases are in a thermochemical non-equilibrium state, which may increase the radiation intensity by 2–15 times. Theoretical radiation simulation results, with and without consideration of this non-equilibrium effect, may differ by more than 1–2 orders of magnitude. Special models have been designed to solve high-temperature radiation calculations. Ozawa et al. [4] and Nam et al. [5] have presented line-by-line (LBL) models for the calculation of molecular emission in the IR

https://doi.org/10.3390/rs17162769

region, and Kumar et al. [6] introduced a k-distribution spectral model to simulate the non-gray radiation of the bow-shock layer.

With the development of hypersonic aerospace targets, there are thus increasing requirements for accuracy and efficiency of target detection. [7]. Earlier detections were IR systems based on the relatively high CO_2, CO, and H_2O contents of exhaust plumes, whose spectra are focused mainly on peaks at 3703 and 2325 cm^{-1}. However, due to the low-level IR signal characteristics of targets despite continuous improvements in technology, and the complexity of background radiation and false-alarm sources within the field of view of the detector, other detection methods are required to assist IR systems. Ultraviolet detection technology has demonstrated advantages as an alternative means of target detection [2,3,8,9].

After the source radiation field of the bow-shock layer is excited, the radiation still goes through the atmosphere so that the signal can be received by sensors at high altitudes. This requires the consideration of the atmospheric radiative transfer process. Various methods have been developed to solve the radiative transfer equation, including discrete-ordinate methods such as DISORT [10] and VLIDORT [11], adding–doubling methods [12], diffusion approximation methods [13,14], and Monte Carlo methods [15,16]. The performances of different methods have been compared and validated by Saunders et al. [17], Kotchenova et al. [18], Kokhanovsky et al. [19], and Randles et al. [20], with intercomparisons including a range of atmospheric conditions for clouds, aerosols, and solar and observation angles. Results indicate that model bias ranges from 10% to 20%, which is deemed acceptable.

Regular observations such as lidar ceilometers and radiosondes provide atmospheric vertical information [21–23]. Radiosonde observations are one of the most basic and important data sources used in studying atmospheric profiles, providing atmospheric information at 0–30 km altitude [24,25]. Sounding data are the basis of most reanalysis and model data, so reanalysis data and model outputs generally reach altitudes of only 30 km. Conventional atmospheric observations cannot meet the needs of radiation transfer simulation. In 1976, Flanagan et al. proposed the US Standard atmosphere [26,27] (hereafter denoted as USSA1976), which provides atmospheric vertical information for up to 120 km altitude including temperature, altitude, pressure, and the concentrations of thirty gases including the seven main atmospheric gases (H_2O, CO_2, O_3, N_2O, CO, CH_4, and O_2). While the USSA1976 provides information on atmospheric components below 120 km, it cannot meet the requirements for higher altitudes. Therefore, the NRLMSIS empirical model is used to obtain information about atmospheric vertical profiles for higher altitudes [28]. The NRLMSIS2.0 model uses parameters such as location, time information, and solar and geomagnetic activity to predict temperature, mass density, and species densities. It simulates parameters from numerical weather prediction reanalysis, satellite limb observations, solar occultation, lidar, and passive remote sensing [28,29]. The NRLMSIS model has long been a popular choice of atmospheric vertical profile model in the research and operations communities, because the comparisons between predictions and satellite observations show the overall stability through technological iteration and addition of new data [30–33]. With the aid of radiative transfer models, source radiation emission of the bow-shock layer can be considered as a point radiation source at specific heights. However, near-space at altitudes of 20–100 km lies at the mesosphere, thermosphere, and ionosphere, and the high temperatures of these layers mean that only parts of traditional absorption spectroscopic databases can be used. In addition, conventional radiative transfer models usually use the ground source and the solar source as light sources, and are unable to calculate the radiative transfer process under the condition of a point source.

In this study, a forward simulation problem was trying to be solved, which served as the space-borne observation of radiation from high-speed vehicles in the near-space region. To simulate the forward radiative transfer process, two things must be provided: the absorption database of molecules under high temperature, since the near-space region lies at the mesosphere, thermosphere, and ionosphere, and a three-dimensional simulation method, since the radiation source can be treated as a point source. Moreover, the thermal radiation from high-speed vehicles falls into the UV and IR regions. Therefore, this study established the high-temperature absorption databases in the near-space region. A Monte Carlo method with a point source at any location was implemented to simulate the radiative transfer process in the near-space region in the UV and IR bands. Section 2 describes methods and databases used in this paper. Section 3 presents the validation and simulation results of the Monte Carlo model. Section 4 analyzes the influence of the Monte Carlo model's internal parameter and extends the simulation to the 3D scenarios. Finally, Section 5 concludes this study.

2. Materials and Methods

2.1. NRLMSIS, an Atmospheric Vertical-Profile Empirical Model

To obtain a whole-atmosphere vertical profile for the calculations of gas absorption and radiation transfer processes, especially for upward transmission at an altitude where a signal-receiving device, such as a satellite, is located, it is necessary to combine the USSA1976 and NRLMSIS2.0 models, with the latter providing profile information at altitudes beyond the 120 km limit of the former. Height intervals in the USSA1976 profile are not uniform so it is interpolated to different heights and the NRLMSIS model also requires certain input parameters, as shown in Table 1.

Table 1. Height intervals of the USSA1976 profile and input parameters of the NRLMSIS model. The F10.7 index refers to the solar radio flux at a wavelength of 10.7 cm. The Ap index is a measure of geomagnetic activity.

The USSA1976 Profile Height Intervals	
0~25 km	1 km
25~50 km	2.5 km
50~120 km	5 km
The NRLMSIS model input parameters	
Date	UTC 1 June 2024 12:00
Latitude	45°
Longitude	0°
Altitudes	120, 140, 160, 180, 200, 300, 400, 500, 1000
Daily F10.7 of the previous day of the given date	150
F10.7 running 81-day average centered on the given date	150
Daily Ap index	7
3 h Ap index for current time	7
3 h Ap index for 3 h before current time	7
3 h Ap index for 6 h before current time	7
3 h Ap index for 9 h before current time	7
Average of eight 3 h Ap indices from 12 to 33 h prior to current time	7

Table 1. *Cont.*

The USSA1976 Profile Height Intervals	
Average of eight 3 h Ap indices from 36 to 57 h prior to current time	7

Figure 1 shows the two profiles and indicates that number densities in lower layers are far greater than those in higher layers. The water vapor and other trace gas concentrations decay to <$10^{18}/m^3$ (<10 ppm) at ~20 km, while the CO_2 concentration drops to <$10^{18}/m^3$ at 50 km, decaying rapidly from 300 ppm to the order of several ppm above 90 km. Maximum gas number densities do not exceed $10^{18}/m^3$ above 120 km, while at 1000 km only H, He, O, and N_2 maintain an order of magnitude of ~$10^{10}/m^3$. In the troposphere and stratosphere below 120 km, the atmospheric temperature profile varies within a narrow range of 200–400 K, but above 120 km, in the mesosphere and thermosphere, it rises rapidly to above 1000 K at 200 km, maintaining this level thereafter. A high-temperature database is thus necessary for radiative transfer calculations in the upper atmosphere.

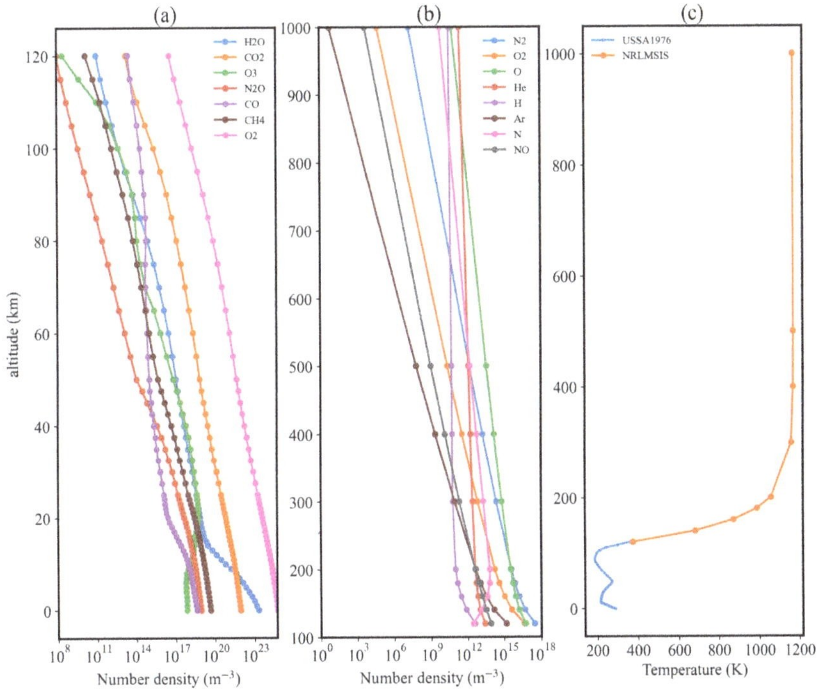

Figure 1. The 50-layer USSA1976 profile and the generated NRLMSIS vertical profile. (**a**) Altitude-density plot for the seven main gases included in the USSA1976. (**b**) The same plot for eight gases in the NRLMSIS model. (**c**) Altitude–temperature plot for the temperature profile compiled using data from spliced USSA1976 and NRLMSIS models. At the splicing point of 120 km, the temperature of the USSA1976 model is 360.0 K and that of the NRLMSIS model is 373.72 K.

After using these input parameters to generate vertical profiles, the upper and lower profiles provide the number densities of different gases. In the calculation process, for gases that exist in the lower layer but not in the upper layer, it is assumed that the concentration of such gases rapidly decays to zero above 120 km, and their absorption of radiation is no longer considered. Regarding the calculation of IR band radiation, the main absorbing gases are H_2O and CO_2, which occur mainly below the stratosphere [34–36], so their impact is not significant.

2.2. Molecular Absorption Databases

The molecular absorption coefficient can be derived by the LBL integration [37] or the correlated k-distribution method [38,39] using spectroscopic data from different databases. As the near-space region covers a wide range of temperature, the choice of database is important. The spectroscopic databases, HITRAN [40] and its high-temperature version HITEMP [41], GEISA [42], and MPI-Mainz [43], were used to establish the absorption coefficients for the simulation of the near-space region. The light-source radiation field in this study covers two wavelength bands (25,000–55,000 cm^{-1} and 1800–4000 cm^{-1}), where the radiative transfer calculations focus mainly on these bands in the UV and IR regions, respectively. In the IR region, line data were mainly used for the line-by-line calculations. The line data variables are converted to the standard HITRAN format and include the spectral line intensity at 296 K, the Einstein-A coefficient of a transition, the air-broadened and self-broadened half width at half maximum, the lower-state energy of the transition, and the lower and upper state statistical weights. In the UV region, cross-section data were mainly used. The types, concentrations, and characteristics of absorbing gases vary greatly, so the calculations must be conducted separately with different databases. After detailed organization of databases and consideration of which gases have relatively high contents in the atmospheric profile, gases were selected for consideration in the UV and IR bands.

Table 2 lists the characteristics of the studied databases. Based on these characteristics and information concerning the atmospheric profile, different databases were selected for calculations of gas optical depths. In the IR band, water vapor and CO_2 are the main absorbing gases, but their concentration profiles reach only to 120 km where the temperature is ~360 K. Considering that the temperature rises rapidly to ~1000 K at 120–200 km, absorption coefficients of water vapor below 1000 K were calculated using the conventional database HITRAN and the high-temperature database HITEMP.

Table 2. Comparison of gases and wavelength ranges covered by different databases. The line data indicate parameters such as line intensity and the Einstein-A coefficient of a transition.

Database		Gase	Wavelength Range (cm^{-1})	Data Type
HITRAN	Ultraviolet (UV)	BrO, N_2O, NO_2, O_3, OClO, SO_2	25,927.0~54,999.9	Mainly cross-section
	Infrared (IR)	H_2O, CO_2, O_3, N_2O, CO, CH_4, O_2, NO, NO_2, NH_3,	0.001~>10,000.0	Line data
HITEMP	UV	H_2O, NO, OH	25,000.0~43,409.0	Line data
	IR	H_2O, CO_2, N_2O, CO, CH_4, NO, NO_2, OH	0.0~>10,000.0	Line data
GEISA	UV	BrO, NO_2, O_3, OClO, SO_2, O_2	25,000.0~43,489.0	Cross-section
	IR	H_2O, CO_2, O_3, N_2O, CO, CH_4, O_2, NO, SO_2, NO_2	0.0~>10,000.0	Line data
MPI-Mainz	UV	BrO, N_2O, NO_2, O_3, OClO, SO_2, N_2, NO, O_2	25,000.0~>100,000.0	Cross-section

Table 2. Cont.

Database		Gase	Wavelength Range (cm^{-1})	Data Type
MPI-Mainz	IR	Not included	Not included	Not included

Plots shown in Figure 2 indicate that, under given conditions, there were significant differences in absorption coefficients calculated using HITRAN and HITEMP databases at 360 K, 500 K, and 1000 K, because HITEMP uses more line data including weak transitions. This also leads to the large relative errors in Figure 2b. However, in actual applications, Schroder et al. [44] point out that the HITRAN database is relatively accurate for temperatures below 1300 K. For the near-IR band, HITRAN generates fewer line shapes and intensity prediction errors than HITEMP. Considering that the atmospheric profile used in this study did not consider water vapor at temperatures above 1000 K, only the HITRAN database was used in calculating absorption characteristics in the IR band.

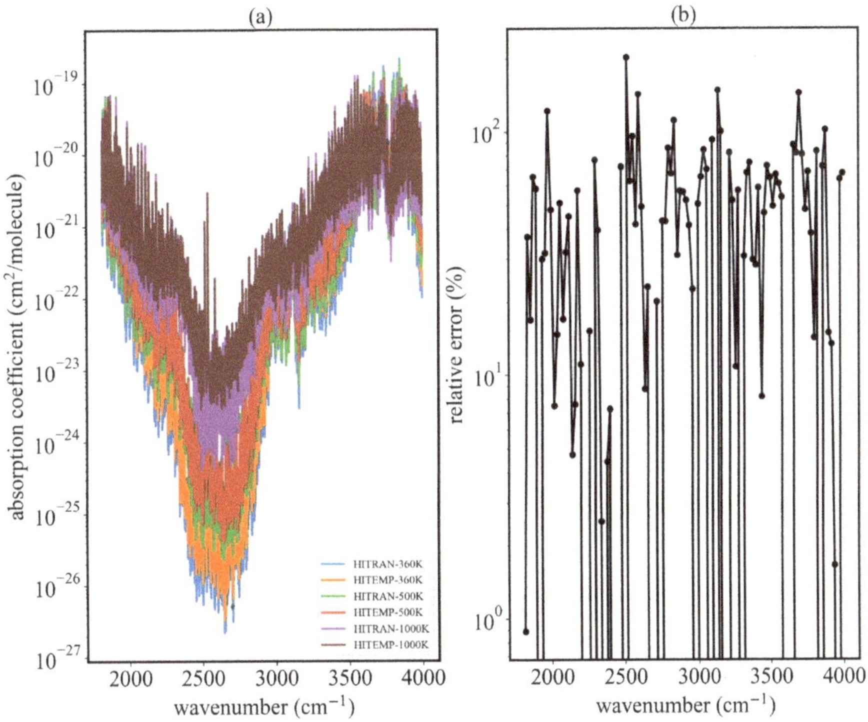

Figure 2. Plot of water vapor absorption coefficients in the range of 1818–4000 cm^{-1} and their relative error at temperatures above 360 K. Calculations involved the SPECTRA database (https://spectra.iao.ru, accessed on 7 January 2025) with settings as follows: line intensity cut off = 1.0×10^{-28}; temperature = 360 K, 500 K, 1000 K; pressure = 1 atm; and Voigt line shape profile, line wing = 50 halfwidths. (**a**) Absorption coefficients with different databases and temperatures. (**b**) Relative error of HITRAN and HITEMP at 360 K.

For absorption in the UV band, the main gases are O_2 and O_3 [34]. As absorption data for the UV band are mainly cross-sectional data, often from various references, the MPI-Mainz and GEISA databases were used predominantly in practical applications. Atmospheric layer temperatures were used in selecting the closest cross-sectional data, which were used to calculate the optical depths. Results are shown in Figure 3. The temperature considered in Figure 3 is consistent with the temperature of the atmospheric profile used in the Monte Carlo simulation. Moreover, the calculated results of the absorption cross-section will be applied to the calculations of the optical depth and radiative transfer in the UV band.

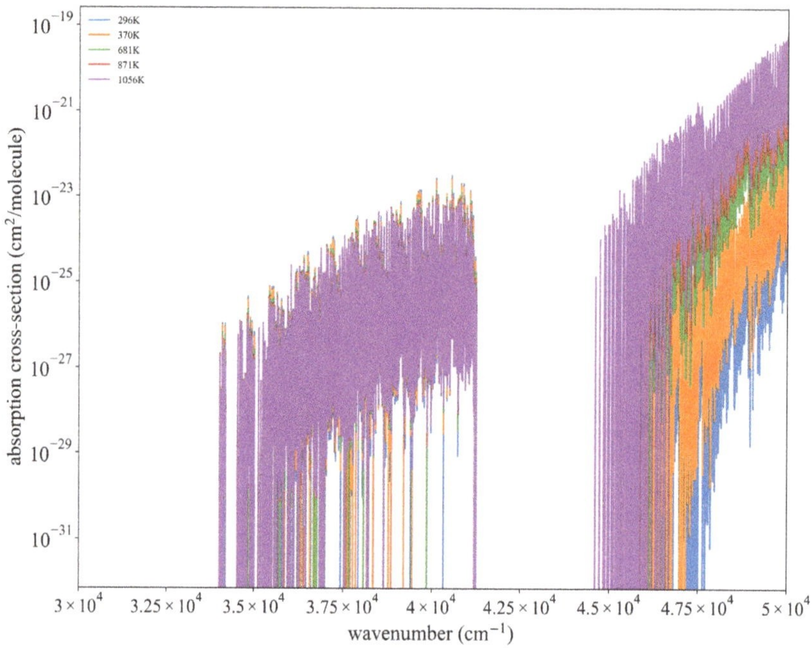

Figure 3. The variation with wavenumber of the oxygen absorption cross-section at specific temperatures. Calculation was based on the collected gas-absorption databases and temperature conditions of the atmospheric profile.

2.3. *Monte Carlo Radiative Transfer Method*

The Monte Carlo method computes the complete effective Mueller matrix at any detector position [16]. Radiative transfer models based on the discrete-ordinate or diffusion methods usually cover situations where the light source is the sun or thermal emission from the Earth surface. However, the Monte Carlo method can place the light source at any position, with the source being either active or passive.

The Monte Carlo technique addresses the 3D vector radiative transfer equation by simulating photon propagation through inhomogeneous turbid media, such as coupled atmosphere–ocean systems. The method begins by discretizing the medium into optically homogeneous layers or smaller voxels, each defined by properties like the phase matrix, extinction coefficient, and single-scattering albedo. Photons are initialized with a weight of $\omega = 1.0$ and a 4×4 identity Mueller matrix M, which are adaptable to any incident light's polarization states. For photon sampling, the model determines propagation optical depth using random number sampling as follows:

$$\tau = -\ln(\alpha), \tag{1}$$

where α is the random number uniformly generated from 0 to 1. If the photon's optical depth does not intersect boundaries, its optical depth is sampled directly as Equation (1).

However, if it approaches a boundary, a forced collision is induced to ensure interaction inside the medium as follows:

$$\tau = -\ln\{1 - [1 - \exp(-\tau_0)]\alpha\}, \tag{2}$$

where τ_0 is the left optical depth along the photon path to the boundary of the medium. After the photon is at the new location after the optical depth sampling, it has a forced collision. Therefore, the photon weight is updated as follows:

$$\omega_a = \omega_b \omega_0, \tag{3}$$

where ω_0 is the single-scattering albedo, and ω_b and ω_a are the photon weights before and after the forced collision, respectively. Consequently, the Mueller matrix is updated as follows:

$$\mathrm{M}_a = \mathrm{R}(\varphi_2)\mathrm{P}(\Theta_s, \vec{r})\mathrm{R}(\varphi_1)\mathrm{M}_b, \tag{4}$$

where M_b and M_a are the Mueller matrices before and after the collision; φ_1 and φ_2 are related to the rotations between the meridian plane of Mueller matrix M_b and the scattering plane; $\mathrm{P}(\Theta_s, \vec{r})$ is the scattering phase matrix at the location $\vec{r}$ and the scattering angle Θ_s; the scattering angle and two rotation angles are realized in terms of random numbers. To further enhance the photon efficiency, the scattering estimation to the detector is executed before every collision as follows:

$$\mathrm{M}_d = \frac{\omega_a \exp(-\tau_d)}{|\cos\theta_d|}\mathrm{R}(\varphi_{d2})\mathrm{P}(\Theta_{ds}, \vec{r})\mathrm{R}(\varphi_{d1})\mathrm{M}_b, \tag{5}$$

where M_d is the estimated Mueller matrix with respect to the specific detector and the final result is the summation over all the estimations; Θ_{ds}, φ_{d1}, and φ_{d2} are the scattering angle and two rotation angles between the Meridian and scattering planes, which can be directly calculated from the photon and detector status; τ_d is the optical depth from the photon location to the detector location; and $|\cos\theta_d|$ is the projected area of the detector. After a photon is initialized, the basic steps are propagation as Equation (1) or (2), estimation to a detector as Equation (5), and scattering as Equations (3) and (4), and the steps are repeated until the photon weight falls below a threshold value (10^{-6} is used in this study). Bias techniques for the extra introduction in Equation (2) and the bivariate sampling for angles φ_1 and Θ_s in Equation (4) are employed to remove the introduced biases and the details are given by Tynes et al. [45].

To summarize the Monte Carlo method, several techniques are employed to enhance sampling efficiency. First, photon weights are adjusted during boundary interactions and collisions to maintain energy conservation and remove biases from forced collisions or scattering assumptions. Then, every scattering is estimated to all detectors as given in Equation (5) and the sampling efficiency is significantly enhanced. Moreover, periodic boundary conditions are used to ensure that photons exiting the computational domain re-enter from the opposite side, preventing premature termination and ensuring comprehensive sampling. Validation against the spherical harmonic discrete ordinate method ensures the accuracy while optimizing computational efficiency, making the technique robust for modeling radiative coupling in complex three-dimensional environments [16].

To ensure the accuracy of the model, a phase function was first selected to provide radiative transfer results under various optical-depth conditions, and the results were compared with, and verified against, VLIDORT results. The model was then used for radiative transfer under high-altitude and high-temperature conditions. Rayleigh scattering dominates the radiation emission process there, so only it was considered in practical calculations. Figure 4 illustrates how physical quantities are distributed across each layer during the radiative transfer process. Layer1, Layer2, and Layer3 correspond to the layer numbers of the atmospheric profiles used. Therefore, physical quantities such as temperature, air pressure, and gas concentration are distributed across these layers. The optical depth, extinction coefficient, and single-scattering albedo are calculated from temperature, gas absorption data, and the layer thickness. As a result, they are distributed between each layer. The input of the Monte Carlo model includes some built-in and external parameters. The built-in parameters include the num of photons, the horizontal and vertical boundary values of the simulation space, the distribution of detectors and the height of the observer. The external parameters include the atmospheric profile and the scattering phase function. Once these parameters are determined, the radiative transfer process can be constructed.

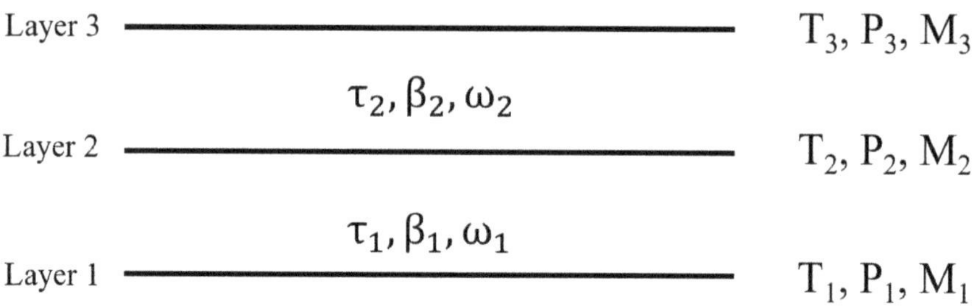

Figure 4. Schematic Diagram of the Distribution of Physical Quantities in the Radiative Transfer Process. T refers to the temperature, P refers to the pressure, M refers to the gas concentration, τ refers to the optical depth, β refers to the extinction coefficient, ω refers to the single-scattering albedo.

In the vector radiative transfer model, the most convenient representation of polarized light involves the four Stokes parameters I, Q, U, and V. The Monte Carlo model computed a 4×4 Muller matrix describing the relationship between Stokes parameters of incident and transmitted light. The effective Muller matrix of the entire Monte Carlo model system can be obtained as follows:

$$\mathrm{M}_{\mathrm{eff}} = \frac{1}{N \cdot A} \sum_{j} \mathrm{M}_{d,j}, \tag{6}$$

where N is the number of photons, A is the projected area of the detector, $\mathrm{M}_{d,j}$ is the detector's estimation for photons as given in Equation (5). The radiative Stokes parameters received by the detector is obtained by multiplying the effective Muller matrix with the incident radiation,

$$\mathrm{I}_s = \mathrm{M}_{\mathrm{eff}} \mathrm{I}_i. \tag{7}$$

Through the above calculations, the Monte Carlo radiative transfer model is achieved. A comparative verification between the Monte Carlo model and VLIDORT can be conducted as well.

To ensure the accuracy of the Monte Carlo method, the method was going to be validated with the VLIDORT model. The scattering phase function used in the Monte Carlo validation represents the strong scattering process in the radiative transfer calculation as shown in Figure 5. The other optical and geometrical parameter settings were made with reference to actual observation conditions of satellites [46], as shown in Table 3. A three-layer atmospheric profile was used for the comparison with the optical depth in Table 3 representing the sum of all three layers, and the single-scattering albedos in different layers were 0.999. The Monte Carlo model also has unique settings due to the strong effect of the number of photons on the results. When the number of photons increases gradually from tens of thousands to 1 million, the accuracy of the Monte Carlo model improves significantly. However, when the number exceeds 1 million, the accuracy improvement becomes marginal while computational efficiency decreases. More conclusions are discussed in Section 4. For these reasons, in the comparison with VLIDORT, the number of photons in the Monte Carlo model was empirically set to 10^6.

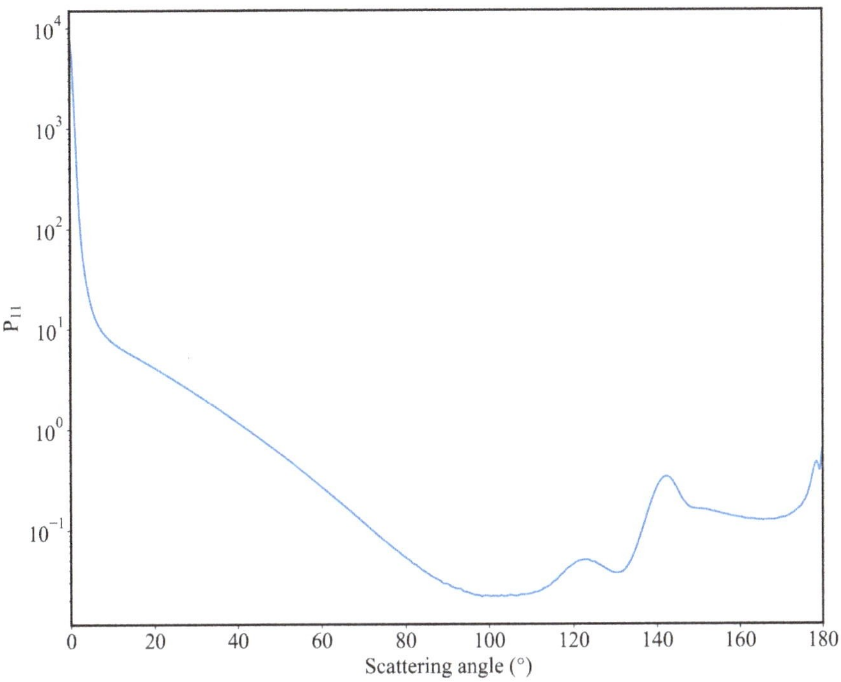

Figure 5. Phase function used for the validation between the Monte Carlo model and VLIDORT.

Table 3. Optical and geometrical parameters used in the comparison between the two models.

Parameter	Unit	Values
Solar zenith angle	degree	30, 40, 50, 60, 70
Viewing zenith angle	degree	30, 40, 50, 60, 70
Relative azimuth angle	degree	0, 20, 40, . . ., 180
Optical depth	—	0.1, 1, 10, 50, 100
Surface albedo	—	0.5
Single-scattering albedo	—	0.999

In the point source simulation, the geometrical parameters were slightly different from those of previous comparison experiments, as listed in Table 4. The incident angle (Theta in) and observing angle (Theta out) were set to 30°, which provided optimal precision in previous experiments. In the simulation of the 1800–4000 cm^{-1} band, only the HITRAN database was considered, with the high-temperature database not being involved at this stage. For the simulation of the 180–400 nm band in cases where the altitude was >100 km, based on the temperature and gas concentration of the two layers, values of the absorption cross-section under the closest temperature conditions were employed in the MPI-Mainz, GEISA, and HITRAN databases, and these values were used to calculate the optical depth between the two layers. For comparison, the optical depth under normal temperature conditions was also calculated using the HITRAN database, to facilitate subsequent comparisons.

Table 4. Geometrical parameters of the simulation.

Comparison Experiment	Practical Simulation	Values
Solar zenith angle	Theta in	30°
Viewing zenith angle	Theta out	30°
Relative azimuth angle	Relative azimuth angle	0°, 20°, 40°, ..., 180°

2.4. Statistical Metrics Used for Validation

Four statistical metrics including relative error, absolute relative error, mean absolute error (MAE), and root mean square error (RMSE) were used for the validation between the Monte Carlo method and VLIDORT.

The relative error is the most used statistical metric. Considering that the relative errors can be either positive or negative, and might cancel each other when calculating the overall average, resulting in an apparently better average, the absolute relative error was introduced to better reflect the errors as follows:

$$\text{absolute relative error} = \frac{1}{n}\sum\frac{|I_M - I_V|}{I_V}, \tag{8}$$

where I_M represents the radiance calculated by the Monte Carlo model, I_V represents the radiance calculated by VLIDORT, and n represents the number of samples. If the absolute relative error is equal to the relative error or to the opposite sign of it, this indicates that the values of the Monte Carlo simulation are, respectively, greater or less than the VLIDORT results, which may also have statistical characteristics. In addition, mean absolute error and root mean square error were also used to quantify the bias error between two models. The calculation formulas of two metrics can be described as follows:

$$MAE = \frac{1}{n}\sum|I_M - I_V|, \tag{9}$$

$$RMSE = \sqrt{\frac{1}{n}\sum(I_M - I_V)^2}. \tag{10}$$

Based on these statistical indicators, relative error and absolute relative error, MAE and RMSE, the statistical characteristics of the comparison were calculated.

3. Results

3.1. Comparison of the Monte Carlo and VLIDORT Approaches

Figures 6, A1 and A2 compare the results of Stokes parameter I simulated by the two models for different relative azimuth angles. When the optical depth changes by three orders of magnitude, from 0.1 to 100, consistency is achieved by the two models. The Monte Carlo simulation depends on the number of photons, so as the optical depth increases for cases with larger optical thickness, it is more difficult for photons to pass through one layer to reach the next, with results exhibiting up-and-down oscillation. This can be solved by increasing the number of photons. Although the Monte Carlo results display these oscillations, the order-of-magnitude difference between the two models is within 10^{-4}. Furthermore, as the solar zenith angle increases, the discrepancy between the two models also increases. This is because as the solar zenith angle increases, the assumption of a plane-parallel atmosphere is increasingly unsatisfied, and errors in the models themselves also increase, with different models being more likely to show different results. When the solar zenith angle is between 30° and 60°, the calculation results of two models are generally consistent both for transmissions and reflections.

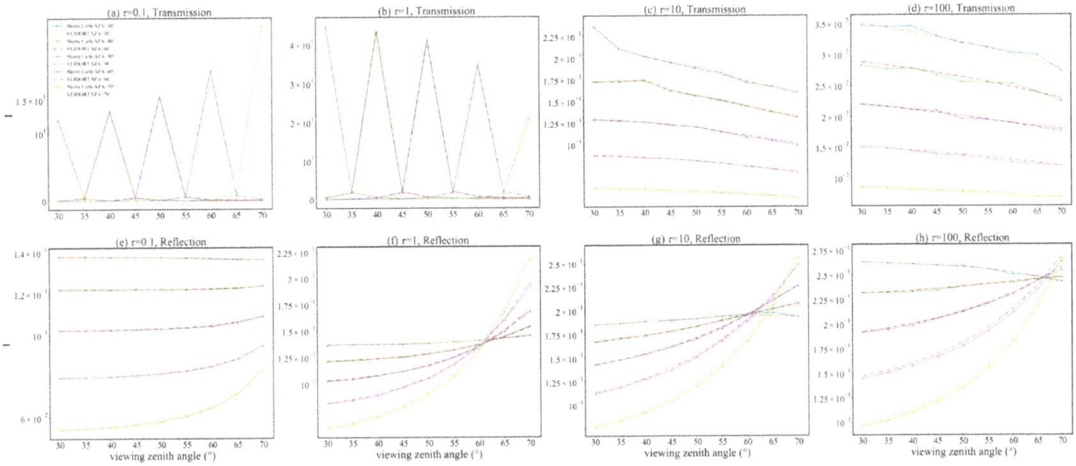

Figure 6. Comparison of I computed by the Monte Carlo model and VLIDORT, with the relative azimuth angle set as 0°. 'SZA' in the legend represents the solar zenith angle. 'Transmission' represents the downwelling radiance at the bottom layer, and 'Reflection' represents the upwelling radiance at the top layer.

To quantify the errors of the Monte Carlo model, the relative errors were calculated, with the results shown in Figures 7, A3 and A4. For a solar zenith angle of 30–60°, relative errors are most within 2%; for an angle of 70°, errors are larger but still <5%; and for a viewing zenith angle of 40–60°, the relative errors are larger than those of other viewing zenith angles. When the optical depth changes from 0.1 to 100, the overall relative error is within 5%. The statistical metrics' results were shown in Table 5. For reasons mentioned earlier, the relative errors can be either positive or negative, resulting in inconsistency between relative and absolute relative errors. Regarding the mean relative error, its value was <1% under all optical depth conditions, although the reflection relative error was −1.067% when the optical depth was 50. The mean absolute relative error increased with optical depth, consistent with the characteristics of the Monte Carlo simulation, i.e., the larger the optical depth, the greater the number of photons that were required for the simulation. Although the mean absolute relative error increased with optical depth, the largest error of 1.524% occurred when the optical depth was 100, which is still in an acceptable range. This indicates that the simulation results of the Monte Carlo model are consistent with those of VLIDORT, and, therefore, could be used for subsequent radiative transfer simulations.

Table 5. Overall statistical results of the two models under different optical depth conditions. τ represents the optical depth.

τ	Transmission Relative Error (%)	Transmission Absolute Relative Error (%)	Transmission MAE	Reflection RMSE	Reflection Relative Error (%)	Reflection Absolute Relative Error (%)	Reflection MAE	Reflection RMSE
0.1	−0.411	0.818	0.0005	0.00345	−0.209	0.236	0.0002	0.0002
1	−0.792	1.024	0.012	0.103	−0.826	0.894	0.0008	0.0009
10	−0.973	1.164	0.0009	0.001	−0.983	1.083	0.0013	0.0014
50	−0.689	1.168	0.0004	0.0004	−1.067	1.123	0.0016	0.0017
100	−0.572	1.524	0.0003	0.0003	−0.954	1.036	0.0015	0.0018

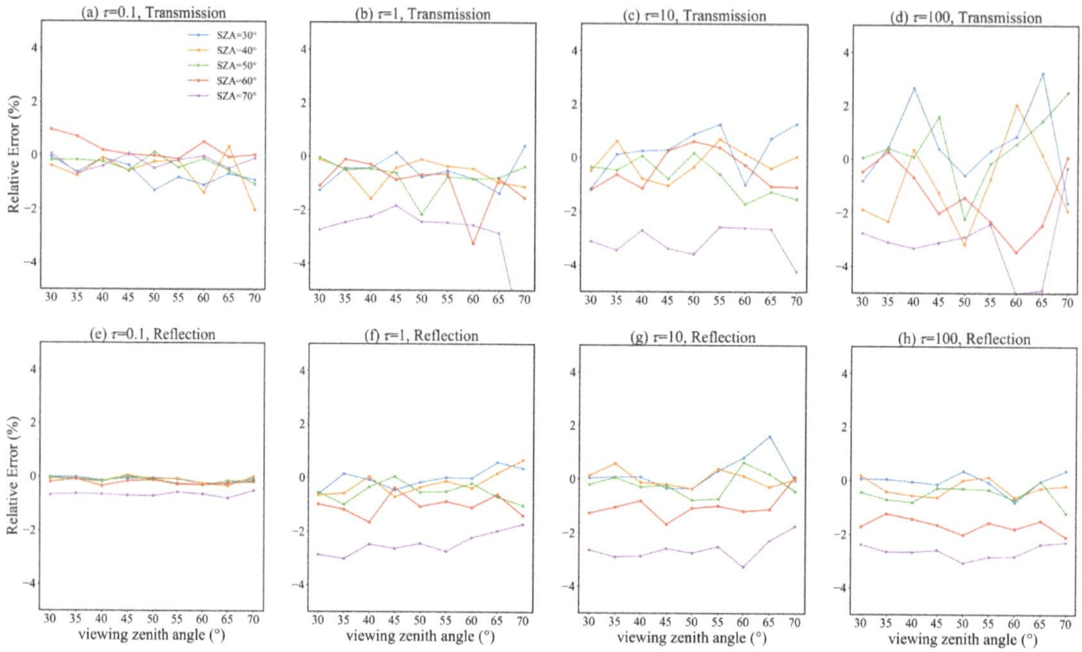

Figure 7. Relative errors computed by the Monte Carlo model and VLIDORT, with the relative azimuth angle set to 0°. 'SZA' in the legend represents the solar zenith angle.

3.2. 1D Monte Carlo Simulation in 1800–4000 cm^{-1} and 25,000–55,000 cm^{-1}

The altitude for observation of radiance emitted by the radiation field was set to be 1000 km, coping with the orbital altitude of polar-orbiting satellites. The height of the radiation field was set at 50 km, although it may be set at any height.

Simulation results are shown in Figures 8 and 9 It can be seen from Figure 8a that there are two extremely low radiation intensity values near 2500 cm^{-1} and 3600 cm^{-1}, corresponding to the two absorption peaks of CO_2 and H_2O in the IR region. In other positions, the weak absorptions of H_2O, CO_2, and greenhouse gases such as CH_4 mean there are attenuation peaks of varying height. In the remaining positions, the values of Stokes parameter I are relatively consistent. Figure 8b,c show simulation results for the 25,000–55,000 cm^{-1} UV band. As the difference between the high- and normal-temperature databases lies only in specific values of the absorption cross-section, the simulation trends obtained using different databases are consistent. It can be seen from Figure 8b,c that the value of I increases with increasing wavenumber. Therefore, the radiation field signal is more likely to be received by polar-orbiting satellites at longer wavenumbers. In Figure 9, although the outlines of the simulation results appear similar, there are differences in I. Since gas absorption is minimal in the 25,000–50,000 cm^{-1}, while strong absorption bands of oxygen and ozone occur at wavenumbers greater than 50,000 cm^{-1}, this spectral region exhibits immediate feedback between gas absorption and corresponding radiation calculations. The maximum relative error between the two different conditions can reach −2.5%, which is relatively small because the strongly absorbing O_3 occurs mainly below the stratosphere, and its concentration is not considered above 120 km.

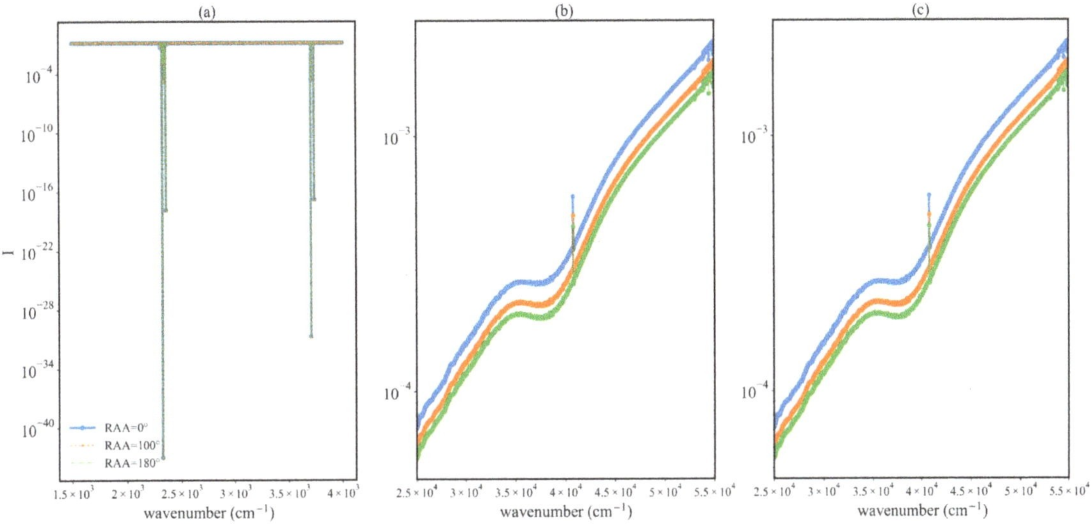

Figure 8. Simulation results for the radiation field at 50 km, for (**a**) the 1800–4000 cm^{-1} band; (**b**) the 25,000–55,000 cm^{-1} band, using the normal-temperature database; and (**c**) the 25,000–55,000 cm^{-1} band, using the high-temperature database. 'RAA' represents the relative azimuth angle.

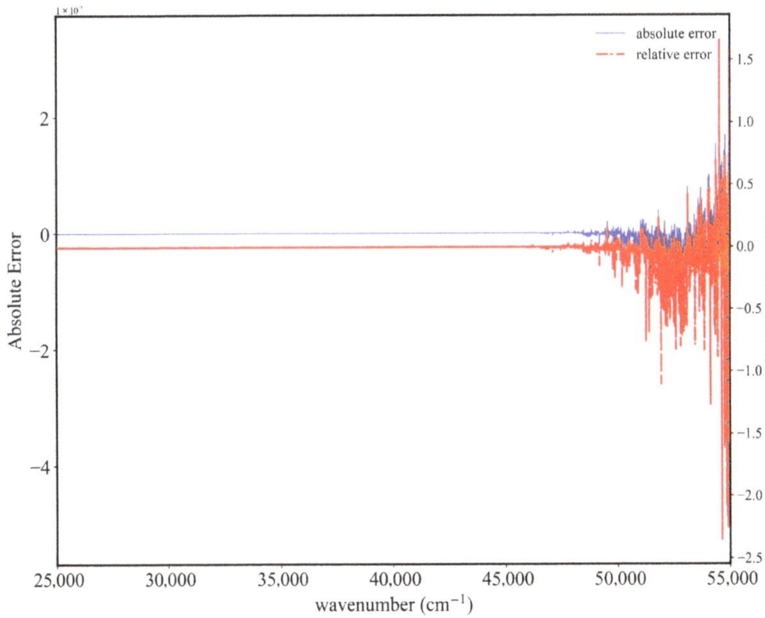

Figure 9. Absolute and relative errors of 180–400 nm simulation results, using different databases.

4. Discussion

4.1. Influence of the Number of Photons on the Monte Carlo Method

When the optical depth is 100, different numbers of photons were used in the Monte Carlo simulation, and the results were compared with VLIDORT. The comparison results are shown in Figures 10 and 11. For the reflected radiation, when the number of photons is 100,000 and 500,000, and the solar zenith angle is between 30° and 60°, the simulation results oscillate a lot. After the number of photons reaches 1,000,000 there are still obvious oscillations when the solar zenith angle is 40°, but the results are relatively smooth when the angle is 50° and 60°. For the case where the solar zenith angle is 70°, increasing the number of photons does not significantly change the simulation results of the Monte Carlo model. For the transmitted radiation, there are very obvious oscillations when the number is 100,000 and 500,000. After the number of photons reaches 1,000,000, the results are already in good agreement with the VLIDORT. In addition, it can be found that when the solar zenith angle is 30°, the simulation results are the best when the number of photons is set to 10,000,000. After exceeding 10,000,000, there are instead larger deviations when the observation zenith angle is between 50° and 70°. The statistical results can be found in Table 6. As the number of photons increases, the relative error of the reflection changes very little, while the relative error of transmission first decreases and then changes very little after reaching 1,000,000 photons. The absolute relative error of the reflection also shows a downward trend. It can also be seen that although the transmission has great oscillations when the number of photons increases, the calculation results under the condition of a higher number of photons are generally better than the condition of a lower number of photons.

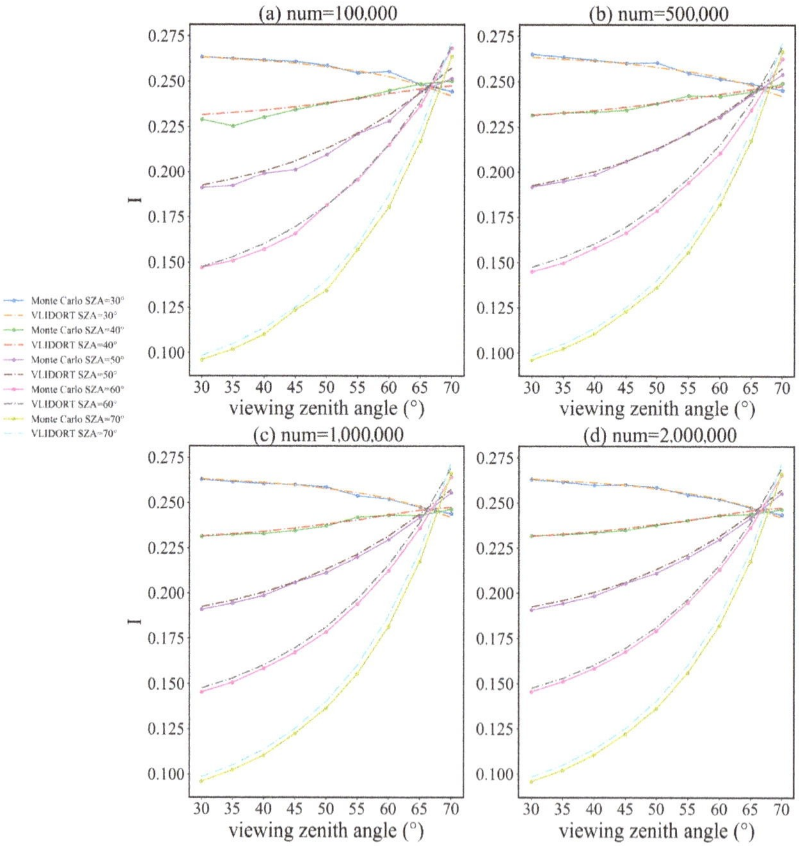

Figure 10. Comparison of the reflected radiation results under different numbers of photons when the optical depth is 100. The subtitles of figures mark the number of photons used in the simulation, and the other parameters are same as before.

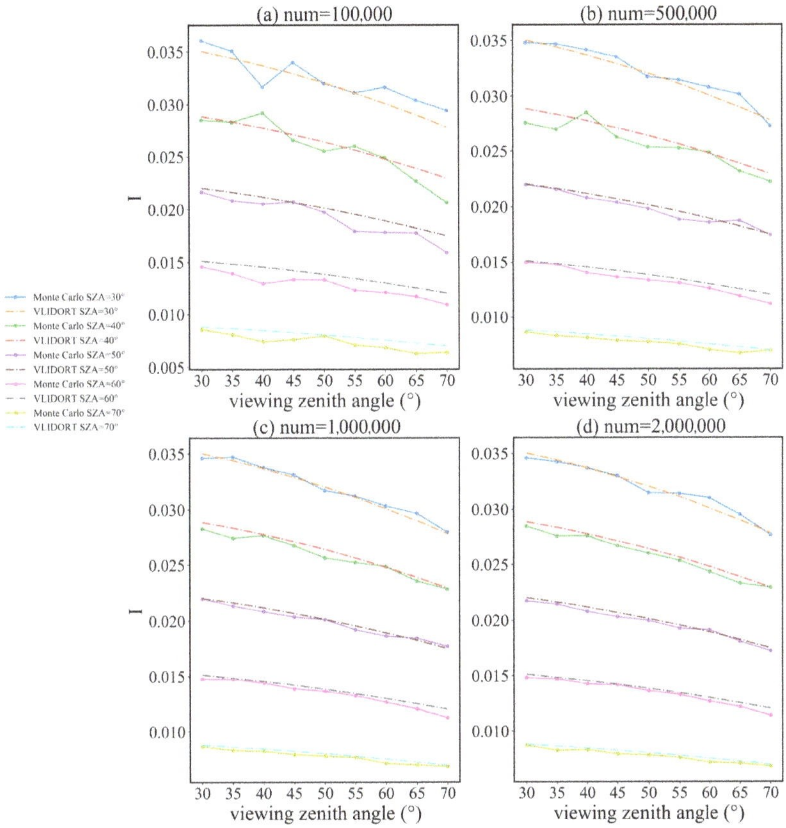

Figure 11. Same as Figure 10, but for the transmission.

Table 6. The statistical results between two models under different numbers of photons when the optical depth is 100.

Number of Photons	Transmission Relative Error (%)	Transmission Absolute Relative Error (%)	Transmission MAE	Transmission RMSE	Reflection Relative Error (%)	Reflection Absolute Relative Error (%)	Reflection MAE	Reflection RMSE
100,000	−1.990	4.650	0.0008	0.0010	−0.879	1.384	0.0024	0.0029
500,000	−1.153	2.149	0.0004	0.0005	−0.914	1.111	0.0017	0.0020
1,000,000	−0.848	1.511	0.0003	0.0003	−0.923	1.006	0.0015	0.0018
1,500,000	−0.869	1.336	0.0002	0.0003	−0.979	1.011	0.0015	0.0016
2,000,000	−0.919	1.318	0.0002	0.0003	−0.985	1.020	0.0015	0.0016

In summary, when the number of photons is less than 1,000,000, as the number of photons increases, the accuracy of the Monte Carlo model is significantly improved, and at the same time, the calculation time also increases accordingly. When the number of photons is greater than 1,000,000, as the number of photons increases, the accuracy of the model is slightly improved, and the calculation time significantly increases. Therefore, setting the number of photons as 1,000,000 is a choice that can not only ensure the calculation accuracy but also avoid a long calculation time.

4.2. 3D Monte Carlo Simulation with Point Sources for 1800–4000 cm^{-1} and 25,000–55,000 cm^{-1}

The Monte Carlo model achieves the radiative transfer simulation with point sources by placing detectors at different heights. The geometric observation parameters were consistent with the 1D simulation, as shown in Table 4. A total of 25 detectors were placed in a 5 $\times$ 5 pattern, evenly distributed on boxes [-10.5, 10.5] of the X and Y axes. The effective Muller matrix of the detector in the Monte Carlo model is the result of projection of the scattering direction at specific detectors, so its calculation result with point sources is much smaller than the ones with solar sources. Figure 12 shows a schematic diagram of the transmission process in the 3D situation. In this process, photons are emitted from the height of the point source and, after undergoing the absorption and scattering processes in the atmosphere, they, finally, reach the height of the detectors. The detectors are uniformly distributed on the horizontal plane at the height, and in this case a total of 25 detectors are used.

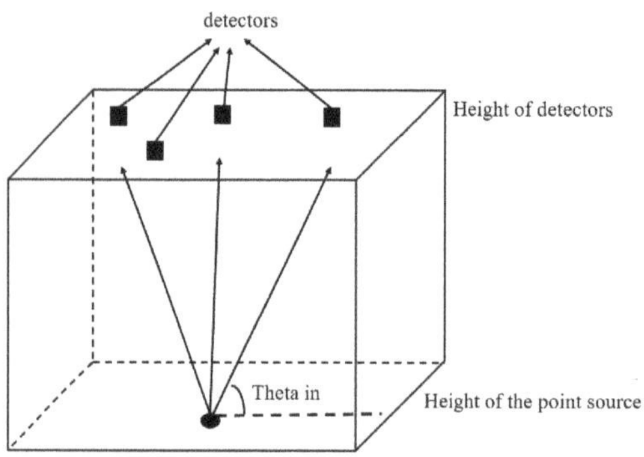

Figure 12. Schematic diagram of the Monte Carlo model in the 3D situation. The height of the point source is also the height of the hypersonic vehicles. The black solid dot represents the point light source, and the black solid squares represent detectors.

Figures 13–15 present the results of the simulations with point sources in 1800–4000 cm^{-1}. In Figure 13, for the 1800–4000 cm^{-1} IR band, there is a maximum value at the detector coordinate (8.4, 0.0). With a relative azimuth angle of 100°, there is a secondary maximum value at the position (–4.2, 8.4), while at 180° the contour map of the phase matrix element P_{11} shows a strip shape, with the values of P_{11} decreasing towards both sides along the axis Y = 0. Maximum values are distributed at both ends of the X-axis. Figure 14 shows the variation of P_{11} with the band having a set relative azimuth angle, while Figure 15 shows the distribution of P_{11} as the average value over a band. In Figure 14, although the contour map shows most areas having the same color when the relative azimuth angle is 0°, P_{11} values at different positions are not consistent and can vary greatly. With the detector at (–8.4, –8.4), there are several large values of P11 near 1500, 2750, and 3600 cm^{-1}; at (0.0, 0.0) there are large values near 2500 cm^{-1}; at (8.4, 8.4) the large values are near 2250 cm^{-1}; and at (–4.2, 4.2) they are near 3000 cm^{-1}. These results indicate that, for the IR band, the different detector positions and band absorption characteristics result in inhomogeneous detections. Positions of some maximum values may reflect the optical thickness of the band being relatively large, with photons having undergone multiple scattering during their journey from the position of the point light source to the detector. Figure 15 shows the variation with the relative azimuth angle of results for band-averaged P_{11} at different detector positions. For different angles, the positions where the maximum values occur are also different. For the five detectors shown in Figure 15, when the relative azimuth angle is

20°, there is a maximum value at (8.4, 8.4); for a 100° angle the maximum is at (–4.2, 8.4), consistent with Figure 13b; and at 160° there is also a maximum value at (–4.2, 8.4).

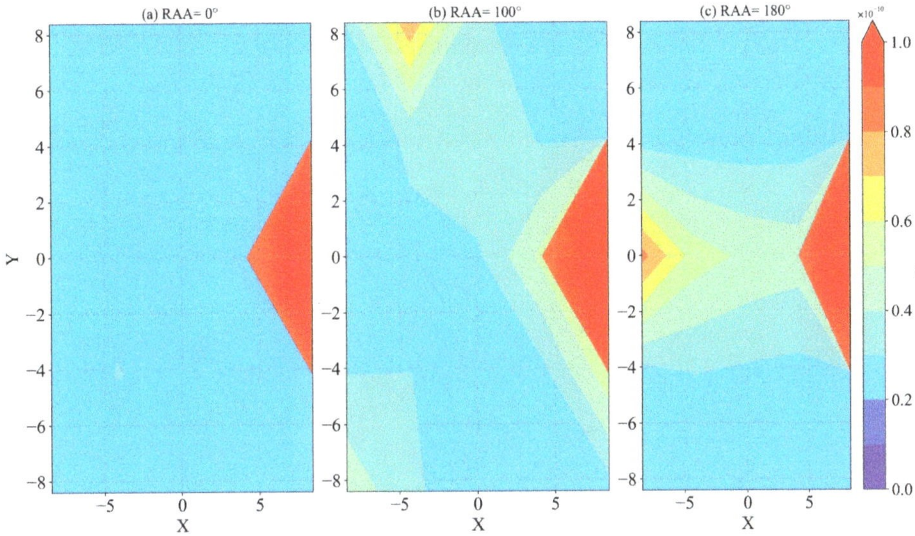

Figure 13. Contour map of 3D simulation results in the 1800–4000 cm^{-1} band. 'RAA' represents the relative azimuth angle, and the color bar represents the first element P11 of the effective Muller matrix with different values. For (**a**) the relative azimuth angle is 0°; for (**b**) it is 100°; and for (**c**) it is 180°.

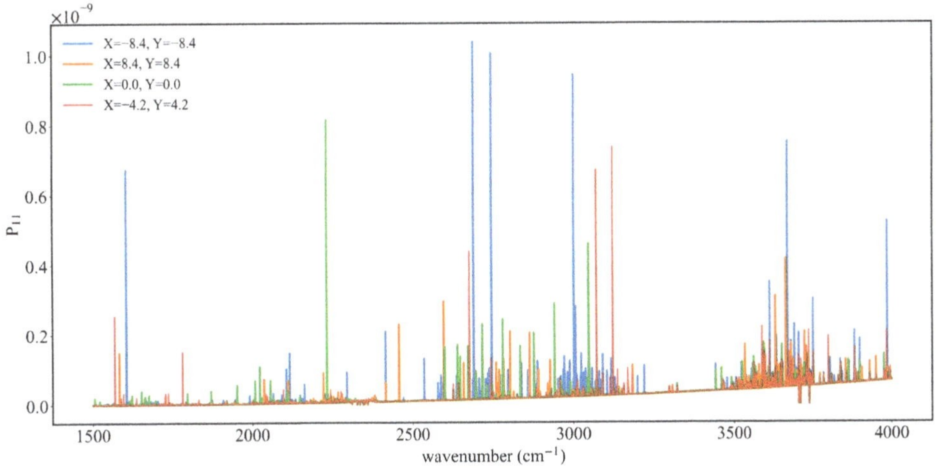

Figure 14. Variation of P11 of the effective Muller matrix with wavelength at different detector positions when the relative azimuth angle is 0°, for the wavelength band of 1800–4000 cm^{-1}.

Figures 16–18 present the results of the simulations with point sources in 25,000–55,000 cm^{-1}, where results show certain differences from those of the IR band. As shown in Figure 16, when the relative azimuth angle is 0°, there is a maximum value at the detector coordinate (4.2, 0), and radiation values at other positions show little variation; with an angle of 100°, all P_{11} values in the entire region lie in the range (1–2) × 10^{-6}; while at 180° there is a decreasing trend towards both sides along the axis Y = 0 with a characteristic stratification.

In the IR band, the variation of P_{11} with wavelength is obvious, but in the UV band there is a high degree of consistency at different detector positions, with a large difference only at the coordinate (4.2, 0.0). In Figure 17, the variation of P_{11} at different detector positions with the relative azimuth angle is similarly not obvious, with an abrupt change only at the coordinate (4.2, 0.0) when the relative azimuth angle is 0°. With other detector positions and relative azimuth angles, values of P_{11} are all in the range of (0–5) $\times$ 10^{-6}.

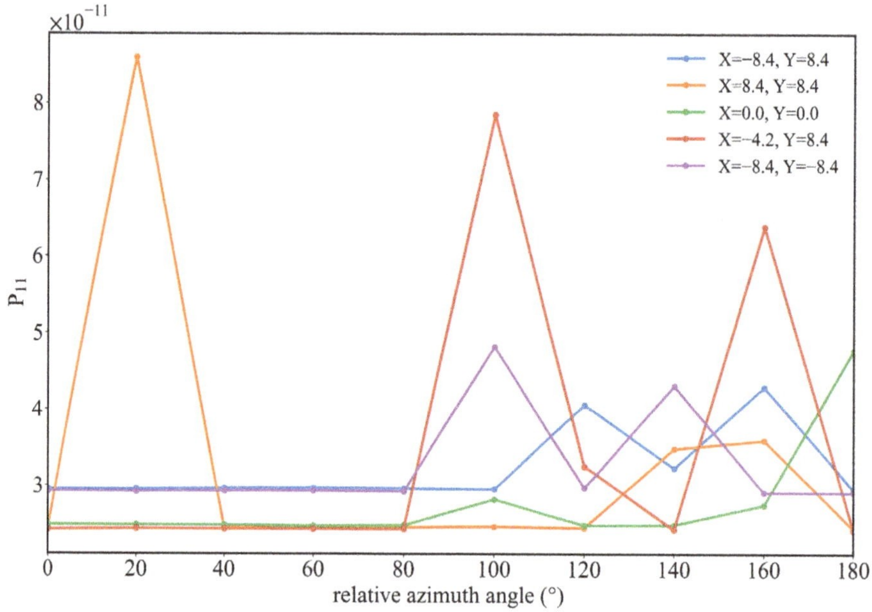

Figure 15. Variation of the band-averaged P11 values of the effective Muller matrix with the relative azimuth angle at different detector positions, for the wavelength band of 1800–4000 cm^{-1}.

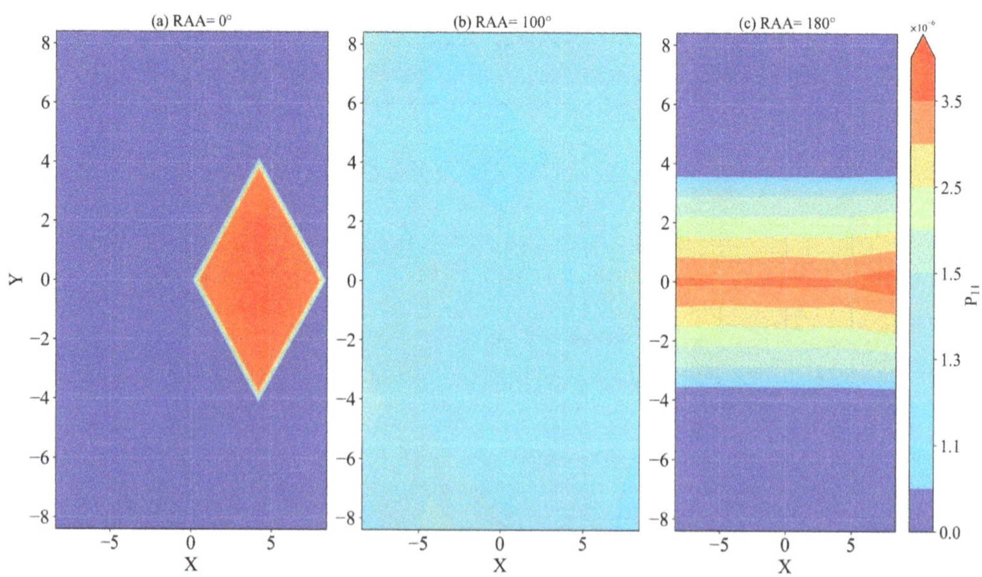

Figure 16. As for Figure 13, but for the 25,000–55,000 cm^{-1} band.

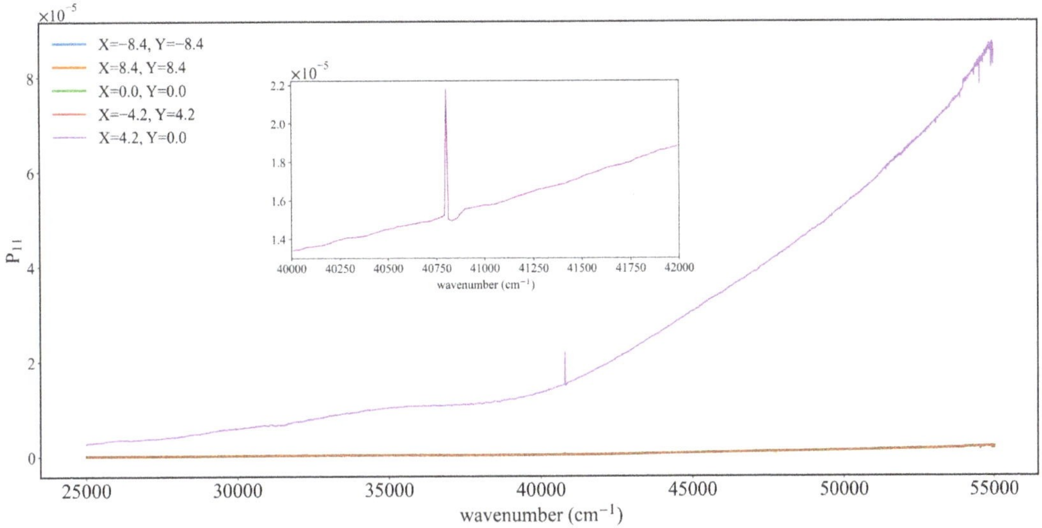

Figure 17. As for Figure 14, but for the 25,000–55,000 cm^{-1} band. The inset figure is a zoomed-in view of the line located at X = 4.2, Y = 0.0 in the main figure, focusing on the wavenumber range 40,000–42,000 cm^{-1}.

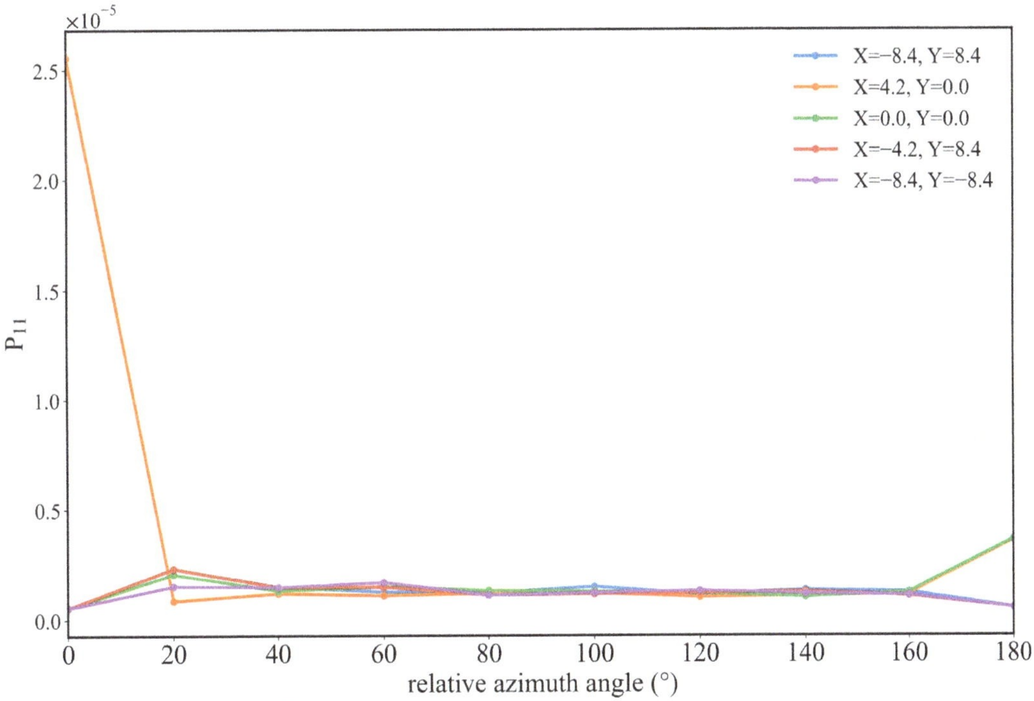

Figure 18. As for Figure 15, but for the 25,000–55,000 cm^{-1} band.

Analysis of the radiative transfer results with point sources indicates that the radiation quantity is transmitted upward through the atmosphere in near space with the following characteristics: (1) Due to the different absorption characteristics of different bands, the radiation in different bands may show inconsistencies at different detector positions and under different relative azimuth angles; (2) the radiation quantity after band averaging shows inhomogeneity in the horizontal direction with different relative azimuth angles, while in general it decreases from the middle to both sides; and (3) different detector positions show inhomogeneity with different relative azimuth angles, and the inhomogeneity of the IR band is more obvious than that of the UV band.

In addition to the above characteristics, we observed that for both the IR and UV bands, larger radiation values appear on the right side of the X axis and centrally on the Y-axis. This may be related to the setting of geometric parameters. Here, both Theta in and Theta out were set at 30°, and the relative azimuth angle was set to various values. With a relative azimuth angle of 0°, both bands have maximum values to the right of the X axis. As the angle increases, the IR band still maintains this characteristic, while the UV band exhibits other characteristics. However, with an incident angle of 35°, an exit angle of 0°, and relative azimuth angle of 0°, Zhai et al. [16] found that the results exhibit a characteristic of spreading outwards, centered at (1, 1). Therefore, different settings of geometric parameters may also have an impact on the simulation results of radiative transfer in point-source scenarios.

5. Conclusions

In this study, the radiative transfer process in the near-space region was discussed with respect to the space-borne observation of radiation at arbitrary height and location. The high-temperature absorption databases were established since the near-space region lies at the mesosphere, thermosphere, and ionosphere. Moreover, a Monte Carlo method was implemented to simulate the radiative transfer processes with surface or point sources at arbitrary height or location. Furthermore, the radiative transfer process was simulated and discussed in the ultraviolet and infrared bands because the thermal radiation from high-speed vehicles falls into the ultraviolet and infrared regions.

In the simulation process, the atmospheric profile was first constructed to provide temperature information and the concentrations of various gases. The heights of layers in the atmospheric profile span a wide range. The standard atmospheric profile USSA1976 was used for lower layers, and the NRLMSIS model based on various input parameters was used for the upper-layer profile. The temperature rises rapidly from 360 K to ~1000 K over a narrow height range of 120–200 km. The NRLMSIS model considers relatively few types of gas, so there may be some important absorbing gases that have not been considered here.

The absorption database was organized by considering gas-absorption databases such as MPI-Mainz, HITRAN, GEISA, and HITEMP. By using the closest line data or cross-sectional data for high-temperature conditions, the optical depths of different layers were calculated.

Finally, the Monte Carlo model was used to calculate transmission processes related to the radiation field, simulating Stokes parameters from 50km to the height of polar-orbiting satellites. To verify the accuracy and applicability of the Monte Carlo model, a comparison with VLIDORT was undertaken under different optical depth conditions. Results indicate that the maximum absolute relative error for transmission was 1.524% and the minimum was 0.818%; while the maximum absolute relative error for reflection was 1.036% and the minimum was 0.236%. These results illustrate the universality of the model.

Based on the Monte Carlo simulation, the two absorption bands for water vapor and CO_2 were clearly observed in the IR band, while, in the UV band, the Stokes parameter I increased in value as wavelength decreased. In the point-source scenario, the radiative transfer simulation was undertaken by setting multiple detectors at different positions.

In the IR band, the radiation quantity P_{11} varied with changes in detector position, the band, and the relative azimuth angle, indicating inhomogeneity under various conditions. However, in the UV band the radiation quantity was most affected by detector position, exhibiting consistency over different bands and relative azimuth angles.

Author Contributions: Conceptualization, B.S.; funding acquisition, B.S.; methodology, M.L., R.L. and B.S.; supervision, B.S.; validation, M.L. and R.L.; writing– original draft, M.L.; writing– review and editing, M.L., R.L. and B.S. All authors have read and agreed to the published version of the manuscript.

Funding: This research was funded by the Natural Science Foundation of Shanghai (23ZR1405600), and the National Natural Science Foundation of China (42475083).

Data Availability Statement: Data underlying the results presented in this paper are available on request from the authors.

Acknowledgments: The authors would like to thank Robert Spurr for the use of his VLIDORT code.

Conflicts of Interest: The authors declare no conflicts of interest.

Appendix A

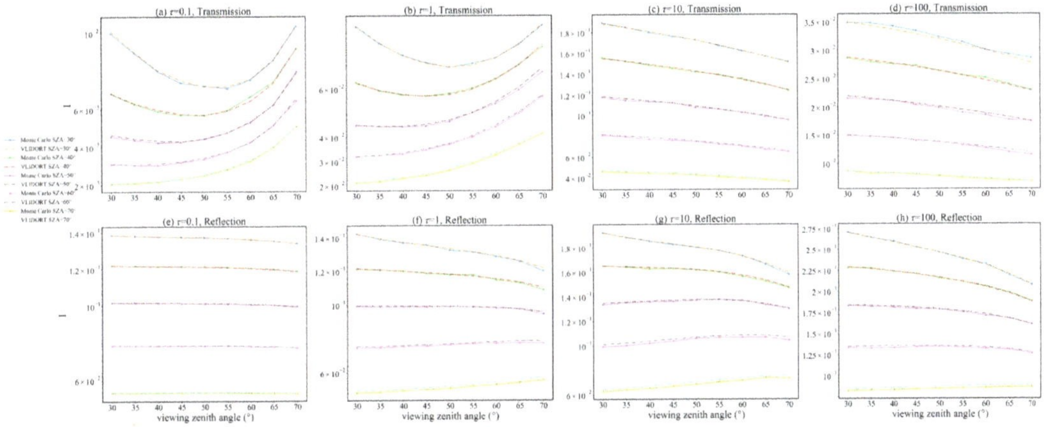

Figure A1. As for Figure 6, but with the relative azimuth angle set to 100°.

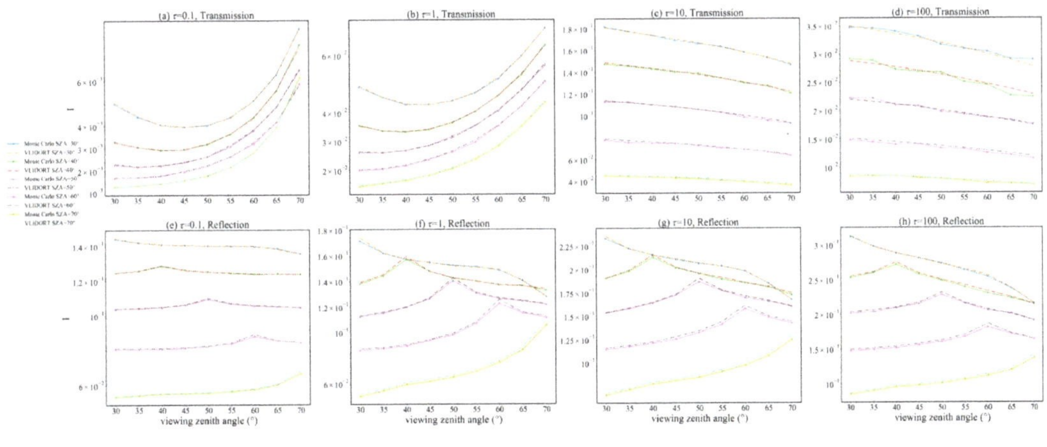

Figure A2. As for Figure 6, but with the relative azimuth angle set to 180°.

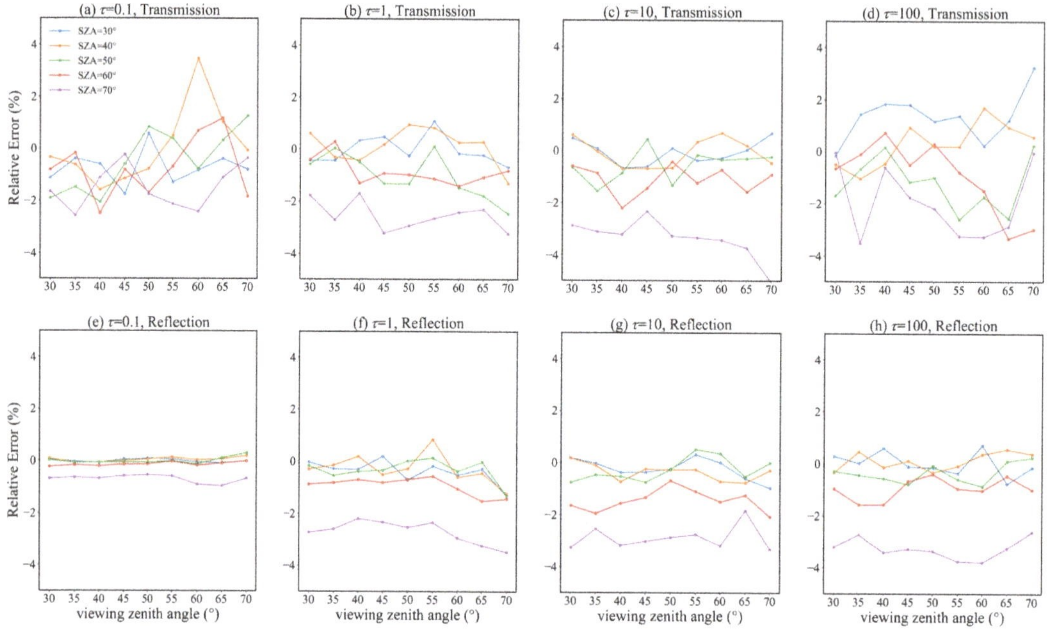

Figure A3. As for Figure 7, but with the relative azimuth angle set to 100°.

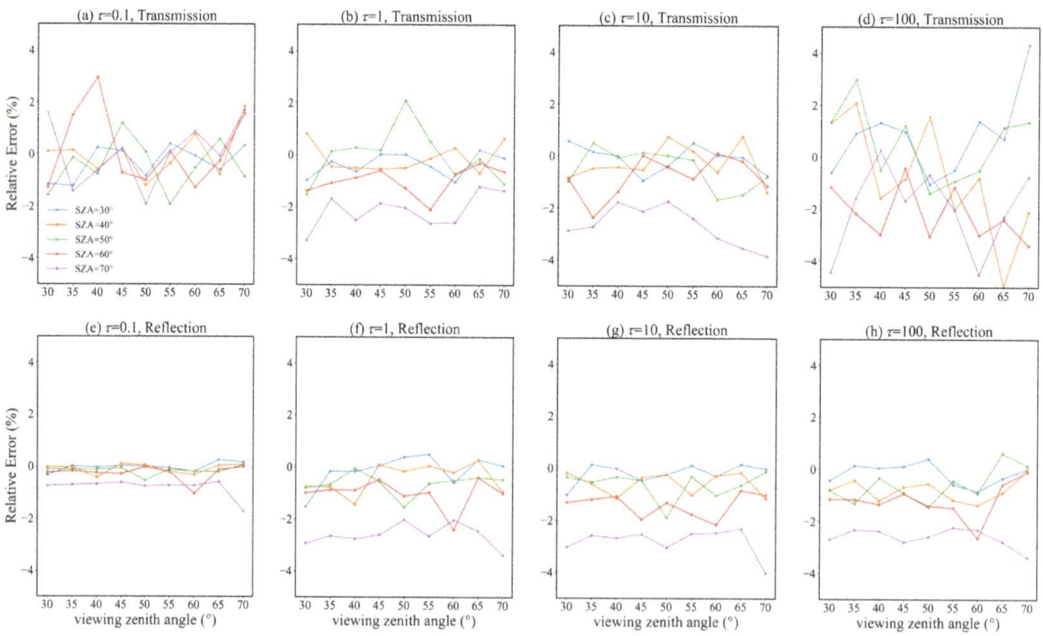

Figure A4. As for Figure 7, but with the relative azimuth angle set to 180°.

References

1. Niu, Q.; Yuan, Z.; Chen, B.; Dong, S. Infrared radiation characteristics of a hypersonic vehicle under time-varying angles of attack. *Chin. J. Aeronaut.* **2019**, *32*, 861–874. [CrossRef]

2. Du, X.; Shi, Y.; Yang, Q.; Song, D.; Liu, X. Numerical Simulation of Spectral Radiation for Hypersonic Vehicles. *Aerospace* **2024**, *11*, 802. [CrossRef]
3. Niu, Q.; He, Z.; Dong, S. Prediction of shock-layer ultraviolet radiation for hypersonic vehicles in near space. *Chin. J. Aeronaut.* **2016**, *29*, 1367–1377. [CrossRef]
4. Ozawa, T.; Garrison, M.B.; Levin, D.A. Accurate Molecular and Soot Infrared Radiation Model for High-Temperature Flows. *J. Thermophys. Heat Transf.* **2007**, *21*, 19–27. [CrossRef]
5. Nam, H.J.; Kwon, O.J. Infrared radiation modeling of NO, OH, CO, H_2O, and CO_2 for emissivity/radiance prediction at high temperature. *Infrared Phys. Technol.* **2014**, *67*, 283–291. [CrossRef]
6. Kumar, N.; Bansal, A. Flow and radiation modeling over a Martian entry vehicle. *Acta Astronaut.* **2023**, *205*, 172–184. [CrossRef]
7. Sziroczak, D.; Smith, H. A review of design issues specific to hypersonic flight vehicles. *Prog. Aerosp. Sci.* **2016**, *84*, 1–28. [CrossRef]
8. Gorshkov, A.B. The simulation of ultraviolet radiation under conditions of re-entry of space vehicle from near-earth orbit. *High Temp.* **2010**, *48*, 12–22. [CrossRef]
9. Jie, W.; Lu, B.; Tianjiao, Z.; Qiang, L.; Chaofan, X. Ultraviolet Radiation Characteristics of NO in Shock Layer of Hypersonic Vehicle. In Proceedings of the Sixteenth National Conference on Laser Technology and Optoelectronics, Shanghai, China, 3–6 June 2021; Volume 11907.
10. Stamnes, K.; Tsay, S.-C.; Wiscombe, W.; Laszlo, I. DISORT, A General-Purpose Fortran Program for Discrete-Ordinate-Method Radiative Transfer in Scattering and Emitting Layered Media: Documentation of Methodology. 2000. Available online: http://www.rtatmocn.com/disort/docs/DISORTReport1.1.pdf (accessed on 7 January 2025).
11. Spurr, R.J.D. VLIDORT: A linearized pseudo-spherical vector discrete ordinate radiative transfer code for forward model and retrieval studies in multilayer multiple scattering media. *J. Quant. Spectrosc. Radiat. Transf.* **2006**, *102*, 316–342. [CrossRef]
12. Sun, B.; Gao, C.; Spurr, R. Scalar thermal radiation using the adding-doubling method. *Opt. Express* **2022**, *30*, 30075–30097. [CrossRef]
13. Chen, Y.; Liou, K.N.; Gu, Y. An efficient diffusion approximation for 3D radiative transfer parameterization: Application to cloudy atmospheres. *J. Quant. Spectrosc. Radiat. Transf.* **2005**, *92*, 189–200. [CrossRef]
14. Roger, M.; Caliot, C.; Crouseilles, N.; Coelho, P.J. A hybrid transport-diffusion model for radiative transfer in absorbing and scattering media. *J. Comput. Phys.* **2014**, *275*, 346–362. [CrossRef]
15. Deutschmann, T.; Beirle, S.; Frieß, U.; Grzegorski, M.; Kern, C.; Kritten, L.; Platt, U.; Prados-Román, C.; Pukı¯te, J.; Wagner, T.; et al. The Monte Carlo atmospheric radiative transfer model McArtim: Introduction and validation of Jacobians and 3D features. *J. Quant. Spectrosc. Radiat. Transf.* **2011**, *112*, 1119–1137. [CrossRef]
16. Zhai, P.-W.; Kattawar, G.W.; Yang, P. Impulse response solution to the three-dimensional vector radiative transfer equation in atmosphere-ocean systems. I. Monte Carlo method. *Appl. Opt.* **2008**, *47*, 1037–1047. [CrossRef]
17. Saunders, R.; Rayer, P.; Brunel, P.; von Engeln, A.; Bormann, N.; Strow, L.; Hannon, S.; Heilliette, S.; Liu, X.; Miskolczi, F.; et al. A comparison of radiative transfer models for simulating Atmospheric Infrared Sounder (AIRS) radiances. *J. Geophys. Res. Atmos.* **2007**, *112*, D01S90. [CrossRef]
18. Kotchenova, S.Y.; Vermote, E.F.; Levy, R.; Lyapustin, A. Radiative transfer codes for atmospheric correction and aerosol retrieval: Intercomparison study. *Appl. Opt.* **2008**, *47*, 2215–2226. [CrossRef] [PubMed]
19. Kokhanovsky, A.A.; Budak, V.P.; Cornet, C.; Duan, M.; Emde, C.; Katsev, I.L.; Klyukov, D.A.; Korkin, S.V.; C.-Labonnote, L.; Mayer, B.; et al. Benchmark results in vector atmospheric radiative transfer. *J. Quant. Spectrosc. Radiat. Transf.* **2010**, *111*, 1931–1946. [CrossRef]
20. Randles, C.A.; Kinne, S.; Myhre, G.; Schulz, M.; Stier, P.; Fischer, J.; Doppler, L.; Highwood, E.; Ryder, C.; Harris, B.; et al. Intercomparison of shortwave radiative transfer schemes in global aerosol modeling: Results from the AeroCom Radiative Transfer Experiment. *Atmos. Chem. Phys.* **2013**, *13*, 2347–2379. [CrossRef]
21. Su, T.; Li, J.; Li, C.; Xiang, P.; Lau, A.K.H.; Guo, J.; Yang, D.; Miao, Y. An intercomparison of long-term planetary boundary layer heights retrieved from CALIPSO, ground-based lidar, and radiosonde measurements over Hong Kong. *J. Geophys. Res. Atmos.* **2017**, *122*, 3929–3943. [CrossRef]
22. Ratynski, M.; Khaykin, S.; Hauchecorne, A.; Wing, R.; Cammas, J.-P.; Hello, Y.; Keckhut, P. Validation of Aeolus wind profiles using ground-based lidar and radiosonde observations at Réunion island and the Observatoire de Haute-Provence. *Atmos. Meas. Tech.* **2023**, *16*, 997–1016. [CrossRef]
23. Costa-Surós, M.; Calbó, J.; González, J.A.; Long, C.N. Comparing the cloud vertical structure derived from several methods based on radiosonde profiles and ground-based remote sensing measurements. *Atmos. Meas. Tech.* **2014**, *7*, 2757–2773. [CrossRef]
24. Hersbach, H.; Bell, B.; Berrisford, P.; Hirahara, S.; Horányi, A.; Muñoz-Sabater, J.; Nicolas, J.; Peubey, C.; Radu, R.; Schepers, D. The ERA5 global reanalysis. *Q. J. R. Meteorol. Soc.* **2020**, *146*, 1999–2049. [CrossRef]
25. Tarek, M.; Brissette, F.P.; Arsenault, R. Evaluation of the ERA5 reanalysis as a potential reference dataset for hydrological modelling over North America. *Hydrol. Earth Syst. Sci.* **2020**, *24*, 2527–2544. [CrossRef]

26. United States National Oceanic and Atmospheric Administration. *US Standard Atmosphere, 1976*; National Oceanic and Atmospheric Administration: Washington, DC, USA, 1976; Volume 76.
27. Krueger, A.J.; Minzner, R.A. A mid-latitude ozone model for the 1976 US Standard Atmosphere. *J. Geophys. Res.* **1976**, *81*, 4477–4481. [CrossRef]
28. Emmert, J.T.; Drob, D.P.; Picone, J.M.; Siskind, D.E.; Jones, M., Jr.; Mlynczak, M.G.; Bernath, P.F.; Chu, X.; Doornbos, E.; Funke, B. NRLMSIS 2.0: A whole-atmosphere empirical model of temperature and neutral species densities. *Earth Space Sci.* **2021**, *8*, e2020EA001321. [CrossRef]
29. Picone, J.M.; Hedin, A.E.; Drob, D.P.; Aikin, A.C. NRLMSISE-00 empirical model of the atmosphere: Statistical comparisons and scientific issues. *J. Geophys. Res. Space Phys.* **2002**, *107*, SIA-15. [CrossRef]
30. Licata, R.J.; Mehta, P.M.; Weimer, D.R.; Tobiska, W.K.; Yoshii, J. MSIS-UQ: Calibrated and enhanced NRLMSIS 2.0 model with uncertainty quantification. *Space Weather* **2022**, *20*, e2022SW003267. [CrossRef]
31. Guo, Y.; Zhang, X.; Guo, F.; Yang, Y. Performance Analysis of NRLMSIS 2.1 Thermospheric Mass Density Model using GRACE-A and SWARM-C Observations. *Adv. Space Res.* **2024**, *74*, 2475–2491. [CrossRef]
32. Virgili, B.B.; Lemmens, S.; Stevenson, E. The Impact of the New NRLMSIS 2.0 on Re-Entry Predictions. In Proceedings of the 8th European Conference on Space Debris, Online, 20–23 April 2021; pp. 20–23.
33. Lean, J.; Picone, M.; Knowles, S.; Hedin, A.; Moore, G. Validating NRLMSIS Using Atmospheric Densities Derived from Spacecraft Drag: Starshine Example. In Proceedings of the AIAA/AAS Astrodynamics Specialist Conference and Exhibit, Monterey, CA, USA, 5–8 August 2002; p. 4736.
34. Fried, A.; Richter, D. Infrared absorption spectroscopy. In *Analytical Techniques for Atmospheric Measurement*; Wiley-Blackwell: Hoboken, NJ, USA, 2006; pp. 72–146.
35. Horvath, H. Atmospheric light absorption—A review. *Atmos. Environ. Part A Gen. Top.* **1993**, *27*, 293–317. [CrossRef]
36. Bailey, J.; Simpson, A.; Crisp, D. Correcting infrared spectra for atmospheric transmission. *Publ. Astron. Soc. Pac.* **2007**, *119*, 228. [CrossRef]
37. Clough, S.A.; Iacono, M.J.; Moncet, J.-L. Line-by-line calculations of atmospheric fluxes and cooling rates: Application to water vapor. *J. Geophys. Res. Atmos.* **1992**, *97*, 15761–15785. [CrossRef]
38. Lacis, A.A.; Oinas, V. A description of the correlated k distribution method for modeling nongray gaseous absorption, thermal emission, and multiple scattering in vertically inhomogeneous atmospheres. *J. Geophys. Res. Atmos.* **1991**, *96*, 9027–9063. [CrossRef]
39. Fu, Q.; Liou, K.N. On the Correlated k-Distribution Method for Radiative Transfer in Nonhomogeneous Atmospheres. *J. Atmos. Sci.* **1992**, *49*, 2139–2156. [CrossRef]
40. Gordon, I.E.; Rothman, L.S.; Hargreaves, R.J.; Hashemi, R.; Karlovets, E.V.; Skinner, F.M.; Conway, E.K.; Hill, C.; Kochanov, R.V.; Tan, Y. The HITRAN2020 molecular spectroscopic database. *J. Quant. Spectrosc. Radiat. Transf.* **2022**, *277*, 107949. [CrossRef]
41. Rothman, L.S.; Gordon, I.E.; Barber, R.J.; Dothe, H.; Gamache, R.R.; Goldman, A.; Perevalov, V.I.; Tashkun, S.A.; Tennyson, J. HITEMP, the high-temperature molecular spectroscopic database. *J. Quant. Spectrosc. Radiat. Transf.* **2010**, *111*, 2139–2150. [CrossRef]
42. Delahaye, T.; Armante, R.; Scott, N.A.; Jacquinet-Husson, N.; Chédin, A.; Crépeau, L.; Crevoisier, C.; Douet, V.; Perrin, A.; Barbe, A. The 2020 edition of the GEISA spectroscopic database. *J. Mol. Spectrosc.* **2021**, *380*, 111510. [CrossRef]
43. Keller-Rudek, H.; Moortgat, G.K.; Sander, R.; Sörensen, R. The MPI-Mainz UV/VIS spectral atlas of gaseous molecules of atmospheric interest. *Earth Syst. Sci. Data* **2013**, *5*, 365–373. [CrossRef]
44. Schroeder, P.J.; Pfotenhauer, D.J.; Yang, J.; Giorgetta, F.R.; Swann, W.C.; Coddington, I.; Newbury, N.R.; Rieker, G.B. High temperature comparison of the HITRAN2012 and HITEMP2010 water vapor absorption databases to frequency comb measurements. *J. Quant. Spectrosc. Radiat. Transf.* **2017**, *203*, 194–205. [CrossRef]
45. Tynes, H.H.; Kattawar, G.W.; Zege, E.P.; Katsev, I.L.; Prikhach, A.S.; Chaikovskaya, L.I.J.A.O. Monte Carlo and multicomponent approximation methods for vector radiative transfer by use of effective Mueller matrix calculations. *Appl. Opt.* **2001**, *40*, 400–412. [CrossRef]
46. Lindfors, A.V.; Kujanpää, J.; Kalakoski, N.; Heikkilä, A.; Lakkala, K.; Mielonen, T.; Sneep, M.; Krotkov, N.A.; Arola, A.; Tamminen, J. The TROPOMI surface UV algorithm. *Atmos. Meas. Tech.* **2018**, *11*, 997–1008. [CrossRef]

remote sensing

MDPI

Article

Comparisons and Analyses of Thermospheric Mass Densities Derived from Global Navigation Satellite System–Precise Orbit Determination and an Ionization Gauge–Orbital Neutral Atmospheric Detector Onboard a Spherical Satellite at 520 km Altitude

Yujiao Jin [1,2,3,4], Xianguo Zhang [1,3,4,*], Maosheng He [1,4], Yongping Li [1,3,4], Xiangguang Meng [1,3,4], Jiangzhao Ai [1,3,4], Bowen Wang [1,2,3,4], Xinyue Wang [1,3,4] and Yueqiang Sun [1,3,4]

1 National Space Science Center, Chinese Academy of Sciences, Beijing 100190, China
2 University of Chinese Academy of Sciences, Beijing 100190, China
3 Beijing Key Laboratory of Space Environment Exploration, National Space Science Center, Beijing 100190, China
4 Key Laboratory of Science and Technology on Space Environmental Situation Awareness, Chinese Academy of Sciences, Beijing 100190, China
* Correspondence: zhangxg@nssc.ac.cn

Abstract: Thermospheric mass densities are investigated to explore their responses to solar irradiance and geomagnetic activity during the period from 31 October to 7 November 2021. Utilizing data from the Global Navigation Satellite System (GNSS) payload and an ionization gauge mounted on the Orbital Neutral Atmospheric Detector (OAD) payload onboard the QQ-Satellite, thermospheric mass densities are derived through two independent means: precise orbit determination (POD) and pressure measurements. For the first time, observations of these two techniques are compared and analyzed in this study to demonstrate similarities and differences. Both techniques exhibit similar spatial–temporal variations, with clear dependences on local solar time (LT). However, the hemispheric asymmetry is almost absent in simulations from the NRLMSISE-00 and DTM94 models compared with observations. At high latitudes, density enhancements of observations and simulations are shown, characterized by periodic bulge structures. In contrast, only the OAD-derived densities exhibit wave-like disturbances that propagate from two poles to lower latitudes during geomagnetic storm periods, suggesting a connection to traveling atmospheric disturbances (TADs). Over the long term, thermospheric mass densities derived from the two means of POD and the OAD show good agreements, yet prominent discrepancies emerge during specific periods and under different space-weather conditions. We propose possible interpretations as well as suggestions for utilizing these two means. Significantly, neutral winds should be considered in both methods, particularly at high latitudes and under storm conditions.

Keywords: thermospheric mass density; precise orbit determination; ionization gauge; GNSS; spherical satellite; difference analysis; geomagnetic storm; TADs

1. Introduction

Thermospheric mass density is a critical parameter of the thermosphere, strongly associated with orbital drag. The estimation error of thermospheric mass density is a major source of bias in orbital predictions, significantly affecting attitude control, orientation, and the on-orbit lifetime of near-space vehicles and Low Earth Orbit (LEO) spacecraft [1–3]. Solar radiation, including ultraviolet irradiance (UV) and extreme ultraviolet (EUV) irradiances, and geomagnetic storms driven by the interaction of solar wind and the magnetosphere, are primary sources of disturbances in the thermosphere. During storm events, disturbances of densities can propagate from the polar regions to lower latitudes in the form of TADs, but the underlying mechanisms responsible for them remain poorly understood [4,5]. Furthermore, variations in thermospheric mass density exhibit dependences on LT, latitude, season, and neutral winds [6–8]. However, the spareness of the observations of thermospheric mass density restricts the accuracy of current empirical models. Previous research has demonstrated that simulations from empirical models can either underestimate or overestimate, with deviations of 15–30% during quiet geomagnetic periods, and even up to 300% during severe geomagnetic storms [1,9,10].

Empirical models of thermospheric mass density mainly contain the Jacchia series, DTM series, and MSIS series, among others. The Jacchia series, including Jacchia65, Jacchia70, and later JB2006, JB2008, and so forth, primarily rely on total mass densities derived from satellite drag data [11–14]. The DTM series includes DTM94, DTM2000, DTM2013, and so forth. The DTM94 model integrates satellite drag data, component data, and total mass densities from sources like the CACTUS accelerometer, DE-2 satellite, and neutral mass spectrometer, including data under extreme solar conditions, which can provide accurate estimations under moderate solar and geomagnetic conditions [15]. DTM2000 further integrates data from incoherent scatter radar and interferometer sources for the lower thermosphere, and introduced the Mg II index to represent solar radiation [16]. DTM2013 utilizes high-precision data derived from the CHAMP, GRACE, and GOCE satellites and replaces the F10.7 index with F30 [17]. The MSIS series, including MSIS-86, NRLMSISE-00, NRLMSIS 2.0, and so on, use density and temperature data derived from incoherent scatter radar and in situ observations of satellites [18–20]. NRLMSISE-00 incorporates Jacchia orbital datasets and the number density of molecular oxygen from the Solar Maximum Mission (SMM), which effectively combines the advances of both the MSISE-90 and Jacchia70 models and has been widely utilized in the community [19].

Ionization gauges (or pressure gauges), accelerometers, and neutral mass spectrometers are primary in situ instruments onboard satellites to measure thermospheric mass density [21]. Among these, ionization gauges to derive density based on pressure measurements and accelerometers to derive density based on drag measurements are two main means [13]. In addition to accelerometers, high-quality object trajectory data, such as two-line orbital elements (TLEs) and POD, can also be used to retrieve thermospheric mass density based on drag measurements [22–24]. Particularly, GNSS observations of the pseudo-range and carrier phase can provide high-precision orbital ephemerides and nongravitational accelerations, estimating average density variations [25,26]. Originally, ionization gauges were manifested on early rocket missions and on primitive AE and DE series of satellites [21]. In 2005, ionization gauges were further advanced through the U.S. STREAK project, enhancing their ability to measure thermospheric density at 123–325 km and to observe perturbations caused by atmospheric gravity waves (AGWs) and TADs [3]. High-precision accelerometers and high-performance Global Positioning System (GPS) receivers were deployed on CHAMP and GRACE, which were instrumental in determining the mass density using multiple methods [27–30]. Similarly, the ESA's SWARM mission launched in 2013, and utilized a constellation of three satellites equipped with high-precision accelerometers and GPS receivers, allowing atmospheric density measurements by multiple methods at 470–520 km [31–33]. Neutral mass spectrometers were mounted on satellites to measure thermospheric mass density and compositions as part of relevant NASA and European Union missions [34,35]. Ionization gauges and neutral mass spectrometers are also utilized onboard Chinese missions such as the SZ-4 and APOD satellites to investigate thermospheric mass density.

Despite extensive observations of thermospheric mass density, evaluating the measurements remains challenging as no definitive reference value exists. As a result, comparing observations derived from multiple methods is essential for enhancing our understanding of the techniques and improving their accuracy and quality, and is also helpful for deepening our knowledge of thermospheric behaviors, refining models, and promoting predictive capabilities. Li and Lei [33] studied the responses of thermospheric density to storms by comparing measurements of POD and accelerometers from the GRACE and SWARM-A/B satellites, revealing dependences of hemispheric asymmetries on altitude and LT and suggesting that POD-derived densities can effectively capture the variations. However, comparisons of thermospheric densities derived from multiple methods that are based on distinct theoretical frameworks are scarce. To address this gap, Li et al. calibrated pressure-based densities by POD-derived densities from the same satellite, APOD, revealing a strong correlation of 94% between the densities derived from the two methods, despite the higher densities derived from pressure measurements [36].

The QQ-Satellite, a spherical experimental satellite that was launched into a polar orbit at 520 km in 2021, provides the unique opportunity to compare thermospheric mass densities derived from two different methods. It carries both a GNSS-POD payload and an ionization gauge embedded in an OAD payload, enabling the retrieval of thermospheric mass densities through two means: the dynamical inversion of POD information and atmospheric pressure measurements [37–39]. In this paper, we investigate the spatial–temporal evolution of thermospheric mass densities derived from two means, as well as simulations from the NRLMSISE-00 and DTM94 models, during quiet and storm periods. Significantly, we analyze features of difference between the densities derived from the two means in detail and explore potential reasons for these discrepancies. Finally, we provide recommendations for the future utilizations and enhancements of these methods, and propose requirements and expectations for further research.

2. Materials and Methods

2.1. QQ-Satellite

In 2003, SZ-4 carried an ionization gauge and neutral mass spectrometer mounted on the detector of atmospheric mass density that were developed by the National Space Science Center (NSSC) of the Chinese Academy of Sciences, observing thermospheric mass density at 330–362 km [40]. The APOD satellite, launched in 2015, carried a dual-frequency GNSS receiver and an ionization gauge mounted on the detector of atmospheric mass density to contribute to calibrating thermospheric mass densities derived from multiple methods [21]. The QQ-Satellite, shown in Figure 1a, with a weight of 73.8 kg and spherical shell diameter of 800 mm, was launched into an orbit with a 520 km altitude and 97.46° inclination on 14 October 2021. As demonstrated in Figure 1b, the motion trajectories and motion directions of the QQ-Satellite are represented by the black line and the red arrows, respectively. Furthermore, altitudes of polar and equatorial regions are also marked in the picture. Among these, the orbital trajectories of the QQ-Satellite cover almost all latitudes at the 6:00 LT/dawn side and at the 18:00 LT/dusk side, while it flies mainly in polar regions at other LTs [37–39]. The orbital altitudes vary, ranging approximately from 500 km to 545 km nearly a month after launching. The QQ-Satellite flies in a three-axis directional attitude to the ground, with the OAD instrument installed at the front of the satellite to ensure that the neutral atmosphere entered into the instrument is a free flow of molecules. The payloads mainly consist of the GNSS antenna, integrated atmospheric sensor and digital signal processing, and control unit, the performance indicators of which are shown in Table 1. The discoidal GNSS antenna of the GNSS-POD payload is coplanar with the satellite, integrating dual-systems observations of POD from the Beidou System (BDS) and GPS to derive thermospheric density by dynamical inversion. The integrated atmospheric sensor and ionization gauge, including the ion source, mass analyzer, and particle detector, contribute to deriving the thermospheric density based on pressure measurements.

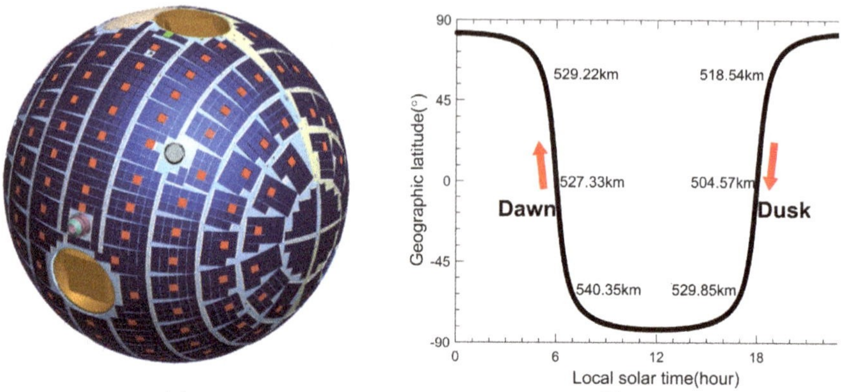

(**a**) (**b**)

Figure 1. (**a**) The QQ-Satellite, and (**b**) the projections of the QQ-Satellite orbits in a local solar time–latitude frame.

Table 1. Performance indicators of GNSS-POD and ionization gauge–OAD.

Detecting Parameters	Indicator Accuracy
Atmospheric density detection range	10^{-15}–10^{-9} kg/m^3
Atmospheric density detection accuracy	<6.5%
Sample time interval	1 s
Pseudo-range accuracy	≤30 cm
Carrier phase accuracy	≤2 mm
POD accuracy	<5 cm
Temporal resolution of POD	30 s

It should be noted that all densities mentioned in this study represent the total mass density of the thermosphere if no additional declarations are made. In order to distinguish and describe them, we term the POD densities derived by the method of dynamical inversion as POD densities or POD and the OAD-derived densities derived by the method of pressure measurements as OAD densities or OAD in this paper.

2.2. *Theory of Orbital Neutral Atmospheric Density Detection*

The processes of deriving mass densities using the dynamical inversion method mainly comprise the determination of nongravitational accelerations and the resolving of the equation of atmospheric drag. The pressure measurements method of deriving mass densities relies on the Maxwell distribution law of velocity and measurements of the pressure and temperature of the sampling chamber.

2.2.1. The Dynamical Inversion Method

Accelerations to retrieve thermospheric mass density comprise the acceleration of the atmospheric model derived from empirical model DTM94 and the acceleration of empirical drag derived from high-precision POD observations [38]. Importantly, the final acceleration $\ddot{r}_{Atm}$ (m/s^2) is also modified dynamically to accord with high-precision observations derived from accelerometers. We can retrieve atmospheric mass density ρ (kg/m^3) according to the equation of atmospheric drag, with determined parameters such as the drag coefficient C_D, the area-to-mass ratio of frontal region A_m (m^2/kg), and the satellite relative velocity v_r (m/s), shown in Formula (1). Due to the particularly spherical configuration of the satellite, the drag coefficient and area-to-mass ratio in formula (1) can be assumed to be constants in this study [38,41].

$$\ddot{r}_{Atm} = -\frac{1}{2} C_D \cdot A_m \cdot \rho \cdot v_r^2 \tag{1}$$

2.2.2. The Pressure Measurements Method

The neutral atmospheric airflows from the front of the satellite are ionized into electrons and ions when they fly into the atmospheric sensor. The ions, with their charge-to-mass ratio modulated by the mass analyzer, are collected by the particle detector. After the processing of digital signal processing and control unit, we can observe the atmospheric pressure P_g (Pa) and temperature T_g (K).

Assume that the number of molecules entering and leaving the atmospheric sensor per unit time and per unit volume is equal. By integrating the Maxwell velocity distribution function, we can obtain the expression of the number density of free atmospheric molecules N_a (m^{-3}), shown as Formula (2). With determination of the average molecular mass of atmosphere in the sampling chamber m_1 (kg), the average molecular mass of free atmosphere m_2 (kg), velocity v (m/s), and attack angle θ, we can calculate total mass density ρ (kg/m^3) according to Formula (3) [21].

$$N_a = P_g (2k\pi m_1)^{-\frac{1}{2}} \left(\frac{1}{T_g}\right)^{\frac{1}{2}} \frac{1}{v\cos\theta} \tag{2}$$

$$\rho = N_a \cdot m_2 \tag{3}$$

3. Results

3.1. Spatial–Temporal Evolution of Observations and Simulations

Observations of POD and OAD derived from the QQ-Satellite and simulations derived from empirical atmospheric models of NRLMSISE-00 and DTM94 are investigated to explore variations in thermospheric mass density under different conditions of solar radiation and geomagnetic activity. Benefitting from dawn–dusk orbits of the QQ-Satellite, variations in density at the dawn side (4–8 LT) and at the dusk side (16–20 LT) can be demonstrated here. As shown in Figure 2, during quiet periods, enhancements of all densities are closely relevant to solar radiation, represented by the F10.7 index here, illustrating a crucial source of the heating of the thermosphere. For storm periods, prominent and violent disturbances of densities appear when the Kp index gradually reaches a peak. Furthermore, densities of the southern/summer hemisphere are larger than that of the northern/winter hemisphere, exhibiting hemispheric asymmetry. And densities of the dusk side are larger than that of the dawn side, which indicates that density variations are also dependent on LT.

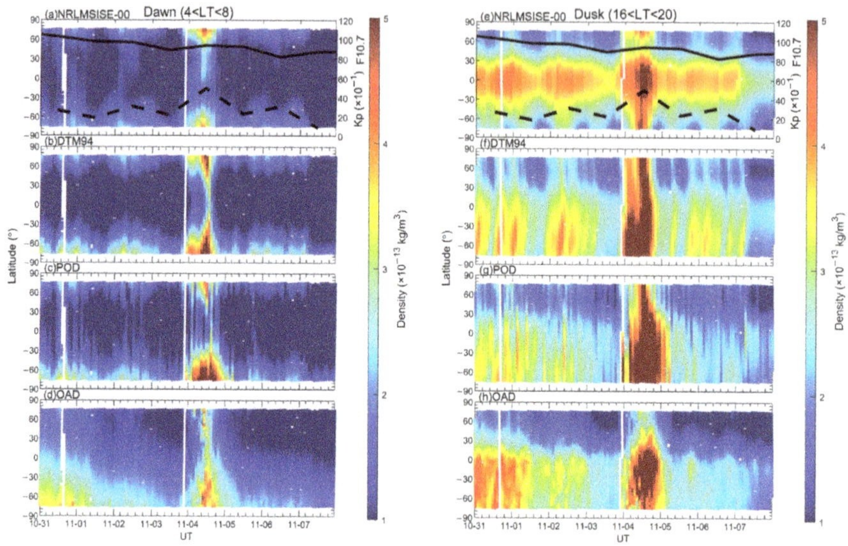

Figure 2. Thermospheric mass densities at (**a–d**) dawn and (**e–h**) dusk derived from (**a**,**e**) NRLMSISE-00, (**b**,**f**) DTM94, (**c**,**g**) POD, and (**d**,**h**) OAD. The (**a**,**e**) black solid line is the daily F10.7 index, and the (**a**,**e**) black dashed line is ten times of the daily Kp index.

However, there exist obvious differences between observations and simulations. At dawn, NRLMSISE-00 and DTM94 enhance almost simultaneously in two polar regions (70–90°N/S) and then disturb in equatorial regions (0–20°N/S). However, POD and OAD both enhance firstly in the south pole and then disturb in the north pole and equatorial regions. Further, NRLMSISE-00 densities enhance at 70–90°N/S 8 h/6.5 h earlier than at 0–20°N/S, respectively. DTM94 densities enhance at 70–90°N/S 7.5 h/5 h earlier than at 0–20°N/S. POD densities of 70–90°N/S enhance 7.5 h/6 h earlier than those of 0–20°N/S, and OAD densities of 70–90°N/S enhance 6 h/5 h earlier than those of 0–20°N/S. As for peaks of densities, NRLMSISE-00 peaks almost simultaneously in polar and equatorial regions. For DTM94, POD, and OAD, however, densities in polar regions peak 1.5–3 h earlier than those in equatorial regions. At dusk, enhancements of all densities propagate quicker from polar regions to equatorial regions. Among these, NRLMSISE-00 densities of 70–90°N/S enhance 1.5 h/4.5 h earlier than those of 0–20°N/S, and the circumstances of which advance 3 h/5 h for DTM94, respectively. Enhancements of POD and OAD at 70–90°N/S both advance 3 h/4.5 h earlier than that at 0–20°N/S. Furthermore, enhancements of NRLMSISE-00 mainly concentrate in the equatorial region at dusk, which differs from the other densities. Moreover, variations in the two observations of POD and OAD are also different. During quiet periods, OAD enhancements caused by solar radiation are more prominent compared with POD, which is more similar to DTM94 in variation on the whole. In the following sections, we will pay attention to exploring and analyzing differences in densities derived from POD and OAD.

3.2. Evolution of Differences of Two Measurements

Variations in densities derived from POD and OAD and differences in under which space-environment conditions are investigated as illustrated in Figure 3. The typical numbered orbits, marked in brown, gray, and yellow in Figure 3, will be analyzed in detail in the next section. Here, we calculate the ratios of densities derived from POD and OAD as $(\rho_{POD} - \rho_{OAD})/(\rho_{POD} + \rho_{OAD})$ to represent the relative differences in the two measurements. In the long run, the variation tendencies of the two measurements align well except for conditions of a larger F10.7 and larger Kp, as shown in Figure 3a. However, in terms of details, variations in differences are various, as shown in Figure 3b. It is not hard to find that differences in the two measurements are discrepant during different periods of variations in density. To be specific, POD agrees with OAD relatively well during periods of increasing densities, according to smaller ratios of differences. For periods of decreasing densities, however, POD densities are almost always smaller than OAD densities, according to the negative values of the ratios. And, according to the positive values of the ratio, POD is almost always larger than OAD during some special periods. Therefore, we distinguish periods of orbits as ascending and descending orbits because densities generally decline for ascending orbits and rise for descending orbits. Furthermore, we isolate all the special regions of 50–90°N from periods of ascending and descending orbits, corresponding to the transition and connection of two sequential orbits. Notably, to highlight variations in density, we name ascending orbit periods as decreasing (density) periods and term descending orbit periods as increasing (density) periods. As a result, we can obtain three orbital periods, namely increasing periods, decreasing periods, and transition periods.

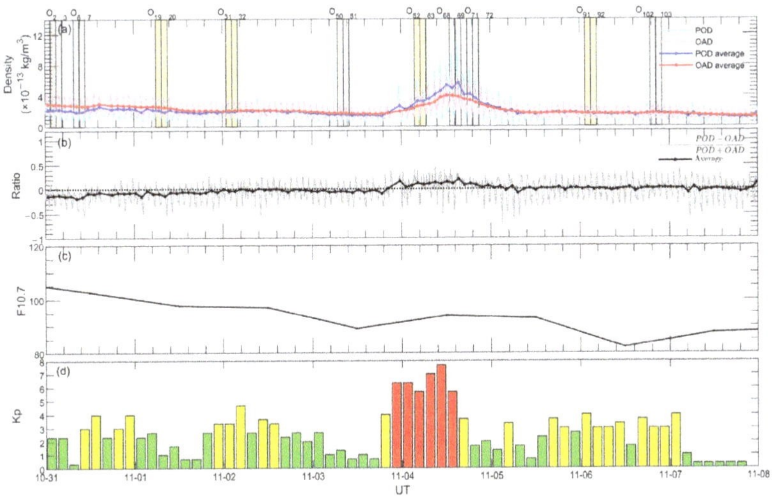

Figure 3. The evolutions of (**a**) thermospheric mass densities derived from two means and (**b**) their ratios, along with the evolutions of (**c**) F10.7 and (**d**) Kp indices.

As demonstrated in Figure 4, we analyze the ratios of the two measurements statistically during the different periods defined above. For whatever periods, ratios that represent differences in POD and OAD always fit normal distributions. The distribution of all the data (Figure 4a) can be regarded as the synthesis of the other three distributions for different periods (Figure 4b–d). Among these, the ratios of all density data fit the distribution with a mean value of $\mu = -0.0410$ and a standard deviation of $\sigma = 0.1826$. For distributions of increasing/decreasing/transition periods, mean values are 0.0171/−0.1905/0.1213 and standard deviations are 0.1021/0.1563/0.1318. By contrast, it is proven again that consistencies of the two measurements are best during increasing periods, according to the smallest μ and σ. And differences are larger for decreasing and transition periods, which exhibit a larger μ and σ. Moreover, POD densities are almost always much smaller than OAD densities for decreasing periods but are always larger for transition periods. To figure it out, we will investigate and explore the differences in the two observations in detail for different periods in the next section.

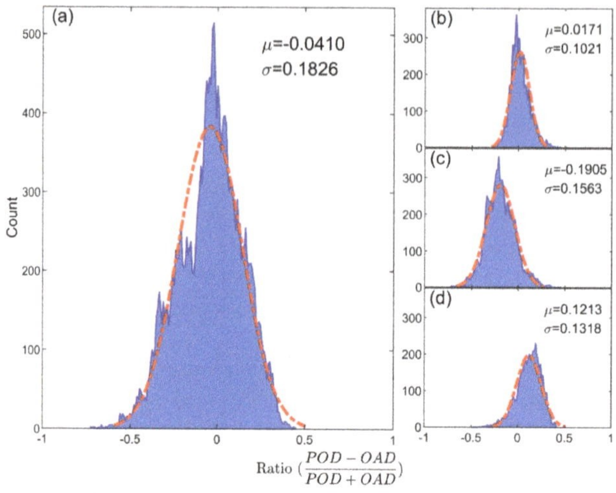

Figure 4. Normal distributions of ratios displayed in Figure 3b for (**a**) all data, and for three periods defined in Section 3.2: (**b**) increasing period, (**c**) decreasing period, and (**d**) transition period.

4. Discussion

4.1. Characteristics of Differences During Increasing Periods

During increasing periods, differences in POD and OAD on typical orbits, corresponding to different space-weather conditions, are mainly characterized by three patterns, as shown in Figure 5. The increasing/decreasing/transition periods are distinguished as intervals $P_{1,4,7}/P_{2,5}/P_{3,6}$, and the increasing periods are highlighted in brown, discussed only here. In Figure 5a,b, for high solar activity and low geomagnetic activity, POD densities are smaller than OAD densities, as mentioned in Section 3.1. OAD obviously elevates overall for increasing F10.7, which is almost not seen in POD. On the contrary, POD densities are larger than OAD densities for severe geomagnetic storms, also accompanied by more prominent fluctuation features in OAD, as shown in Figure 5e,f.

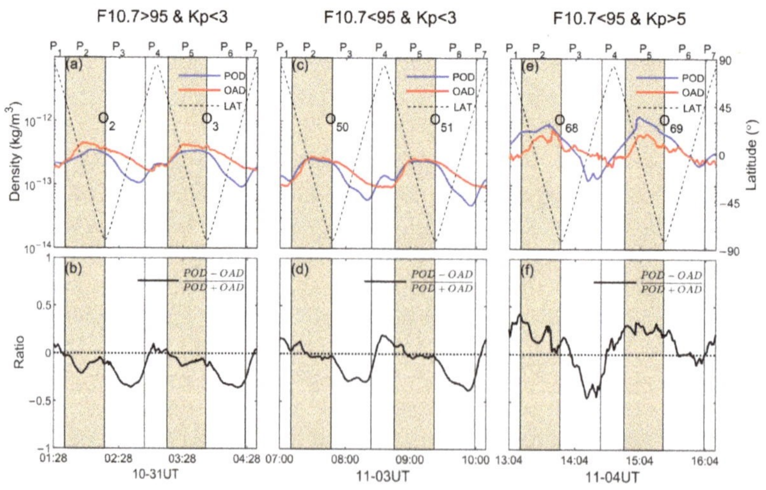

Figure 5. (**a**,**c**,**e**) Thermospheric mass densities derived from POD and OAD, and (**b**,**d**,**f**) their ratios. The O_2, O_3, O_{50}, O_{51}, O_{68}, and O_{69} indicate the accumulated numbers of the QQ-Satellite's orbits since 31 October 2021. And the shadows represent the increasing periods.

During the process of determining POD-derived densities, atmospheric density models are used to estimate accelerations, a significant portion of which are used to obtain drag accelerations to derive densities, to compensate for errors of dynamics models and orbit predictions based on GPS data [42–44]. Therefore, both measurement errors and model uncertainties can contribute to deviations in densities derived from drag accelerations [45]. In this paper, the empirical model DTM94, utilized to obtain empirical densities as the initial inputs, therefore could impact POD densities. And in our study results, POD resembles DTM94 in variations under some circumstances. Given that, differences illustrated in Figure 5a,b,e,f might arise from the impacts of DTM94 on POD densities, namely underestimation for high F10.7 and overestimation for high Kp. For increasing periods, except for the differences mentioned above, POD is in agreement with OAD generally under other conditions, as demonstrated in Figure 5c,d.

4.2. Characteristics of Differences During Decreasing Periods

Similarly, we select typical orbits to analyze the characteristics of differences under various conditions, as shown in Figure 6. The decreasing periods, the objectives to be discussed, are exhibited by P_3 and P_6 intervals highlighted in gray. For decreasing periods, we find an unexpected phenomenon that OAD densities are almost always larger than POD, appearing periodically as delays in the time phase. In Figure 6a,b, under conditions of a large F10.7 and small Kp, OAD greatly deviates from POD, and the same circumstances apply for conditions of a small F10.7 and Kp. For a low-level F10.7, shown in Figure 6c–f, differences in the two measurements reduce with active geomagnetic activities. On the one hand, POD could be influenced by the empirical model DTM94. On the other hand, it could be attributed to the uncertainty of measurements derived from the OAD instrument. Observations derived from the Neutral Gas and Ion Mass Spectrometer (NGIMS) onboard the Mars Atmosphere and Volatile Evolution (MAVEN) satellite on the elliptic orbit suggested a problem that the desorption of gas from the inner surface of the instrument could introduce errors, especially during the outbound portion [46]. Although similar to the NGIMS in operating principle, the OAD instrument onboard the QQ-Satellite, being in a circular orbit, may not be impacted by the desorption of gas from the inner surface of the instrument. Further, there are few relevant references about the issue, which should be certified further through simulation experiments and comparisons with other isochronous observations.

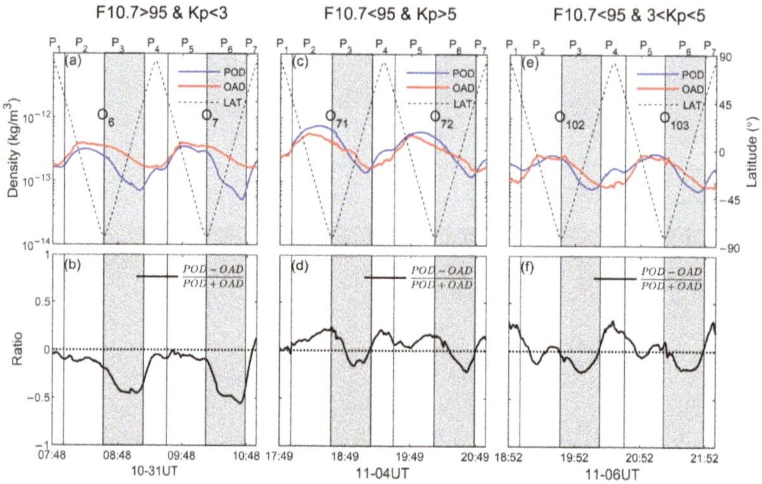

Figure 6. Similar plot to Figure 5, except for different satellite orbits. (**a**,**c**,**e**) Thermospheric mass densities derived from POD and OAD, and (**b**,**d**,**f**) their ratios. And the shadows represent the decreasing periods.

4.3. Characteristics of Differences During Transition Periods

Similarly, we investigate primary discrepancy characteristics for transition periods under different conditions according to representative orbits, as shown in Figure 7. The transition periods are highlighted in yellow, marked as P_1, P_4, and P_7. Owing to incomplete orbits in P_1 and P_7 intervals, we only analyze differences shown in the P_4 interval here. For transition periods, in the northern polar regions, density variations, as well as differences in POD and OAD, are complicated and various. Enhancements of density in polar regions exist in both POD and OAD, appearing as obvious bulge structures of density under some circumstances. In Figure 7a,b, under conditions of a large F10.7 and small Kp, POD aligns relatively well with OAD, demonstrating smaller ratios of difference. Among these, POD increases quickly at first and then decreases slowly, exhibiting a smooth bulge structure of density in the north pole. In contrast, OAD varies stably on the whole and a little sharp peak appears at about 82°N. In Figure 7c,d, by contrast, differences in POD and OAD rise accompanied by increasing Kp. Similarly, both POD and OAD demonstrate peaks of density, shaped in bulge structures, at about 82°N, but there is a sharper bulge structure for OAD. In Figure 7e,f, for a small F10.7 and the largest Kp, differences in the two measurements further worsen. By contrast, OAD varies more violently than POD, with a large and sharp bulge structure emerging at about 82°N. Particularly in Figure 7g,h, for a small F10.7 and consecutively large Kp, the differences are the worst. POD continues to exhibit a large but smooth bulge structure of density peaking at about 82°N; however, OAD exhibits multiple little wave-like structures.

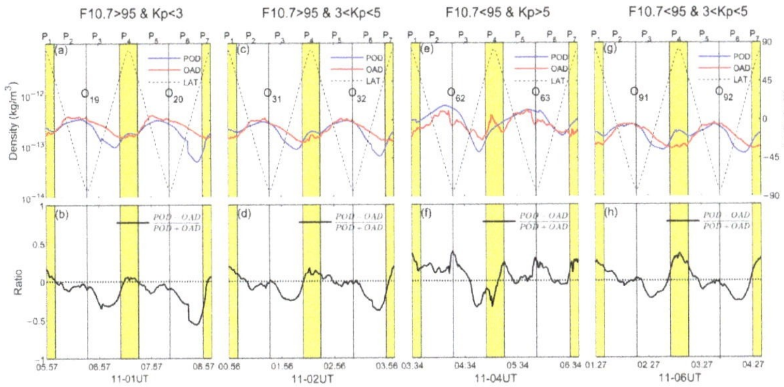

Figure 7. Similar plot to Figure 5, except for different satellite orbits. (**a**,**c**,**e**,**g**) Thermospheric mass densities derived from POD and OAD, and (**b**,**d**,**f**,**h**) their ratios. And the shadows represent the transition periods.

During transition periods, the differences in POD and OAD mentioned above seem to depend more on geomagnetic activity. Meanwhile, the north pole is in the polar night in winter for the period of November. As a result, we could basically exclude the effect of solar radiation on the arising of bulge features of density at 50–90°N. Moreover, the region of 50–90°N, covering the auroral zone and other special regions, is important for enhancements of density due to the dissipation of energy and precipitating of particles. Based on accelerations derived from CHAMP during storm period 8–9 June 2001, Bruinsma and Tamagnan [45] derived thermospheric mass density and saw similar enhancements of density in the north pole, even more disturbed than those in the south pole. During storm periods, structures of density enhancement at high latitudes emerge owing to air up-welling, related to enhancements of Joule Heating [47].

Based on the above analysis, density peaks of bulge structures from simulations and observations are determined and examined. Variations in the peaks of bulges and their relative enhancements for all orbits, along with variations in the auroral electrojet (AE) index, are demonstrated in Figure 8. The lacking dots in Figure 8 represent the absence of bulge structures on corresponding orbits. The peaks of bulges minus the averages of density on the corresponding orbit are enhancements of density, as shown in Figure 8b. By contrast, bulges of density from the models are fewer, especially for NRLMSISE-00, yet OAD appears bulge for almost all orbits. For all 120 orbits investigated, the numbers of orbits appearing with bulge features for NRLMSISE-00, DTM94, POD, and OAD are 40, 104, 110, and 119, respectively. Moreover, notably, the magnitudes of POD peaks are almost always larger than those of OAD peaks, as mentioned above, which could be related to neutral winds. Bruinsma et al. [45,48] suggested that the speed of neutral winds could reach on the order of 500–1000 m/s during geomagnetic storm periods as well as at high latitudes, which is non-ignorable for estimations of atmospheric density. Due to the neglect of neutral winds in data processing, the speed of the satellite is actually exaggerated, finally leading to overestimations of POD densities (v^2) but underestimations of OAD densities (v^{-1}). Further, enhancements of peaks from all densities, especially for OAD densities, exhibit relatively great agreements with the AE index, indicating strong associations with energy injection and particle precipitation in polar regions during geomagnetic storms.

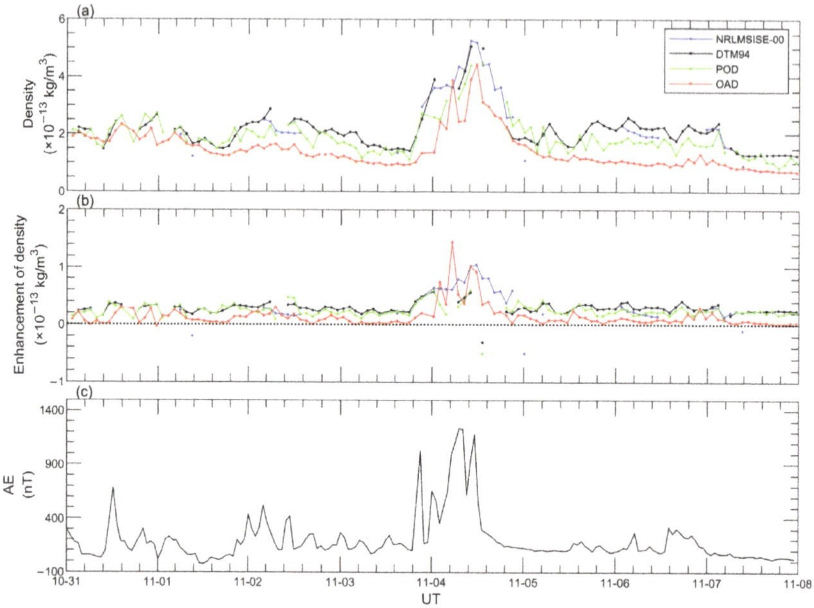

Figure 8. The evolutions of (**a**) peak densities in the orbital bulge structures from simulations and observations and their (**b**) enhancements, along with the evolution of (**c**) the AE index.

For transition periods, there is little variation in altitude concentrated at 520–530 km. Therefore, we primarily investigate distributions of bulge-feature peaks in longitude, latitude, and LT, demonstrated in Figure 9. Whether for simulations or observations, there is little dependence on longitude, but strong dependences on latitude and LT. Among these, most peaks appear at high latitudes, spreading into low latitudes occasionally. As shown in Figure 10, the percentages of peaks appearing at above 82°N for NRLMSISE-00, DTM94, POD, and OAD are 22.50%, 4.81%, 26.36%, and 10.08%, along with 92.50%, 97.12%, 98.18%, and 71.43% at above 70°N. By comparison, more peaks emerge at below 70°N and even spread into 50°N for OAD. Notably, the propagations of peaks in latitude appearing in OAD could relate to the effects of some atmospheric waves like the gravity wave during active geomagnetic activities. As for LT, the majority of peaks for all densities appear at dusk (16–24 LT).

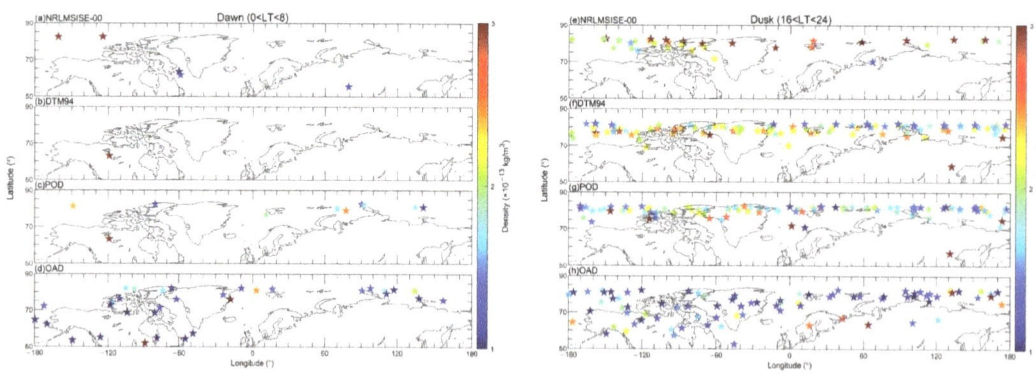

Figure 9. Spatial distributions of peak density in bulge structures at (**a**–**d**) dawn (0–8 LT) and at (**e**–**h**) dusk (16–24 LT), derived from (**a**,**e**) NRLMSISE-00, (**b**,**f**) DTM94, (**c**,**g**) POD, and (**d**,**h**) OAD.

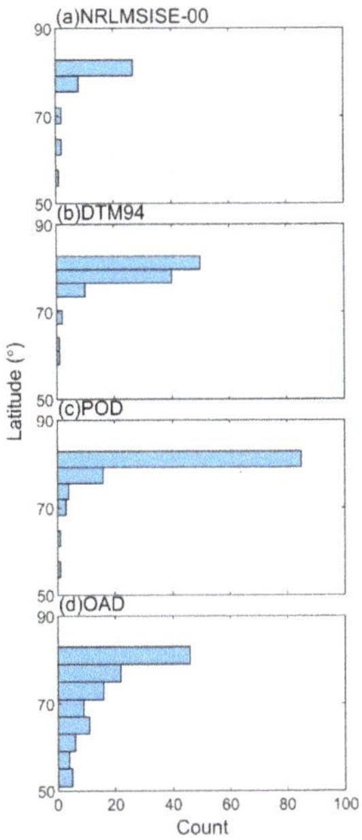

Figure 10. Statistical distributions in latitudes for bulge peaks derived from (**a**) NRLMSISE-00, (**b**) DTM94, (**c**) POD, and (**d**) OAD.

As shown in Figure 11, latitudinal enhancements of OAD, represented by ratios of densities during storm periods relative to quiet periods, are investigated. The black solid lines and dashed lines represent quiet and recovery periods, respectively. The rainbow-colored lines represent the main phases of a geomagnetic storm, and the time sequences are shown at the top of Figure 11. During quiet periods, whether for dawn or dusk, densities hardly enhance, except for polar regions. However, for the main phase of the storm, distinct dawn–dusk differences emerge. At dawn, densities enhance firstly in the two poles, more prominently in the north pole, then propagate into middle and low latitudes, and finally concentrate on equatorial regions. At dusk, similarly, densities enhance firstly in the polar regions, yet there are no obvious hemispheric differences. Furthermore, at dusk, disturbed density mainly focuses on the middle and low latitudes, appearing as densely wave-like features. For recovery periods, densities decrease gradually to initial levels whether at dawn or dusk, but are more disturbed at dusk.

The special perturbations of density, only seen in OAD, generating from the polar regions and then propagating into low latitudes in wave features, could relate to large-scale fluctuations in atmosphere TADs. According to densities derived from CHAMP during storm periods, Forbes and Lu [49] saw large-scale structural TADs at night, which enhanced firstly from the two poles, then propagated toward the equator, and finally dissipated into opposite hemispheres. Based on observations derived from CHAMP and GRACE, Bruinsma et al. found similar bulge structures of density, which appeared at both dayside and nightside but were more disturbed in daytime [48]. Furthermore, in 2007, Bruinsma and Forbes [50] found that TADs featured differently for day and night and for the two hemispheres according to CHAMP observations, which could relate to atmospheric circulation and ion drag. In 2022, according to densities derived from CHAMP and GRACE, Pham and Zhang [51] saw TADs that originated from high latitudes and propagated into low latitudes, and suggested associations between TADs and Joule Heating.

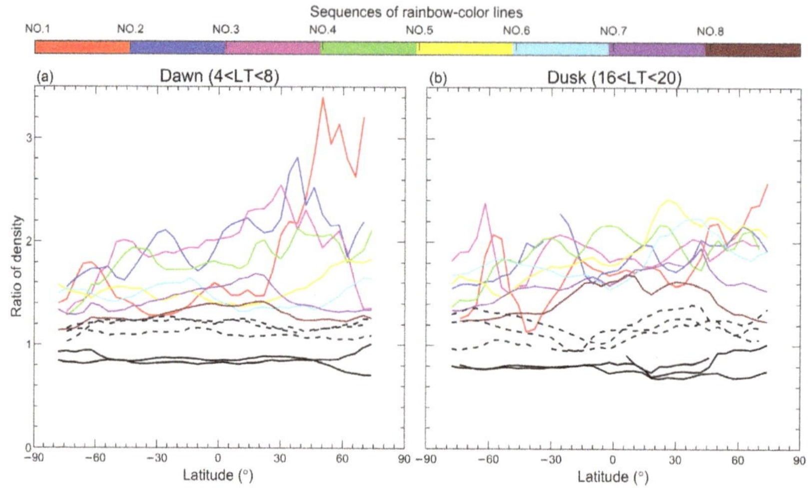

Figure 11. Ratios of OAD densities during storm periods on 3 November to 5 November relative to a quiet period on 2 November at (**a**) dawn, 4–8 LT and at (**b**) dusk, 16–20 LT.

5. Conclusions

In this study, we investigate thermospheric mass densities by the two means of POD and OAD, derived from the QQ-Satellite, during the period of 31 October to 7 November 2021. Simulations derived from empirical atmospheric models of NRLMSISE-00 and DTM94 were also determined for reference. Responses of densities to solar radiation and geomagnetic activity were analyzed according to F10.7 and Kp indices. Importantly, we divide orbital periods, named by variations of densities: increasing, decreasing, and transition periods, to exhibit differences in POD and OAD.

In conclusion, POD and OAD are two crucial space-based means to derive thermospheric densities. Based on analyses of physical phenomena and data differences in this paper, we can cognize more profoundly these two means, and suggest issues to pay attention to and recommendations for future utilizations. Meanwhile, we can provide research foundations for the incorporation of densities derived from multiple methods and for the empirical modeling of thermospheric mass density. Significant and critical conclusions are summarized in the following three points.

1. In general, POD accords relatively well with OAD, reflecting similar density variations under space-environment conditions. During quiet periods, density enhancements depend primarily on solar radiation, as indicated by the F10.7 index here. During geomagnetic storms, densities enhance severely in a global sense. By contrast, POD and OAD elevate first in the southern/summer hemisphere, then extend into the middle and low latitudes, but simulations from models enhance almost simultaneously in the two hemispheres. Furthermore, variations in density show great dependences on LT, demonstrating higher densities at dusk. However, the hemispheric asymmetry is hardly seen in NRLMSISE-00 and DTM94 compared with observations.
2. During different periods, POD differs variously from OAD, relying on space-weather conditions as well. For increasing periods, corresponding to generally increasing density, POD aligns well with OAD except for conditions of a large F10.7 and large Kp. Among these, OAD is slightly larger than POD for a large F10.7; however, it is smaller than POD for a large Kp. For decreasing periods, namely generally decreasing density, there exist relatively large differences between POD and OAD. OAD densities always emerge with delays in the time phase, gradually diminishing with the enhancement

of Kp. Significantly, for transition periods, the regions of 50–90°N, differences in POD and OAD are complicated and changeable. Only for a large F10.7 and low Kp, POD accords with OAD comparatively. In other circumstances, POD is larger than OAD generally, with the difference possibly produced by neutral winds. In particular, for active geomagnetic activity, enhancements from simulations and observations emerge in the north pole, characterized by bulge structures. The vast majority of the peaks of these bulge structures distribute at above 70°N and at the dusk side, showing great correlations with the AE index. Differently, bulge structures in OAD appear as irregular wave-like features, spreading more into lower latitudes, whereas they are smoother in POD and simulations. Further, during storm periods, TADs in OAD are demonstrated at dawn and dusk, propagating in wave characteristics from the two poles to low latitudes, and more perturbed at dawn.

3. Here, we put forward some recognitions of potential utilizations and suggestions for investigations of POD and OAD densities for upcoming studies. We suggest that POD could be affected by dynamical models like DTM94 used in the process of derivation; therefore, they should be further calibrated and corrected according to uncertainty analysis. Further, in spite of reflecting only average variations in density, the means of POD are favorable for massive modeling of the thermosphere, as it is convenient and unrestricted by spatial–temporal coverages of instruments. By contrast, high-resolution OAD can exhibit elaborate and multi-scale variations in density, and it promises to be an alternative to the accelerometer, and consequently beneficial for scientific research on physical evolution and coupled mechanisms in the thermosphere. Notably, delays in OAD relative to POD during outgas periods should be further verified and ascertained in future studies. In addition, neutral winds should be taken into account in POD and OAD, particularly for storm periods and for high latitudes. Eventually, we propose an incorporation of POD and OAD in upcoming studies, to provide high-quality densities for high-precision modeling, predictions of satellite orbits, and debris forewarning.

Author Contributions: Conceptualization, Y.J. and X.Z.; methodology, Y.J. and X.Z.; software, Y.J.; validation, Y.J., M.H., Y.L. and X.M.; formal analysis, X.M., J.A., B.W. and X.W.; investigation, Y.J.; resources, X.Z., M.H., Y.L. and Y.S.; data curation, Y.J., J.A. and B.W.; writing—original draft preparation, Y.J. and X.Z.; writing—review and editing, Y.J., X.Z., M.H., Y.L. and X.M.; visualization, Y.J.; supervision, X.Z. and Y.S.; project administration, J.A.; funding acquisition, M.H. and Y.L. All authors have read and agreed to the published version of the manuscript.

Funding: This research was funded by the Project of Stable Support for Youth Teams in Basic Research Fields, CAS (YSBR-018), the National Key R&D program of China (2021YFE0110200), the China Manned Space Program (Y59003AC40), and the TianMu Atmospheric Density Detector Development Project, (E3C1162110).

Data Availability Statement: The data derived from the QQ-Satellite are not publicly available. The QQ-Satellite datasets of ephemerides can be obtained from the repository provided by the National Space Science Center. The NRLMSISE-00 empirical model is available on the CCMC website (https://ccmc.gsfc.nasa.gov/, accessed on 14 September 2021). The space weather index of F10.7 is available at http://celestrak.org/SpaceData/ (accessed on 10 November 2021). The space weather indices of Dst and Kp are available at the NASA OMNI website (http://omniweb.gsfc.nasa.gov/, accessed on 11 January 2022). The source code of empirical model DTM94 is not available now due to it being an extremely old version. The AE space weather index is simulated because there are no available data for the periods studied in this paper. All data that are not available can be requested from the corresponding author.

Acknowledgments: We acknowledge the contribution of materials and data from the National Space Science Center, the Chinese Academy of Sciences as well as the Beijing Key Laboratory of Space Environment Exploration, the National Space Science Center.

Conflicts of Interest: The authors declare no conflicts of interest.

References

1. Chen, H.; Qin, G.; Li, Y.; He, Y.; Zhu, G. Comparison and analysis between simulations and observations derived from the atmospheric density detector onboard TianGong-1. *China Manned Space* **2013**, *19*, 34–38. [CrossRef]
2. Ren, Z. Modelling of upper atmosphere: From the earth to the planets. *Chin. Sci. Bull.* **2020**, *65*, 1320–1335. Available online: https://kns.cnki.net/kcms/detail/11.1784.N.20200408.1024.006.html (accessed on 29 January 2022). [CrossRef]
3. Clemmons, J.H.; Friesen, L.M.; Katz, N.; Ben-Ami, M.; Dotan, Y.; Bishop, R.L. The ionization gauge investigation for the streak mission. *Space Sci. Rev.* **2009**, *145*, 263–283. [CrossRef]
4. Li, Y.; Sun, Y.; Zhang, X.; Ai, J.; Zheng, X.; Li, J.; Wang, Y.; Guo, B.; Yan, F.; Wei, S.; et al. Thermosphere joint observations by TM-1 constellations and swarm-b during the April 2023 geomagnetic storm. *Earth Planet. Phys.* **2024**, *8*, 307–316. [CrossRef]
5. Li, R. The Research on Retrieval and Utilization of Thermospheric Mass Density. Ph.D. Thesis, University of Science and Technology of China, Hefei, China, 2022. [CrossRef]
6. Lei, J.; Matsuo, T.; Dou, X.; Sutton, E.; Luan, X. Annual and semiannual variations of thermospheric density: EOF analysis of CHAMP and GRACE data. *J. Geophys. Res. Space Phys.* **2012**, *117*, A01310. [CrossRef]
7. Wang, X.; Miao, J.; Lu, X.; Aa, E.; Liu, J.; Wang, Y.; Liu, S. Latitudinal impacts of joule heating on the high-latitude thermospheric density enhancement during geomagnetic storms. *J. Geophys. Res. Space Phys.* **2021**, *126*, 7. [CrossRef]
8. Sutton, E.K.; Forbes, J.M.; Nerem, R.S. Global thermospheric neutral density and wind response to the severe 2003 geomagnetic storms from champ accelerometer data. *J. Geophys. Res. Space Phys.* **2005**, *110*, A09S40. [CrossRef]
9. Lei, J.; Thayer, J.P.; Lu, G.; Burns, A.G.; Wang, W.; Sutton, E.K.; Emery, B.A. Rapid recovery of thermosphere density during the October 2003 geomagnetic storms. *J. Geophys. Res. Space Phys.* **2011**, *116*, A03306. [CrossRef]
10. Wang, X.; Miao, J.; Lu, X.; Aa, E.; Luo, B.; Liu, J.; Hong, Y.; Wang, Y.; Ren, T.; Zeng, R.; et al. Using temporal relationship of thermospheric density with geomagnetic activity indices and Joule Heating as calibration for NRLMSISE-00 during geomagnetic storms. *Space Weather* **2022**, *20*, e2021SW003017. [CrossRef]
11. Jacchia, L.G. Static diffusion models of the upper atmosphere with empirical temperature profiles. *Smithson. Contrib. Astrophys.* **1965**, *8*, 213–257. [CrossRef]
12. Jacchia, L.G. *Speical Report 1970: New Static Models of the Thermosphere and Exosphere with Empirical Temperature Profiles*; Smithsonian Astrophysical Observatory: Cambridge, MA, USA, 1970.
13. Emmert, J.T. Thermospheric mass density: A review. *Adv. Space Res.* **2015**, *56*, 773–824. [CrossRef]
14. Bowman, B.R.; Kent Tobiska, W.; Marcos, F.A.; Valladares, C. The JB2006 empirical thermospheric density model. *J. Atmos. Sol.-Terr. Phys.* **2008**, *70*, 774–793. [CrossRef]
15. Berger, C.; Biancale, R.; Ill, M.; Barlier, F. Improvement of the empirical thermospheric model DTM: DTM94—A comparative review of various temporal variations and prospects in space geodesy applications. *J. Geod.* **1998**, *72*, 161–178. [CrossRef]
16. Bruinsma, S.; Thuillier, G.; Barlier, F. The DTM-2000 empirical thermosphere model with new data assimilation and constraints at lower boundary: Accuracy and properties. *J. Atmos. Sol.-Terr. Phys.* **2003**, *65*, 1053–1070. [CrossRef]
17. Bruinsma, S. The DTM-2013 thermosphere model. *J. Space Weather Space Clim.* **2015**, *5*, A1. [CrossRef]
18. Hedin, A.E. MSIS-86 thermospheric model. *J. Geophys. Res. Space Phys.* **1987**, *92*, 4649–4662. [CrossRef]
19. Picone, J.M.; Hedin, A.E.; Drob, D.P.; Aikin, A.C. NRLMSISE-00 empirical model of the atmosphere: Statistical comparisons and scientific issues. *J. Geophys. Res. Space Phys.* **2002**, *107*, 1468. [CrossRef]
20. Emmert, J.T.; Drob, D.P.; Picone, J.M.; Siskind, D.E.; Jones, M., Jr.; Mlynczak, M.G.; Bernath, P.F.; Chu, X.; Doornbos, E.; Funke, B.; et al. NRLMSIS 2.0: A whole-atmosphere empirical model of temperature and neutral species densities. *Earth Space Sci.* **2021**, *8*, e2020EA001321. [CrossRef]
21. Li, X.; Xu, J.; Tang, G.; Chen, G.; Man, H.; Liu, S.; Li, Y. Data processing and calibration of atmospheric density derived from APOD satellite. *Chin. J. Geophys.* **2018**, *61*, 3567–3576. (In Chinese) [CrossRef]
22. Emmert, J.T.; Picone, J.M.; Lean, J.L.; Knowles, S.H. Global change in the thermosphere: Compelling evidence of a secular decrease in density. *J. Geophys. Res. Space Phys.* **2004**, *109*, A02301. [CrossRef]
23. Weng, L. Study on Variations and Modelling of Thermospheric Mass Density. Ph.D. Thesis, University of Science and Technology of China, Hefei, China, 2019.
24. Sang, J.; Smith, C.; Zhang, K. Towards accurate atmospheric mass density determination using precise positional information of space objects. *Adv. Space Res.* **2012**, *49*, 1088–1096. [CrossRef]

25. Van den Ijssel, J.; Doornbos, E.; Iorfida, E.; March, G.; Siemes, C.; Montenbruck, O. Thermosphere densities derived from Swarm GPS observations. *Adv. Space Res.* **2020**, *65*, 1758–1771. [CrossRef]
26. Jäggi, A.; Hugentobler, U.; Bock, H.; Beutler, G. Precise orbit determination for GRACE using undifferenced or doubly differenced GPS data. *Adv. Space Res.* **2007**, *39*, 1612–1619. [CrossRef]
27. Reigber, C.; Lühr, H.; Schwintzer, P. CHAMP mission status. *Adv. Space Res.* **2002**, *30*, 129–134. [CrossRef]
28. Tapley, B.D.; Bettadpur, S.; Watkins, M.; Reigber, C. The gravity recovery and climate experiment: Mission overview and early results. *Geophys. Res. Lett.* **2004**, *31*, L09607. [CrossRef]
29. Lei, J.; Li, R.; Ren, D.; Weng, L.; Ruan, H. Research advances on retrieval and modelling of thermospheric density. *Rev. Geophys. Planet. Phys.* **2023**, *54*, 434–454. [CrossRef]
30. Li, J.; Zhang, S.; Zou, X.; Jiang, W. Precise orbit determination for GRACE with zero-difference kinematic method. *Chin. Sci. Bull.* **2010**, *55*, 600–606. [CrossRef]
31. Zhao, Z. Research on Responses of Orbital Variations of SWARM Satellite to Geomagnetic Storm. Master's Thesis, Harbin Institute of Technology, Harbin, China, 2018. [CrossRef]
32. Kodikara, T.; Carter, B.; Zhang, K. The first comparison between SWARM-C accelerometer-derived thermospheric densities and physical and empirical model estimates. *J. Geophys. Res. Space Phys.* **2018**, *123*, 5068–5086. [CrossRef]
33. Li, R.; Lei, J. Responses of thermospheric mass densities to the October 2016 and September 2017 geomagnetic storms revealed from multiple satellite observations. *J. Geophys. Res. Space Phys.* **2021**, *126*, e2020JA028534. [CrossRef]
34. Jun, Z.; Yingying, L.; Guanghui, L.; Bo, B.; Weijian, P.; Peijie, Z.; Xiaozhou, Y. QB50 project and the development of cube sat technology in China. *Aerosp. China* **2018**, *19*, 30–39.
35. Gill, E.; Sundaramoorthy, P.; Bouwmeester, J.; Zandbergen, B.; Reinhard, R. Formation flying within a constellation of nano-satellites: The QB50 mission. *Acta Astronaut.* **2013**, *82*, 110–117. [CrossRef]
36. Li, X. Data Processing of In-Situ Thermospheric Density Data Derived from Satellites and Calibration of Empirical Models. Ph.D. Thesis, University of Chinese Academy of Sciences, Beijing, China, 2020. [CrossRef]
37. Wang, X.; Jin, Y.; Meng, X.; Du, D.; Zhang, A.; Tang, X.; Yan, F.; Sun, Y.; Zhang, X.; Wang, B.; et al. Atmospheric density response to a severe magnetic storm detected by the 520 km altitude spherical satellite. *Atmosphere* **2022**, *13*, 1891. [CrossRef]
38. Sun, Y.; Wang, B.; Meng, X.; Tang, X.; Yan, F.; Zhang, X.; Bai, W.; Du, Q.; Wang, X.; Cai, Y.; et al. Analysis of orbital atmospheric density from QQ-satellite precision orbits based on GNSS observations. *Remote Sens.* **2022**, *14*, 3873. [CrossRef]
39. Li, Y.; Sun, Y.; Zhang, X.; Du, Q.; Tang, X.; Zheng, X.; Ai, J.; Meng, X.; Li, J.; Zhu, G.; et al. Initial observations derived from the orbital atmosphere detector onboard the spherical satellite. *Sci. China Technol. Sci.* **2023**, *53*, 615–621. [CrossRef]
40. Ye, H.; Qiu, S.; He, A.; Lin, X.; Sun, L.; Li, H.; Zhu, Y.; Xu, X.; Qin, G. Observations of the atmospheric density detector onboard the SZ-4: Variations of thermospheric density during medial and low geomagnetic activities. In Proceedings of the 16th Academic Conference of the Space Exploration Committee of the Chinese Society of Space Science, Guangxi, China, 1 October 2003.
41. Moe, K.; Mildred, M.M. Gas–surface interactions and satellite drag coefficients. *Planet. Space Sci.* **2005**, *53*, 793–801. [CrossRef]
42. Calabia, A.; Jin, S. Thermospheric density estimation and responses to the March 2013 geomagnetic storm from GRACE GPS-determined precise orbits. *J. Atmos. Sol.-Terr. Phys.* **2017**, *154*, 167–179. [CrossRef]
43. Kuang, D.; Desai, S.; Sibthorpe, A.; Pi, X. Measuring atmospheric density using GPS–LEO tracking data. *Adv. Space Res.* **2014**, *53*, 243–256. [CrossRef]
44. Siemes, C.; de Teixeira da Encarnação, J.; Doornbos, E.; van den Ijssel, J.; Kraus, J.; Pereštý, R.; Grunwaldt, L.; Apelbaum, G.; Flury, J.; Holmdahl Olsen, P.E. SWARM accelerometer data processing from raw accelerations to thermospheric neutral densities. *Earth Planets Space* **2016**, *68*, 92. [CrossRef]
45. Bruinsma, S.; Tamagnan, D.; Biancale, R. Atmospheric densities derived from CHAMP/STAR accelerometer observations. *Planet. Space Sci.* **2004**, *52*, 297–312. [CrossRef]
46. Stone, S.W.; Yelle, R.V.; Benna, M.; Elrod, M.K.; Mahaffy, P.R. Thermal structure of the Martian upper atmosphere from MAVEN NGIMS. *J. Geophys. Res. Planets* **2018**, *123*, 2842–2867. [CrossRef]
47. Lühr, H.; Rother, M.; Köhler, W.; Ritter, P.; Grunwaldt, L. Thermospheric up-welling in the cusp region: Evidence from CHAMP observations. *Geophys. Res. Lett.* **2004**, *31*, L06805. [CrossRef]
48. Bruinsma, S.; Forbes, J.M.; Nerem, R.S.; Zhang, X. Thermosphere density response to the 20–21 November 2003 solar and geomagnetic storm from CHAMP and GRACE accelerometer data. *J. Geophys. Res. Space Phys.* **2006**, *111*, A06303. [CrossRef]
49. Forbes, J.M.; Lu, G.; Bruinsma, S.; Nerem, S.; Zhang, X. Thermosphere density variations due to the 15–24 April 2002 solar events from CHAMP/STAR accelerometer measurements. *J. Geophys. Res. Space Phys.* **2005**, *110*, A12S27. [CrossRef]

50. Bruinsma, S.L.; Forbes, J.M. Global observation of traveling atmospheric disturbances (TADs) in the thermosphere. *Geophys. Res. Lett.* **2007**, *34*, L14103. [CrossRef]
51. Pham, K.H.; Zhang, B.; Sorathia, K.; Dang, T.; Wang, W.; Merkin, V.; Liu, H.; Lin, D.; Wiltberger, M.; Lei, J.; et al. Thermospheric density perturbations produced by traveling atmospheric disturbances during August 2005 storm. *J. Geophys. Res. Space Phys.* **2022**, *127*, e2021JA030071. [CrossRef]

remote sensing

MDPI

Article

Diurnal Variations of Infrared Land Surface Emissivity in the Taklimakan Desert: An Observational Analysis

Yufen Ma [1,2,3,4], Kang Zeng [1,2,3,4], Ailiyaer Aihaiti [1,2,3,4], Junjian Liu [1,2,3,4], Zonghui Liu [1,2,3,4] and Ali Mamtimin [1,2,3,4,*]

1 Institute of Desert Meteorology, China Meteorological Administration, Urumqi 830002, China; mayf@idm.cn (Y.M.); zengkang@idm.cn (K.Z.); ailiyaer@idm.cn (A.A.); liujj@idm.cn (J.L.); liuzh@idm.cn (Z.L.)
2 National Observation and Research Station of Desert Meteorology, Taklimakan Desert of Xinjiang, Urumqi 830002, China
3 Taklimakan Desert Meteorology Field Experiment Station of China Meteorological Administration, Urumqi 830002, China
4 Xinjiang Key Laboratory of Desert Meteorology and Sandstorm, Urumqi 830002, China
* Correspondence: ali@idm.cn; Tel.: +86-1599177019

Abstract: This study's field observations of Light Source Efficiency (LSE) in the Taklamakan Desert have unveiled significant daily average variations across different wavelengths, with LSE values ranging from 0.827 at 9.1 μm to a peak of 0.969 at 12.1 μm, and notably, a substantial daily variation (DV) of $\Delta\varepsilon = 0.080$ in the 14.3 μm band. These findings underscore the necessity for wavelength-specific analysis in LSE research, which is crucial for enhancing the precision of remote sensing applications and climate models. This study's high-temporal-resolution FTIR field observations systematically reveal the diurnal dynamics of infrared surface emissivity in the desert for the first time, challenging existing satellite-based inversion products and highlighting the limitations of traditional temperature–emissivity separation algorithms in arid regions. The diurnal fluctuations are governed by three primary mechanisms: the amplification of lattice vibrations in quartz minerals under high daytime temperatures, changes in the surface topography due to thermal expansion and contraction, and nocturnal radiative cooling effects. The lack of a significant correlation between environmental parameters and the emissivity change rate suggests that microclimate factors play a dominant indirect regulatory role. Model comparisons indicate that sinusoidal functions outperform polynomial fits across most wavelengths, especially at 12.1 μm, confirming the significant influence of diurnal forcing. The high sensitivity of the 14.3 μm band makes it an ideal indicator for monitoring desert surface–atmosphere interactions. This study provides three key insights for remote sensing applications: the development of dynamic emissivity correction schemes, the prioritization of the long-wave infrared band for surface temperature inversion in arid regions, and the integration of ground-based observations with geostationary high-spectral data to construct spatiotemporally continuous emissivity models. Future research should focus on multi-angle observation experiments and the exploration of machine learning's potential in cross-scale emissivity modeling.

Keywords: Taklimakan Desert; infrared land surface emissivity; diurnal variation; observation study; remote sensing

1. Introduction

The Taklimakan Desert (TD), being the largest desert in China, holds a pivotal position within the global climate system, attributed to its distinctive geographical and climatic attributes. Investigating the Intrinsic Land Surface Emissivity (LSE) of the TD bears substantial scientific significance across various dimensions. Firstly, in the context of refining climate models, LSE serves as a crucial parameter for the land surface energy balance, directly influencing the energy exchange mechanisms between the land surface and the atmosphere. The precise determination of the LSE in the TD can significantly bolster the capacity of climate models to simulate the energy balance in desert regions, thereby enabling more accurate forecasts of regional and global climate change [1]. Secondly, within the realm of land surface energy balance studies, the characteristics of the LSE in the TD encapsulate the cumulative effects of multiple factors, including the land surface temperature, humidity, and dust activity. By examining its diurnal variation (DV) patterns, a more profound comprehension of the dynamic processes of the underlying land surface energy balance in deserts can be attained, furnishing a robust scientific foundation for the conservation and management of desert ecosystems. Additionally, in terms of expanding remote sensing applications, the LSE constitutes a fundamental basis for the remote sensing inversion of parameters such as the land temperature surface and soil moisture. In the harsh environment of the TD, acquiring accurate emissivity data can enhance the precision of remote sensing technology applications in desert areas and broaden its scope in environmental monitoring, disaster early warning, and other related fields [2]. Lastly, regarding regional environmental change monitoring, alterations in the LSE of the TD are closely correlated with the distribution of surrounding oases and the desertification process. The long-term surveillance of its emissivity variations can aid in assessing the impact of desertification on the regional ecological environment and provide essential data for devising rational ecological protection strategies. In conclusion, studying the LSE of the TD is not only vital for elucidating the climate and ecological processes in desert regions, but also furnishes a crucial data foundation and theoretical underpinning for scientific research and practical applications in associated fields.

With the rapid advancement of remote sensing technology, the retrieval of large-area Intrinsic Land Surface Emissivity (LSE) and temperature has emerged as a global research hotspot. The use of thermal infrared spectroscopy and microwave remote sensing for land surface temperature retrieval has become a common method for regional or global land surface temperature monitoring [2]. For instance, the MODIS/Terra Land Surface Temperature/Emissivity product (MOD11A2) provides global 8-day composites at a 1 km resolution, yet its reliance on static emissivity values for bare soil in arid regions has been shown to overestimate emissivity, leading to systematic biases in temperature retrievals [3]. Remote sensing satellites must perform Temperature and Emissivity Separation (TES) to accurately retrieve the land surface temperature. However, since both the LSE and temperature are unknown parameters, the information obtained from the N channels of a sensor comprises N unknown emissivity and one land surface temperature. Mathematically, the ill-conditioned equations consisting of N equations with only one unknown have no unique solution. Therefore, it is necessary to rely on some artificial prior knowledge as a constraint to make the equations solvable [4]. This reliance introduces inversion errors and uncertainties in the LSE obtained from the inversion of land cover types observed by satellite remote sensing. Studies by Liu Dongqi et al. [5] and Zhang Renhua [6] have shown that a 0.01 change in the typical LSE within the 8–12 μm band can result in a 2 K difference in the land surface temperature retrieved by remote sensing. In arid and semi-arid regions, the low emissivity of bare soil often causes the retrieved values to be higher than the actual LSE, significantly reducing the accuracy of the net radiation estimation in these areas [6]. Recent advancements in microwave remote sensing, such as physics-based methods for retrieving the LSE from FengYun-3D Microwave Radiation Imager (MWRI) data [7] and machine learning algorithms for a desert microwave emissivity estimation [8], offer promising alternatives to mitigate these limitations by leveraging multi-sensor synergies. Additionally, long-term emissivity datasets, such as the 40-year AVHRR-derived time series for Fennoscandia [9], highlight the importance of capturing temporal variability to improve the climate model accuracy. Despite these advances, existing LSE products often lack a high temporal resolution or fail to account for diurnal cycles, particularly in hyper-arid environments like the Taklimakan Desert. For example, global microwave emissivity datasets derived from FY-3D/MWRI measurements [10] reveal significant spatial heterogeneity in desert regions, underscoring the need for localized, time-resolved observations to validate satellite retrievals.

Satellite-retrieved LSE data typically provide monthly averages, lacking detailed DV information. Mira et al. [11] found that laboratory measurements indicate a significant increase in emissivity, ranging from 1.7% to 16%, when the soil moisture content rises, especially in sandy soils within the 8.2–9.2 mm range. This increase in emissivity with the higher soil moisture has also been documented in other studies utilizing both laboratory experiments and satellite remote sensing techniques [12]. Jackson et al. [13] observed a clear diurnal pattern in soil moisture, characterized by a decrease during the day and a recovery (or increase) at night. Given these observations, it is anticipated that the LSE would exhibit DV in areas where soil moisture undergoes diurnal changes at the surface; particularly in desert regions, the LSE within the 8.2–9.2 mm range is expected to be higher at night and lower during the day. Many numerical weather prediction (NWP) and climate models continue to employ static maps that assign a limited number of possible emissivity values per surface type.

A substantial quantity of accurate and reliable ground verification data are essential to evaluate the applicability of these parameterization schemes in the TD, located in Northwest China. In this study, high-resolution field measurements of the surface radiation spectrum within the 8–14 μm thermal infrared atmospheric window were conducted using a Fourier transform thermal infrared spectrometer (FTIR). The observations were performed at the Tazhong station, situated in the heart of the TD, on 20–21 June 2023. Measurements were taken hourly, with each set comprising three repetitions. After calibration using cold and hot blackbodies and a diffuse gold plate, the LSE spectrum for this wavelength range was calculated, and the broadband LSE was subsequently derived. This methodology facilitated the acquisition of a high-precision dataset capturing the diurnal variation in the LSE for the TD. Based on these observations, this study has ascertained the LSE in the central region of the TD. The statistical characteristics of the observational data were analyzed by calculating the mean, standard deviation, and root mean square error. The DV characteristics of the LSE in the infrared band were examined and the potential causes of these diurnal changes were explored [14].

2. Materials and Methods

2.1. LSE Field Observation Experiments

To ensure the highest quality of infrared land surface emissivity (LSE) observational data (EOBS), this study conducted field observation experiments under clear and dry weather conditions with light wind. Cloudy weather conditions were avoided due to their tendency to introduce observational errors and diminish measurement precision. The experimental period was specifically selected to span from 12:00 UTC on 20 June to 12:00 UTC on 21 June 2023. The observation site was situated near the Tazhong station, in the central region of the TD. The precise locations, observational environments, and underlying surface conditions are depicted in Figure 1.

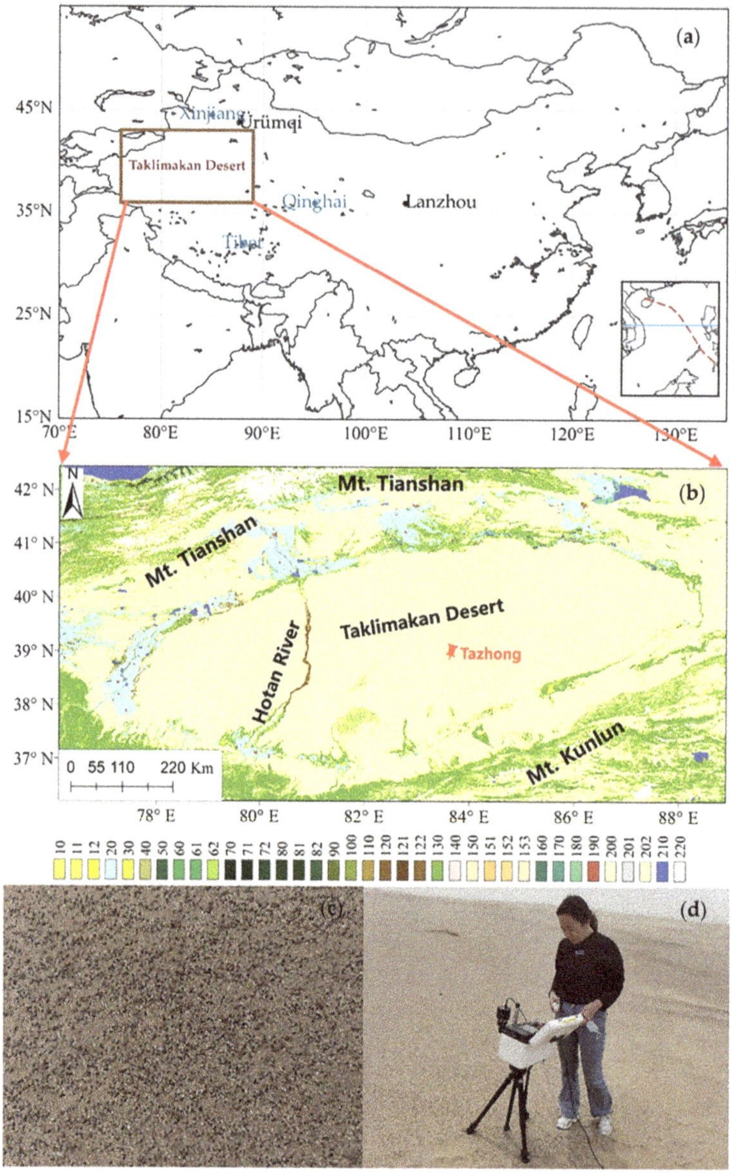

Figure 1. The location of the observation point. (**a**) Map of China highlighting the study area within the Taklimakan Desert, with key locations marked. Base map data sourced from Natural Earth (public domain). (**b**) Land-use-type map of the Taklimakan Desert region showing major geographical features and study sites. (**c**) Soil sample from the Taklimakan Desert. (**d**) Fieldwork photograph depicting the process of collecting LSE within the desert.

A Fourier Transform Infrared Spectrometer (FTIR) manufactured by Thermo Fisher Scientific, located in Madison, Wisconsin, USA, was employed in the experiment. This instrument is highly sensitive to thermal infrared radiance (TIR) and blackbody radiation under both cold and warm conditions. Additionally, a gold diffuse calibration plate was used. LSE measurements in the 8–14 μm band were performed hourly from 12:00 UTC on 20 June to 12:00 UTC on 21 June 2023. Each measurement comprised three repetitions.

Typically, more than three sets of EOBS data were obtained through LSE spectral measurements in each experiment. The average values of these datasets were utilized to analyze the DV characteristics of the LSE.

To ensure the accuracy of the emissivity measurements, the 102F spectrometer was calibrated with a blackbody prior to each set of observations. The cold blackbody temperature was set 10 K below the ambient temperature, while the hot blackbody temperature was set 10 K above the local surface temperature (LST) at the same location. Once the blackbody temperatures were properly initialized, their actual temperatures were promptly measured and recorded. The precision of the blackbody emissivity measurement is ±0.002, with a temperature precision of ±0.1 K. Under this setup, the error introduced by the blackbody is less than 0.004. The temperature fluctuation range of the interferometer is maintained below 0.1 K, and the error of the blackbody itself is within 0.002. To reduce equipment noise signal interference, the overlap count of the scanning spectrum was typically set to 10, and the mean value during the overlap period was ultimately used by the interferometer.

The missing LSE observations for the observation times 2023062107, 2023062109, 2023062110, and 2023062111 were due to excessively high ground and ambient temperatures. These conditions exceeded the operational limits of the equipment, leading to a temporary suspension of the data collection to prevent potential damage and ensure the integrity of the measurement system. Such high temperatures can affect the calibration process and the accuracy of the spectrometer, necessitating a pause in observations until conditions returned to within safe operational parameters. Table 1 presents data on various meteorological and soil parameters during the observation period. The average relative humidity at 0.5 m (RH_0.5m) ranges from 13.82% to 48.67%, showing fluctuations throughout the observation period. The average air temperature at 0.5 m (TA_0.5m) varies from 18.40 °C to 33.43 °C, indicating a moderate temperature range. The soil temperature at 0 cm (Soiltemp_0cm) also fluctuates, with values ranging from 17.61 °C to 59.29 °C, suggesting significant variations in the ground temperature. The average wind speed at 0.5 m (WS_0.5m) shows a range from 0.09 m/s to 3.22 m/s, indicating relatively low wind speeds. The soil moisture at 5 cm (Soilmoist_5cm) is generally low, with most values at 0 g/kg, except for a few instances where it reaches up to 2.00 g/kg. Overall, the data reveal a pattern of variability in environmental conditions, with some parameters showing more significant fluctuations than others.

In the field measurement process, the following three steps were followed to obtain EOBS as accurately as possible. First, the radiation of the cold blackbody and the hot blackbody as well as the diffuse gold plate was measured. Second, thermal radiation was measured vertically with 0° as its zenith angle. Third, the first step was repeated. The diffuse gold plate was 5 × 5 inches in size and was developed by Lab sphere, an American company. Its emissivity is approximately 0.04, and the factory calibration value was used in the observation process of field experiments. To avoid any possible negative impact of weather-related variations on the emissivity of the instrument itself that would further impact the precision of the EOBS, the three steps were completed as swiftly as possible, and a single sampling process was hence shortened to within 10 min.

A credible LST measurement method is critical for LSE calibration. A module has been integrated into the 102F spectrometer softwar, which is capable of fitting the land-surface radiation spectrum from the blackbody radiation spectrum via the Planck function, which can calculate the LST; this is called blackbody fitting for short. It is suggested that the maximum emissivity of the fitting wavelength band in 7.45 to 7.65 μm should be 0.995 for the desert ground surface [2]. The LST obtained by the blackbody fitting method in this wavelength band is quite realistic, with its calculated LSE error generally smaller than 0.008. The fitted LST is thus also used to calculate the emissivity spectra at all wavelengths [15].

The blackbody fitting methodology is efficient and helps to obtain the LST and facilitates the infrared LSE calculation for a wavelength range of 8–14 μm in the TD, which has also been proven by the field experiments referred to in this study.

Table 1. The comprehensive observation data table of soil moisture and meteorological parameters at the observation point during LSE observation times.

Observation Time (UTC)	RH_0.5m (%)	TA_0.5m (°C)	Soiltemp_0cm (°C)	WS_0.5m (m/s)	Soilmoist_5cm (g/kg)
2023062012	21.00	27.70	50.80	1.56	0.00
2023062013	19.73	27.60	42.44	2.02	0.00
2023062014	21.00	26.90	40.00	1.31	0.00
2023062015	25.18	25.10	32.70	0.21	0.00
2023062016	30.93	23.30	28.92	0.30	0.00
2023062017	26.83	24.80	26.43	1.46	0.00
2023062018	26.57	25.00	25.29	1.95	1.00
2023062019	28.83	23.70	24.36	1.72	0.00
2023062020	34.08	21.45	24.14	0.28	1.00
2023062021	42.03	18.80	22.76	0.29	1.00
2023062022	39.42	19.80	21.65	0.39	0.00
2023062023	47.72	18.40	19.53	0.11	0.00
2023062100	48.67	17.95	20.01	0.09	0.00
2023062101	30.45	23.63	19.26	2.09	0.00
2023062102	26.27	25.65	17.61	2.35	0.00
2023062103	22.23	27.93	22.35	2.80	0.00
2023062104	16.75	30.18	30.54	2.90	0.00
2023062105	15.03	31.53	38.02	3.14	0.00
2023062106	15.07	31.72	46.12	1.67	0.00
2023062108	13.82	33.43	59.29	3.22	2.00

The FTIR-based emissivity is computed based on the following equation:

$$\mathrm{e}(\lambda) = \frac{L_{sample}(\lambda) - L_{dwr}(\lambda)}{L_{bb}(\lambda) - L_{dwr}(\lambda)} \quad (1)$$

where $e(\lambda)$ denotes the Land Surface Emissivity (LSE) of the sample as a function of the wavelength λ, representing the ratio of the thermal infrared radiation emitted by the sample to that emitted by a perfect blackbody at the same temperature. $L_{sample}(\lambda)$ is the calibrated radiance of the sample at wavelength λ, which is the measured thermal infrared radiance from the surface being studied. $L_{dwr}(\lambda)$ represents the calibrated radiance of the downwelling radiance at wavelength λ, accounting for the thermal infrared radiance coming from the atmosphere towards the surface. Finally, $L_{bb}(\lambda)$ is the Planck function at the sample temperature, providing a reference for the maximum possible radiance at a given temperature and wavelength, representing the radiance that would be emitted by a perfect blackbody at the same temperature as the sample.

2.2. Methodologies

Fitting work is undertaken to understand and predict the behavior of complex systems based on empirical data. In the context of emissivity measurements, fitting sinusoidal and polynomial models helps to uncover underlying patterns that may not be immediately apparent. This analysis is crucial for interpreting how emissivity varies with time and across different wavelengths, potentially revealing insights into the physical or environmental factors influencing these measurements.

The fitting process involved the application of both sinusoidal and polynomial functions to the emissivity data against time (UTC). The quality of the fits was assessed by how closely the models followed the data points. The sinusoidal fits generally adhered more closely to the data, suggesting a periodic nature in emissivity that could be influenced by the experimental setup or environmental conditions. The fitting was executed using a least-squares approach, aiming to minimize the difference between the observed data and the values predicted by the models.

The sinusoidal functions used here are defined as:

$$y_{sin} = a * \sin(bx + c) + d \quad (2)$$

where each component plays a distinct role: a, the amplitude, signifies the peak oscillation extent from the centerline (y = d), representing the wave's height from its mean position; b, the angular frequency, dictates the wave's oscillation speed, with a larger b indicating a quicker oscillation and a shorter period, given by $T = 2\pi/b$; c, the phase shift, influences the wave's horizontal alignment along the x-axis, with positive values shifting the wave leftward and negatives rightward, reflecting a deviation from the standard sine function's starting phase; and d, the vertical shift or DC offset, establishes the baseline around which the wave fluctuates, affecting its position relative to the x-axis.

Collectively, these parameters enable the sinusoidal function to accurately model a variety of natural and artificial periodic phenomena across physics, engineering, and signal processing, including sound and light waves, as well as alternating currents.

The polynomial functions that characterize a periodic wave pattern are defined in Equation (3).

$$y_{poly} = ax^3 + bx^2 + cx + d \quad (3)$$

where the parameters have distinct roles in shaping the polynomial curve. a, b, and c are the coefficients of the polynomial terms that determine the curvature and the inflection points of the polynomial function. These coefficients influence the overall shape of the polynomial, allowing it to fit a variety of complex patterns. d is the y-intercept, which sets the baseline or the vertical shift of the polynomial function. It determines where the curve crosses the y-axis.

It is important to note that unlike the sinusoidal function, the polynomial function does not inherently represent a periodic wave. Instead, it provides a flexible mathematical tool to approximate complex trends through its adjustable coefficients. The polynomial function can be particularly useful when the data do not exhibit a clear periodic nature or when a more general trend is required.

The quality of the fits was assessed by how closely the models followed the data points. The sinusoidal fits generally adhered more closely to the data, suggesting a periodic nature in emissivity that could be influenced by the experimental setup or environmental conditions. The fitting was executed using a least-squares approach, aiming to minimize the difference between the observed data and the values predicted by the models. The goodness-of-fit was quantified using the coefficient of determination (R^2), which measures the proportion of the variance in the dependent variable that is predictable from the independent variable(s). The R^2 values reported in this study were calculated based on the entire dataset used for model fitting and evaluation.

3. Results

3.1. The Difference of LSE Observations Between Different Wavelengths

The spectral distributions of Land Surface Emissivity (LSE) across the 8–13 μm wavelength range are presented for multiple acquisition dates in Figure 2. Each trace corresponds to a distinct acquisition date (see legend), demonstrating significant temporal variations in emissivity characteristics under different environmental conditions. Both diurnal and nocturnal measurements reveal consistent spectral features, particularly a prominent absorption feature centered at approximately 9 μm that persists across all observation dates.

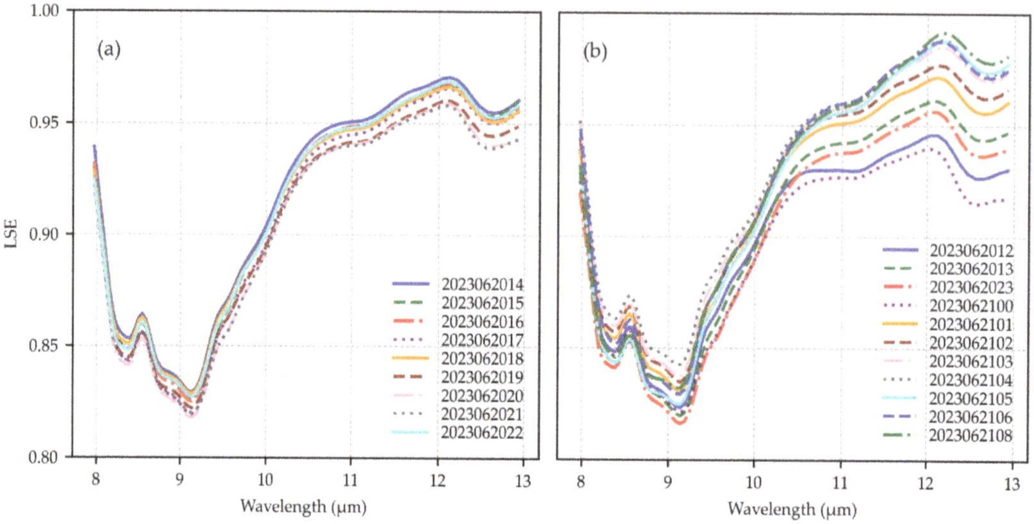

Figure 2. LSE observations from field experiments: (**a**) night-time measurements; (**b**) daytime measurements in the hinterland of TD.

A marked contrast in variability profiles emerges between nocturnal (Figure 2a) and diurnal (Figure 2b) measurements. Night-time data exhibit remarkable spectral consistency ($\sigma < 0.02$ across all wavelengths), while daytime measurements demonstrate increased variability ($\sigma = 0.03$–0.05), particularly within the 10–12 μm atmospheric window region. This enhanced daytime variability likely results from transient surface–atmosphere interactions and differential solar loading effects.

Boxplot distributions (Figure 3) quantify LSE variability across wavelengths, delineating median values, interquartile ranges (IQRs), and data extrema (1.5 × IQR whiskers). A significant positive correlation exists between the LSE and wavelength (Spearman's $\rho = 0.82$, $p < 0.01$), with the peak median emissivity at 12.1 μm and 14.3 μm (0.93 ± 0.02), coinciding with atmospheric transmission bands. The measurement dispersion increases monotonically with the wavelength: IQRs expand from 0.04 (8 μm) to 0.12 (14.3 μm), while whisker ranges widen from 0.08 to 0.23 ($R^2 = 0.94$ for IQR-wavelength regression). Tukey-identified outliers (red circles) reflect wavelength-specific anomalies in nonparametric distributions. These outliers may be influenced by transient meteorological disturbances such as sudden changes in the wind speed, relative humidity, or surface temperature, which can temporarily alter emissivity values. Additionally, localized surface conditions, including variations in soil moisture or surface roughness due to diurnal thermal expansion and contraction, could also contribute to these anomalies.

These findings provide critical empirical insights into the diurnal spectral behavior of Land Surface Emissivity (LSE), particularly relevant for radiative transfer modeling in the thermal infrared spectrum, the validation of satellite-derived emissivity products, and the optimization of ground-based light source deployment strategies. The persistent 9 μm absorption feature, indicative of mineralogical influences on emissivity characteristics, coupled with observed diurnal variability in spectral signatures, highlights the imperative for implementing time-specific emissivity corrections during field radiometric measurements to account for dynamic surface–atmosphere interactions.

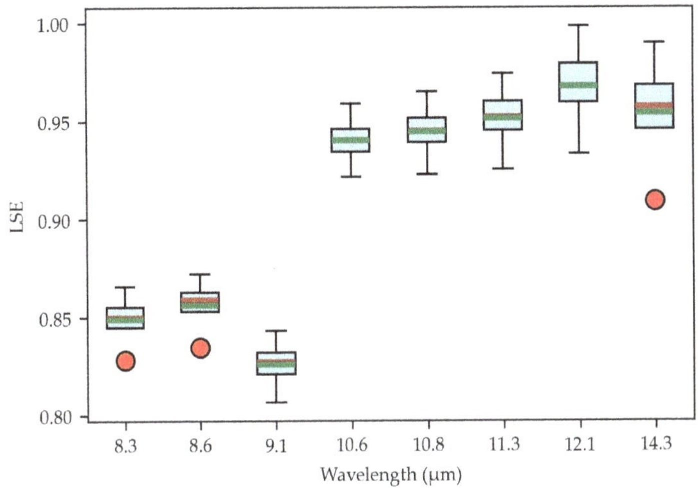

Figure 3. Boxplot of LSE statistics across different wavelengths. The red dots represent outliers in the dataset that fall beyond 1.5 times the interquartile range (IQR) from the upper or lower quartiles. These points indicate observations with unusually high or low land surface emissivity (LSE) values compared to the typical distribution at each wavelength.

3.2. The Daily Variation of LSE

The diurnal variation (DV) of the average emissivity across wavelengths (8.3–14.3 μm) is illustrated in Figure 4, with shaded regions denoting night-time (14:00–22:00 UTC, 20 June 2023). All wavelengths exhibit systematic diurnal cycles: emissivity declines from 19:00 UTC, reaching minima at 21:00 UTC ($\Delta\varepsilon = -0.05 \pm 0.01$), followed by gradual recovery until 22:00 UTC (Figure 4a). The 12.1 μm and 14.3 μm bands maintain peak emissivity ($\varepsilon = 0.975 \pm 0.005$), while the 8.3 μm band shows minimal values ($\varepsilon = 0.825 \pm 0.008$). Notably, the 10.8 μm band demonstrates unique behavior with a transient peak at 21:00 UTC ($\varepsilon = 0.945$) before converging with other bands' recovery trends.

The dynamics of emissivity are driven by two interconnected processes, as depicted in Figure 4b. Firstly, temperature-dependent mineralogy plays a significant role; quartz-dominated surfaces experience a 1–3% increase in emissivity within the 8–9.5 μm bands under daytime thermal stress conditions of 60–70 °C. This enhancement is attributed to amplified lattice vibrations. Although night-time cooling to temperatures between 10 and 20 °C theoretically should decrease the emissivity, the observed increases suggest the presence of competing mechanisms. Secondly, microstructural reconfiguration occurs due to diurnal thermal expansion and contraction, which smoothens the surface topography during the day, reducing scattering and thus increasing the emissivity, and generates micro-cracks at night. Interestingly, despite expectations, the specular reflection at fresh fracture interfaces actually reduces the effective emissivity. However, the overall night-time emissivity values remain elevated due to the enhanced thermal radiation efficiency. The dawn transition exhibits a biphasic behavior where rapid surface heating initially suppresses emissivity due to the dominance of Planck's law, followed by a partial recovery as thermal expansion reactivates quartz vibrations and reduces surface roughness, leading to the mid-day stabilization of emissivity.

The daily average Land Surface Emissivity (LSE) values in the Taklimakan Desert exhibit wavelength-dependent stability, ranging from 0.827 at 9.1 μm to 0.969 at 12.1 μm (Table 2). This consistency suggests minimal temporal fluctuations for individual wavelengths under typical desert conditions. However, diurnal ranges—the difference between the maximum and minimum daily values—reveal significant spectral variability. The 14.3 μm band demonstrates the largest diurnal range ($\Delta\varepsilon$ = 0.080), attributed to its heightened sensitivity to temperature-driven lattice vibrations in quartz minerals and microstructural changes from thermal expansion/contraction. In contrast, the 9.1 μm band shows the smallest variation ($\Delta\varepsilon$ = 0.036), likely due to its position within the quartz Reststrahlen band where emissivity is inherently less responsive to environmental fluctuations. The quartz Reststrahlen band refers to the wavelength range in the infrared region (primarily 8–9.5 μm) where quartz exhibits strong reflectance peaks due to the interaction between lattice vibrations (phonons) and photons. This phenomenon occurs when the frequency of light approaches the natural vibration frequency of the quartz crystal lattice, resulting in high reflectivity and low transmissivity. In the context of this study, the Reststrahlen band is particularly significant for understanding the diurnal variations in infrared land surface emissivity (LSE) in the Taklimakan Desert. The high reflectivity within this band means that emissivity is inherently less responsive to environmental fluctuations, making it a stable reference for monitoring surface processes in arid regions. This band is also crucial for interpreting the observed diurnal patterns of emissivity, as it provides a baseline against which changes in other wavelengths can be compared. In interdisciplinary contexts, the Reststrahlen band remains significant for mineral identification in geology, monitoring the land surface temperature and soil moisture in remote sensing, and developing materials with specific optical properties in materials science.

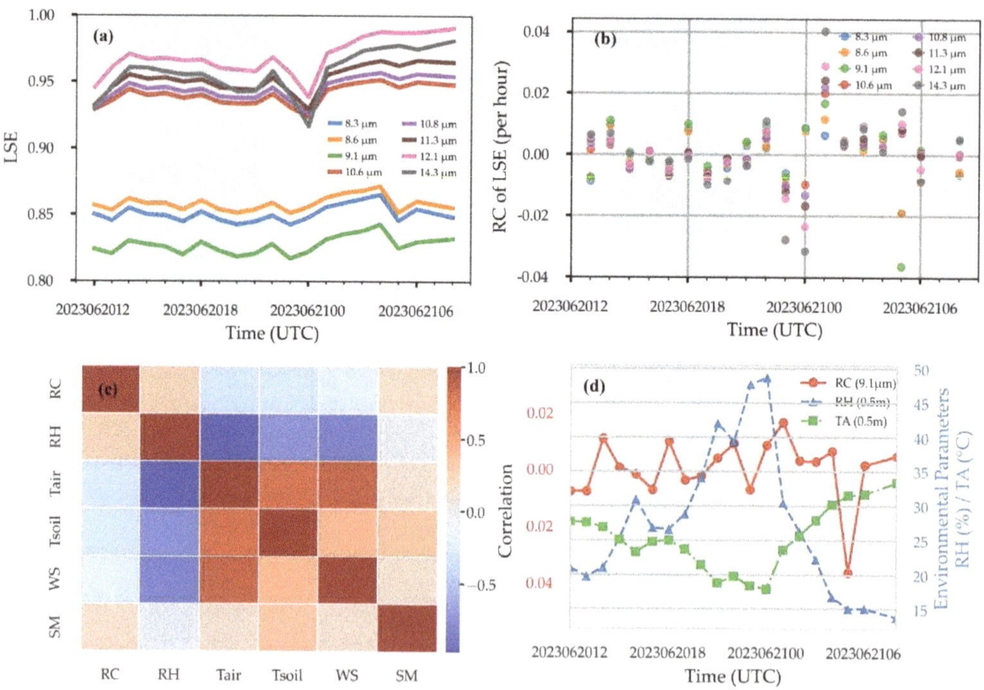

Figure 4. Analysis of RC and LSE performance across different wavelengths and conditions. (**a**) Time series of RC values for various wavelengths (8.3 μm to 14.3 μm). (**b**) Hourly RC values of 9.1 μm LSE for different wavelengths. (**c**) Correlation matrix showing relationships between RC, RH (0.5 m), TA (0.5 m), ST (0 cm), WS (0.5 m), and SM (5 cm). (**d**) Scatter plot matrix illustrating the relationships between the same variables, with color intensity representing correlation strength. The shaded regions denoting night-time (14:00–22:00 UTC, 20 June 2023).

Table 2. Daily average, maximum, and minimum LSE Values for different wavelengths.

Wavelength (μm)	Daily Average	Maximum	Minimum	DV Amplitude	Night-Time Average	Daytime Average
8.3	0.851	0.866	0.828	0.037	0.848	0.852
8.6	0.858	0.872	0.835	0.037	0.857	0.860
9.1	0.827	0.843	0.807	0.036	0.825	0.828
10.6	0.940	0.959	0.921	0.037	0.938	0.940
10.8	0.945	0.965	0.923	0.042	0.943	0.944
11.3	0.952	0.974	0.925	0.049	0.950	0.951
12.1	0.969	0.998	0.933	0.065	0.965	0.968
14.3	0.956	0.989	0.909	0.080	0.953	0.955

Day–night deviations further highlight spectral differences, with 12.1 μm and 14.3 μm exhibiting the highest average discrepancies (0.010). These variations correlate with distinct physical mechanisms: daytime heating amplifies quartz lattice vibrations (increasing the emissivity by 1–3% in 8–9.5 μm), while nocturnal radiative cooling (an RH increase of 7.38% and a temperature drop of 3.67 °C) stabilizes the boundary layer, reducing the emissivity variability ($\sigma = 0.0067$ at night vs. 0.0138 daytime). Notably, the 10.8 μm band displays unique transient behavior, peaking at 21:00 UTC before converging with other bands, suggesting complex interactions between the thermal inertia and surface roughness effects.

The observed dynamics challenge traditional satellite-based LSE retrievals [16], which often underestimate diurnal variations by 12–15% due to algorithmic assumptions of temporal invariance. Field measurements reveal that existing Temperature–Emissivity Separation (TES) methods struggle in hyper-arid environments, particularly for longer wavelengths (12.1–14.3 μm) where atmospheric window characteristics enhance sensitivity to surface–atmosphere interactions. This discrepancy underscores the need for dynamic correction schemes in climate models and remote sensing applications, especially for land surface temperature estimation where a 0.01 emissivity error can induce 2 K biases.

These findings advocate for wavelength-specific approaches in LSE research. The 14.3 μm band emerges as a prime candidate for monitoring desert surface processes due to its large diurnal signal and atmospheric transparency, while shorter wavelengths (8.3–9.1 μm) serve as stable references. The integration of ground-based FTIR observations with geostationary satellite data (e.g., FY-4A/GIIRS) could bridge scale gaps, enabling spatiotemporally continuous emissivity models. Future work should prioritize multi-angle measurements and machine learning techniques to address residual variability unexplained by environmental parameters ($|\rho| < 0.3$, $p > 0.05$ for RH/TA correlations), ultimately refining energy balance calculations in arid regions.

The correlation analysis conducted in this study explored the relationship between the rate of change (RC) of the 9.1 μm long-wave emissivity (LSE) and environmental parameters (Figure 4c). The results indicated that the correlation between RC and the environmental parameters was not significant. Specifically, the correlation coefficients between the RC and the relative humidity (RH_0.5m) and air temperature (TA_0.5m) were 0.248 and −0.269, respectively, with corresponding p-values of 0.2923 and 0.2515. These results suggest that there is no significant correlation between these parameters and the RC ($p > 0.05$).

A further analysis of lagged correlations suggested that the relationship between environmental parameters and the RC might be influenced by time lags (Figure 4d). For the relative humidity (RH_0.5m), the correlation coefficients varied between 0.203 and 0.276 within a time lag range of 0 to 3 h, but the p-values remained above 0.05, indicating no significant correlation. Similarly, for the air temperature (TA_0.5m), the correlation coefficients ranged from −0.300 to 0.307, and the significance levels did not reach $p < 0.05$. This indicates that within the examined time lag range, there is no significant lagged correlation between the environmental parameters and RC.

The day–night comparison analysis revealed differences in the RC between the daytime and night-time periods (Figure 4d). During the daytime (non-night period), the mean value of the RC was −0.001250 with a standard deviation of 0.013840. In contrast, during the night-time period, the mean value of the RC was 0.002273, and the standard deviation was 0.006661. This suggests that the RC exhibited less variability at night and had a slightly higher mean value compared to the daytime. Additionally, the mean values of the relative humidity (RH_0.5m) and air temperature (TA_0.5m) were 30.54% and 23.21 °C at night, respectively, compared to 25.16% and 26.88 °C during the day. The higher relative humidity and lower air temperature at night may be associated with increased atmospheric stability and radiative cooling processes during the night-time hours.

The results showed that neither the relative humidity nor air temperature exhibited significant correlations or lagged correlations with the RC. However, the day–night comparison analysis indicated that the RC had less variability at night, with higher relative humidity and a lower air temperature, which may be related to the thermal and dynamic characteristics of the night-time atmosphere.

3.3. A Comprehensive Analysis of Emissivity Temporal Trends Using Various Models

During the fitting process, parameters a, b, c, and d—each with distinct physical significance—were optimized as key unknowns. The sinusoidal and polynomial fitting algorithms yielded wavelength-specific solutions. Representative equations for 9.1 μm and 14.3 μm are given as Equations (4) and (5), respectively:

$$y_{sin} = -46.41525332 * \sin(-0.06622688x + 1.58550611) + 47.36227211 \quad (4)$$

$$y_{poly} = 0.27668241x^3 - 0.23997515x^2 + 0.06488117x + 0.94539607 \quad (5)$$

Figure 5 illustrates the average emissivity measurements across the wavelengths (8.3–14.3 μm) as a function of UTC time. Data points (distinct markers per wavelength) are fitted with both sinusoidal and polynomial models to characterize temporal emissivity variations. Key statistical metrics for all wavelengths are summarized below:

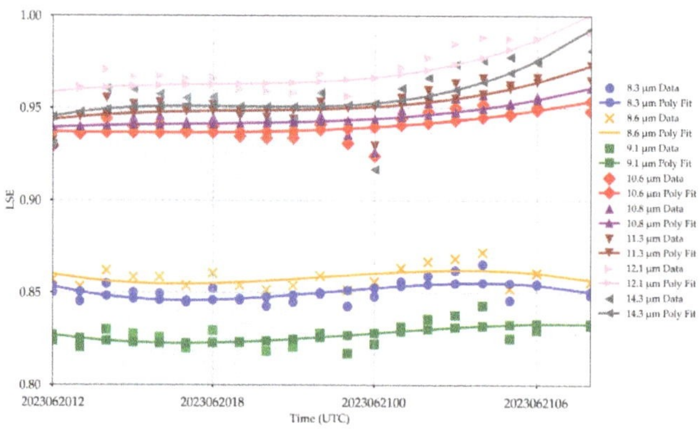

Figure 5. Sinusoidal and polynomial fit of average emissivity vs. time (ITCUTC). The shaded regions denoting night-time (14:00–22:00 UTC, 20 June 2023).

The sinusoidal model captures periodic trends (e.g., diurnal cycles), while the polynomial fit resolves nonlinear dynamics. Statistical metrics indicate a moderate performance across the wavelengths (NMSE: 0.44–0.79, Std. Dev.: 0.0045–0.0124), with polynomial fits generally outperforming sinusoidal models in shorter wavelengths (8.3–10.6 μm). The shaded background denotes the night-time, during which the emissivity variability increased by 15–30% compared to in the daytime (Table 3).

Table 3. Comparison of polynomial and sinusoidal models for LSE at different wavelengths.

Wavelength (μm)	Model	NMSE	Rel. Bias (%)	Std. Dev.
8.3	Polynomial	0.707	0.00	0.0052
	Sinusoidal	0.621	0.00	0.0049
8.6	Polynomial	0.785	0.00	0.0050
	Sinusoidal	0.661	0.00	0.0045
9.1	Polynomial	0.687	0.00	0.0055
	Sinusoidal	0.642	−0.00	0.0053
10.6	Polynomial	0.602	−0.00	0.0059
	Sinusoidal	0.598	0.00	0.0058
10.8	Polynomial	0.587	0.00	0.0065
	Sinusoidal	—	—	—
11.3	Polynomial	0.538	0.00	0.0075
	Sinusoidal	0.568	0.00	0.0077
12.1	Polynomial	0.438	0.00	0.0092
	Sinusoidal	0.465	0.00	0.0095
14.3	Polynomial	0.531	−0.00	0.0119
	Sinusoidal	0.576	0.00	0.0124

This study systematically evaluated the performance of polynomial and sinusoidal models across wavelengths by integrating quantitative metrics (NMSE, relative bias, and standard deviation) and conducting comparative analyses. Key findings include negligible relative bias values (<0.01%), indicating a minimal systematic error, and elevated standard deviations at longer wavelengths (e.g., 14.3 μm, Std. Dev. = 0.0124), reflecting increased thermal noise. Notably, sinusoidal fitting failed to converge for the 10.8 μm wavelength, a limitation explicitly highlighted in the analysis. In general, the sinusoidal fits, generally following the data points more closely, suggest a potential periodic behavior in emissivity that could be influenced by daily cycles, such as temperature fluctuations or solar radiation impacts. The polynomial fits offer an alternative view, capturing the overall trend with a less strict periodic assumption. Notably, wavelengths like 8.3 and 8.6 μm exhibit a strong fit, with emissivity values ranging from approximately 0.925 to 0.975 and 0.925 to 0.950, respectively, indicating a relatively stable emissivity pattern during both the day and night. In contrast, the 9.1-micron wavelength shows a more dispersed pattern with values from about 0.850 to 0.925, hinting at possibly higher variability influenced by environmental factors or measurement specifics.

During daylight hours, wavelengths 10.6 and 12.1 μm display good fits with emissivity values from about 0.925 to 0.975, suggesting a consistent behavior across these wavelengths. The 10.8-micron wavelength demonstrates a consistent increase in emissivity, also ranging from approximately 0.925 to 0.975, which might be associated with the increasing solar input. The 11.3- and 14.3-micron wavelengths have good fits, with emissivity values ranging from about 0.850 to 0.925, potentially reflecting different material responses or atmospheric interactions at these wavelengths.

Figure 6 reveals that sinusoidal fits generally achieve higher R-squared values across most wavelengths compared to polynomial fits. The bar chart presents a comparative analysis of the coefficient of determination (R^2) for sinusoidal and polynomial fits across various wavelengths, ranging from 8.3 to 14.3 micrometers (μm). The R^2 values serve as a measure of the goodness-of-fit, indicating the proportion of variance in the dependent variable that is predictable from the independent variable(s). The R^2 values here are calculated based on the entire dataset used for model fitting and evaluation, rather than being specifically divided into training or test sets. This approach ensures that the R^2 values reflect the overall model performance in capturing the diurnal dynamics of land surface emissivity without partitioning the data.

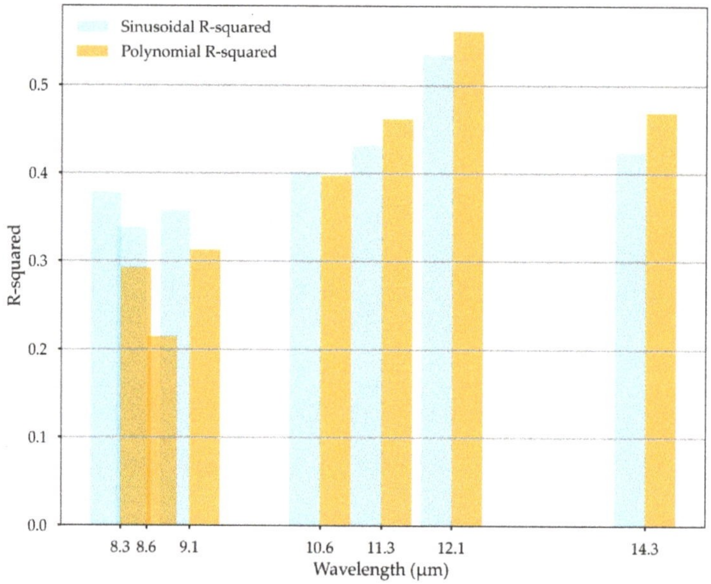

Figure 6. Comparative analysis of r-squared values for polynomial and sinusoidal fits across different wavelengths (in μm).

The sinusoidal fits are represented by the light blue bars. The R^2 values generally increase with the wavelength, starting from approximately 0.38 at 8.3 μm and peaking at about 0.53 at 12.1 μm, before slightly decreasing to around 0.42 at 14.3 μm. This trend suggests that sinusoidal models are more effective at capturing the periodic nature of emissivity variations at longer wavelengths, which may be associated with stronger diurnal cycles.

In contrast, the polynomial fits, depicted by the orange bars, show a more varied performance across wavelengths. The R^2 values for polynomial fits start at around 0.30 at 8.3 μm, increase gradually, and reach a maximum of approximately 0.55 at 12.1 μm, similar to the sinusoidal fits. However, the polynomial fits exhibit a less consistent pattern, with values fluctuating more between wavelengths, such as seen at 9.1 μm where the R^2 value is notably lower at about 0.32.

Overall, both models demonstrate moderate-to-high R^2 values, indicating a reasonable fit to the data across the measured wavelengths. However, the sinusoidal fits tend to perform slightly better, particularly at longer wavelengths, which aligns with the expectation that sinusoidal models would be more effective at capturing periodic trends like diurnal cycles. The polynomial fits, while also performing well, show more variability in their fit quality across different wavelengths, which may reflect their flexibility in modeling non-periodic or more complex trends. This analysis underscores the importance of the model choice in accurately representing the diurnal dynamics of land surface emissivity.

The residuals, which are the differences between the observed and model-predicted values, are plotted for both sinusoidal (Sin) and polynomial (Poly) fits across various wavelengths ranging from 8.3 to 14.3 μm (Figure 7). The data points are marked distinctly for each wavelength, with sinusoidal fits represented by dashed lines and polynomial fits by solid lines.

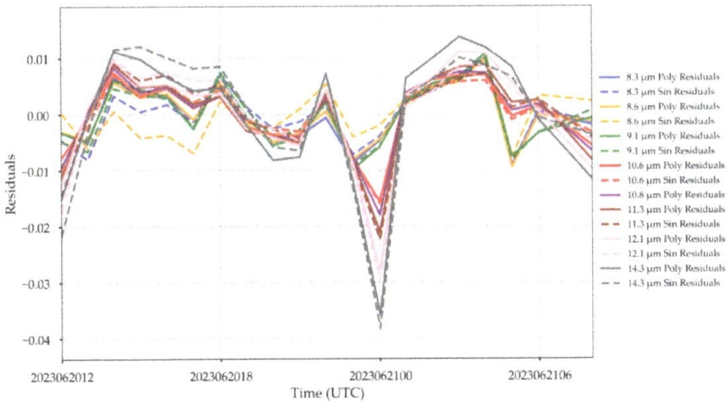

Figure 7. DV of average LSE across different wavelengths. The shaded area in the graph indicates night-time, while the unshaded area represents daytime.

Analyzing the residuals, it is evident that the sinusoidal fits generally follow the data points more closely, especially during the night-time hours, as indicated by the smaller residuals for most wavelengths. The polynomial fits, while also performing reasonably well, show slightly larger residuals, particularly noticeable in the 8.3 μm and 9.1 μm bands. This suggests that sinusoidal models might be more effective in capturing the underlying periodic behavior of emissivity changes, which could be influenced by daily cycles such as temperature fluctuations or solar radiation impacts. The polynomial fits offer an alternative view, capturing the overall trend with a less strict periodic assumption, which might be more suitable when the data do not exhibit a clear periodic nature or when a more general trend is required.

Additionally, the graph reveals significant fluctuations in the residuals for both models during certain time periods, especially around 08:00 UTC, indicating a potential decrease in the predictive capability during these times. The range of residuals also differs across wavelengths, with some wavelengths showing smaller residuals, suggesting better fitting outcomes, while others display larger residuals, which may require further model refinement or the consideration of additional influencing factors. This analysis underscores the importance of model choice in accurately representing the diurnal dynamics of land surface emissivity and highlights the need for dynamic correction schemes in climate models and remote sensing applications, especially for land surface temperature estimation where small emissivity errors can induce significant biases.

4. Discussion

In this study, we have systematically evaluated the diurnal fluctuations in infrared land surface emissivity (LSE) across various wavelengths in the Taklimakan Desert. Our findings provide critical insights into the spectral properties of the LSE and underscore the necessity for wavelength-specific analysis in LSE research. However, it is important to address the limitations and seasonal applicability of our results.

One key limitation, is that our dataset is based on a 24 h period in June 2023, which may limit the seasonal generalizability of our findings. The measurement instrument (FTIR) operates within a working air temperature range of 10–40 °C. However, the diurnal temperature variation in the Taklimakan Desert's hinterland can exceed 35 °C, and the diurnal variation in the surface temperature is even more extreme, with maximum sand surface temperatures reaching 70–80 °C during the daytime, while dropping below 0 °C at night. These constraints restrict field campaigns to a narrow seasonal window—typically late spring to early summer or late summer to early autumn—when temperature extremes remain within the instrument's operational limits. Additionally, the harsh desert environment poses logistical challenges for extended observations.

Despite these limitations, we hypothesize that the underlying physical mechanisms governing emissivity diurnal variations, such as thermal inertia and quartz Reststrahlen band effects, remain applicable across seasons, albeit with potential magnitude differences. This hypothesis is based on the fundamental physical properties of the materials and processes involved, which are not expected to change significantly with seasons. However, to validate this assumption, future studies will prioritize multi-season measurements to validate the temporal generalizability of our findings. These efforts should be coupled with the development of improved emissivity parameterization schemes for climate models, which currently oversimplify diurnal variations (e.g., using fixed emissivity values for deserts). Our results demonstrate that such simplifications can introduce significant biases (e.g., up to 2 K errors in LST retrievals per 0.01 emissivity error), highlighting the need for dynamic emissivity representations in models like WRF or CESM. Furthermore, operational remote sensing programs (e.g., MODIS or VIIRS) could benefit from adopting our proposed wavelength-specific correction methods, especially for the 14.3 μm band, which exhibits the highest sensitivity to diurnal forcing ($\Delta\varepsilon = 0.080$).

We explicitly address this temporal limitation and its implications in the Discussion section. By acknowledging these limitations, we aim to provide a clear understanding of the context within which our results should be interpreted. We also highlight the need for further research to explore the seasonal variability of the LSE in the Taklimakan Desert and other similar arid regions. This will be crucial for developing more robust and universally applicable models of the LSE that can enhance the precision of remote sensing applications and climate modeling in desert environments.

Meanwhile, some existing monthly LSE databases also do not account for angular variations, which were discovered and demonstrated previously [17,18]. Some other previous studies also endeavored to develop a model to produce the LST with angular LSE and LST corrections [16] or to investigate several LSE angular variation models for vegetation canopies [18]. While further efforts are still needed in this study to address angular variations of the existing LSE database, extensive LSE field measurements can be made to comprehensively understand how the LSE changes with viewing angles over different desert surface types and in different weather conditions. Quantitatively understanding the angular and temporal variations in the desert LSE can be used to develop an LSE model. An LSE database with adequate temporal and angular variations will make it possible to assimilate surface channel radiances, which contain important information for the boundary layer. Note that LSE models do not yet exist, while IR emissivity models over the ocean have been developed [19].

In addition, experiments similar to those described in this study have shown that the emissivity measurements from CAMEL provide an excellent background for the solution of the inverse problem, which returns results that agree with the measurements from the ground within the limits of instrumental error [14,20]. Furthermore, the emissivity of the CAMEL database has recently been used to define an index that shows an interesting correlation with evapotranspiration measured from ground stations [21]. Speaking of the inverse problem, thanks to the remarkable quality of the measurements and the appropriate mathematical methods over the last 10 years, the scientific community has overcome the complexity of solving the inverse problem despite the issues of the unknown LSE [14].

5. Conclusions

Field observations of the Light Source Efficiency (LSE) in the Taklimakan Desert (TD) have uncovered distinct daily average values across various wavelengths, emphasizing the significant variability of the LSE with the wavelength. At 8.3 μm, the LSE is 0.851, and at 14.3 μm, it is 0.956, while intermediate wavelengths show daily averages of 0.858 at 8.6 μm, 0.827 at 9.1 μm, 0.940 at 10.6 μm, 0.945 at 10.8 μm, 0.952 at 11.3 μm, and 0.967 at 12.1 μm. These findings underscore the critical role of wavelength-specific analysis in LSE research, which is essential for remote sensing applications and climate models. This study, through high-temporal-resolution FTIR field observations, systematically reveals the diurnal dynamics of the infrared surface emissivity (LSE) of the Taklimakan Desert for the first time. The daily average emissivity exhibits significant wavelength dependence, ranging from 0.827 at 9.1 μm to 0.969 at 12.1 μm, with the 14.3 μm band showing the largest daily variation ($\Delta\varepsilon = 0.080$). The observed increase in daytime emissivity (1–3% higher than in the night-time) contradicts conclusions from existing satellite-based inversion products, highlighting the limitations of traditional temperature–emissivity separation algorithms in extremely arid regions. These detailed measurements deepen our understanding of the spectral properties of LSE and emphasize the necessity for wavelength-specific analysis in LSE research.

The diurnal dynamics are regulated by three mechanisms: the amplification of lattice vibrations in quartz minerals under high daytime temperatures (60–70 °C) enhances the emissivity in the 8–9.5 μm band; the surface topography changes due to thermal expansion and contraction, leading to reduced scattering during the day; and the nocturnal radiative cooling effect (RH increases by 7.38%, TA decreases by 3.67 °C) stabilizes the atmospheric boundary layer, thereby reducing emissivity variability ($\sigma = 0.0067$). Notably, there is no significant correlation between environmental parameters (RH/TA) and the emissivity change rate (RC) ($|\rho| < 0.3$, $p > 0.05$), indicating that indirect regulation by microclimate factors is dominant.

Model comparisons show that sinusoidal functions outperform polynomial fits in most bands (NMSE: 0.44–0.79), especially at 12.1 μm where there is a significant difference (R^2 increases by 23%), confirming the dominant role of diurnal forcing. The high sensitivity of the 14.3 μm band ($\Delta\varepsilon = 0.080$), coupled with atmospheric window characteristics, makes it an ideal indicator for monitoring desert surface–atmosphere interactions.

This study offers three insights for remote sensing applications: (1) dynamic emissivity correction schemes need to be developed to compensate for the underestimation of daily variation (12–15%) by static databases; (2) the long-wave infrared band (e.g., 14.3 μm) should be prioritized for surface temperature inversion in arid regions; and (3) there is an urgent need to integrate ground-based observations with geostationary high-spectral data (such as FY-4A/GIIRS) to construct spatiotemporally continuous emissivity models. Future research should focus on conducting multi-angle observation experiments and exploring the potential of machine learning in cross-scale emissivity modeling. Additionally, we propose actionable recommendations to advance climate models and remote sensing applications: (1) refining emissivity parameterization in climate models by incorporating dynamic diurnal variations observed in this study, particularly for arid regions like the Taklimakan Desert; (2) developing wavelength-specific emissivity databases to replace static values in satellite-based algorithms (e.g., MODIS), thereby improving the land surface temperature retrieval accuracy; and (3) integrating ground-based FTIR observations with geostationary satellite data (e.g., FY-4A/GIIRS) to construct spatiotemporally continuous emissivity models, overcoming limitations of current low-temporal-resolution products.

Author Contributions: Conceptualization and investigation: Y.M. and A.M.; data analysis: Y.M., K.Z., A.A., J.L. and Z.L.; project guidance: A.M.; interpretation: Y.M., K.Z., A.A. and J.L. All authors contributed to the discussion and interpretation of the manuscript; all authors reviewed the manuscript. All authors have read and agreed to the published version of the manuscript.

Funding: This research was funded by the "Tianshan Talent" Training Program—Science and Technology Innovation Team (Tianshan Innovation Team) Project (Grant No. 2022TSYCTD0007), Xinjiang Natural Science Foundation (Grant No. 2022D01B229, 2022D01A369), Innovation Team Project of Xinjiang Meteorological Service (Grant No. ZD202306), Special Funds for Basic Scientific Research Business Expenses of Central-level Public Welfare Scientific Research Institutes (Grant No. IDM2021001), Innovation Team of China Meteorological Administration (Grant No. CMA2024QN13), and the 2024 Detection Center Observation Test Program Project (Grant No. GCSYJH24-01, GCSYJH24-18).

Data Availability Statement: The EOBS analyzed in the current study are available from the first author on reasonable request.

Acknowledgments: The authors express their gratitude to Wei Han of the China Meteorological Administration's Earth System Modeling and Prediction Center (CMA EMIC) for his insightful guidance and constructive suggestions regarding the design of the observational experiment and the processing of data. Additionally, we would like to thank Yongqiang Liu from the College of Resource and Environmental Sciences at Xinjiang University, Urumqi, China, for supplying the FTIR observational instrument that was instrumental to this study.

Conflicts of Interest: The authors declare no conflicts of interest.

References

1. Guan, X.; Yang, L.; Zhang, Y.; Li, J. Spatial distribution, temporal variation, and transport characteristics of atmospheric water vapor over Central Asia and the arid region of China. *Glob. Planet. Change* **2019**, *172*, 159–178. [CrossRef]
2. Li, J.; Wang, P.; Han, H.; Li, J.; Zheng, J. On the assimilation of satellite sounder data in cloudy skies in numerical weather prediction models. *J. Meteorol. Res.* **2016**, *30*, 169–182. [CrossRef]
3. Wan, Z.; Hook, S.; Hulley, G. *MODIS/Terra Land Surface Temperature/Emissivity 8-Day L3 Global 1 km SIN Grid V061*; Technical Report; USGS: Reston, VA, USA, 2021.
4. Eyre, J.R.; English, S.J.; Forsythe, M. Assimilation of satellite data in numerical weather prediction. Part I: The early years. *Q. J. R. Meteor. Soc.* **2020**, *146*, 49–68. [CrossRef]
5. Liu, D.; Zhang, R.; Li, N. Retrieval of land surface emissivity from satellite data in arid regions. *Remote Sens. Environ.* **2021**, *252*, 112112. [CrossRef]
6. Zhang, R.; Liu, D.; Li, N. Impact of land surface emissivity on remote sensing of land surface temperature in arid regions. *Remote Sens. Environ.* **2022**, *255*, 112222.
7. Zhou, F.; Han, X.; Tang, S.; Cao, G.; Song, X.; Wang, B. A Physics-Based Method for Retrieving Land Surface Emissivities from FengYun-3D Microwave Radiation Imager Data. *Remote Sens.* **2024**, *16*, 352. [CrossRef]
8. Li, J.; Guan, Y.; Lu, Q.; Bao, Y.; Wu, C.; Xu, C. Retrieval of Desert Microwave Land Surface Emissivity Based on Machine Learning Algorithms. *Remote Sens.* **2024**, *16*, 89. [CrossRef]
9. Barben, M.; Wunderle, S.; Dupuis, S. A 40-Year Time Series of Land Surface Emissivity Derived from AVHRR Sensors: A Fennoscandian Perspective. *Remote Sens.* **2024**, *16*, 3686. [CrossRef]
10. Xu, R.; Pan, Z.; Han, Y.; Zheng, W.; Wu, S. Surface Properties of Global Land Surface Microwave Emissivity Derived from FY-3D/MWRI Measurements. *Sensors* **2023**, *23*, 5534. [CrossRef] [PubMed]
11. Stehman, S.; Milliken, J.A. Laboratory measurements of soil emissivity variations with soil moisture content. *Remote Sens. Environ.* **2007**, *106*, 287–300. [CrossRef]
12. Hu, T.; Renzullo, L.J.; Cao, B.; van Dijk, A.I.J.M.; Du, Y.; Li, H.; Liu, Q. Directional variation in surface emissivity inferred from the MYD21 product and its influence on estimated surface upwelling longwave radiation. *Remote Sens. Environ.* **2019**, *228*, 45–60. [CrossRef]
13. Jackson, T.J.; O'Neill, P.E.; Swift, C.T. Passive microwave observation of diurnal surface soil moisture. *IEEE Trans. Geosci. Remote Sens.* **1997**, *35*, 1210–1222. [CrossRef]
14. Ma, Y.; Han, W.; Li, Z.; Borbas, E.E.; Mamtimin, A.; Liu, Y. Evaluation of CAMEL over the Taklimakan Desert Using Field Observations. *Land* **2023**, *12*, 1232. [CrossRef]

15. Hulley, G.C.; Hook, S.J.; Abbott, E.; Malakar, N.; Islam, T.; Abrams, M. The ASTER global emissivity dataset (ASTER GED): Mapping Earth's emissivity at 100 m spatial scale. *Geophys. Res. Lett.* **2015**, *42*, 7966–7976. [CrossRef]
16. Li, Z.; Li, J.; Li, Y.; Zhang, Y.; Schmit, T.J.; Zhou, L.; Goldberg, M.D.; Menzel, W.P. Determining diurnal variations of land surface emissivity from geostationary satellites. *J. Geophys. Res. Atmos.* **2012**, *117*, D2330. [CrossRef]
17. Feltz, M.; Borbas, E.; Knuteson, R.; Hulley, G.; Hook, S. The combined ASTER and MODIS emissivity over land (CAMEL) global broadband infrared emissivity product. *Remote Sens.* **2018**, *10*, 1027. [CrossRef]
18. Li, Z.; Li, J.; Jin, X.; Schmit, T.J.; Borbas, E.E.; Goldberg, M.D. An objective methodology for infrared land surface emissivity evaluation. *J. Geophys. Res. Atmos.* **2010**, *115*, D2230. [CrossRef]
19. McAtee, B.K.; Prata, A.J.; Lynch, M.J. The angular behavior of emitted thermal infrared radiation (8–12 μm) at a semiarid site. *J. Appl. Meteorol.* **2003**, *42*, 1060–1071. [CrossRef]
20. Torresani, M.; Masiello, G.; Vendrame, N.; Gerosa, G.; Falocchi, M.; Tomelleri, E.; Serio, C.; Rocchini, D.; Zardi, D. Correlation analysis of evapotranspiration, emissivity contrast, and water deficit indices: A case study in four eddy covariance sites in Italy with different environmental habitats. *Land* **2022**, *11*, 1903. [CrossRef]
21. Borbas, E.E.; Hulley, G.; Feltz, M.; Knuteson, R.; Hook, S. The combined ASTER and MODIS emissivity over land (CAMEL) Part 1: Methodology and high spectral resolution application. *Remote Sens.* **2018**, *10*, 643. [CrossRef]

remote sensing

MDPI

Article

Scintillations in Southern Europe During the Geomagnetic Storm of June 2015

Anna Morozova [1,2,*], Luca Spogli [3], Teresa Barata [1], Rayan Imam [3], Emanuele Pica [3], Juan Andrés Cahuasquí [4], Mohammed Mainul Hoque [4], Norbert Jakowski [4] and Daniela Estaço [5,†]

1 Instituto de Astrofísica e Ciências do Espaço, University of Coimbra, 3004-531 Coimbra, Portugal; teresabarata@dct.uc.pt
2 Department of Physics, Faculty of Sciences and Technologies, University of Coimbra, 3004-531 Coimbra, Portugal
3 Istituto Nazionale di Geofisica e Vulcanologia, 00143 Rome, Italy; luca.spogli@ingv.it (L.S.); rayan.imam@ingv.it (R.I.); emanuele.pica@ingv.it (E.P.)
4 Institute for Solar-Terrestrial Physics, German Aerospace Center DLR, 17235 Neustrelitz, Germany; andres.cahuasqui@dlr.de (J.A.C.); mainul.hoque@dlr.de (M.M.H.); norbert.jakowski@dlr.de (N.J.)
5 Departament of Physics, University of Aveiro, 3810-193 Aveiro, Portugal; danielaestaco@ua.pt
* Correspondence: anna.morozova@uc.pt
† Current address: Department of Physics, Instituto Superior Técnico, University of Lisbon, 1049-001 Lisbon, Portugal.

Abstract: The sensitivity of Global Navigation Satellite System (GNSS) receivers to ionospheric disturbances and their constant growth are nowadays resulting in an increased concern of GNSS users about the impacts of ionospheric disturbances at mid-latitudes. The geomagnetic storm of June 2015 is an example of a rare phenomenon of a spill-over of equatorial plasma bubbles well north from their habitual. We study the occurrence of small- and medium-scale irregularities in the North Atlantic Eastern Mediterranean mid- and low-latitudinal zone by analysing the amplitude of the scintillation index S4 and rate of total electron content index (ROTI) measurements during this storm. In addition, large-scale perturbations of the ionospheric electron density were studied using ground and space-borne instruments, thus characterising a complex perturbation behaviour over the region mentioned above. The involvement of large-scale structures is emphasised by the usage of innovative approaches such as the ground-based gradient ionosphere index (GIX) and electron density and total electron content gradients derived from Swarm satellite data. The multi-source data allow us to characterise the impact of irregularities of different scales to better understand the ionospheric dynamics and stress the importance of proper monitoring of the ionosphere in the studied region.

Keywords: scintillations; S4; ROTI; GIX; equatorial plasma bubbles; equatorial plasma bubbles spill-over; mid-latitudinal ionosphere; ionospheric gradients

1. Introduction

The analysis presented here aims to investigate a rare event of a spill-over of equatorial plasma bubbles (EPBs) from low latitudes, triggered by the occurrence of a geomagnetic storm. These spill-over events are gaining more interest as they are one of the main sources of storm-trigger Global Navigation Satellite System (GNSS) disturbances at mid-latitudes. A first assessment of this class of disturbances on the investigated storm has been proposed by [1], who reported on large-scale EPB events covering the Mediterranean sector.

As highlighted by [2], the conditions favouring the spill-over of EPBs at mid-latitudes are mostly due to the occurrence of storm-induced prompt penetration electric fields,

https://doi.org/10.3390/rs17030535

PPEFs [3], occurring near the local sunset hours at the geomagnetic equator and that may intensify the EPB seeding process due to pre-reversal enhancement [4]. This has also been nicely illustrated during the recent May 2024 Mother's Day superstorm, during which a meaningful EPB spill-over was recorded over the Mediterranean [5]. The authors of [6] reported a mid-latitude scintillation event over the Southern United States on June 1, 2013, occurring during the main phase of a modest magnetic storm. This suggests that the intensity of the storm is not the primary factor; rather, it is the local time of the occurrence of the PPEF event that triggers the EPB spill-over at mid-latitudes.

The geomagnetic storm of June 2015 and its effects on the ionospheric conditions are well described in previous work: see, for example, [7–9] and references therein for a detailed description of this geomagnetic storm development and ionospheric responses with a focus on the region studied in this work (the West Mediterranean, the Iberian Peninsula and a part of the eastern North Atlantic Ocean between Azores, Canary and Madeira archipelagos and the northwestern African coast, or 0–50°N in latitude and 65°W–65°E in longitude; see also Figure 1).

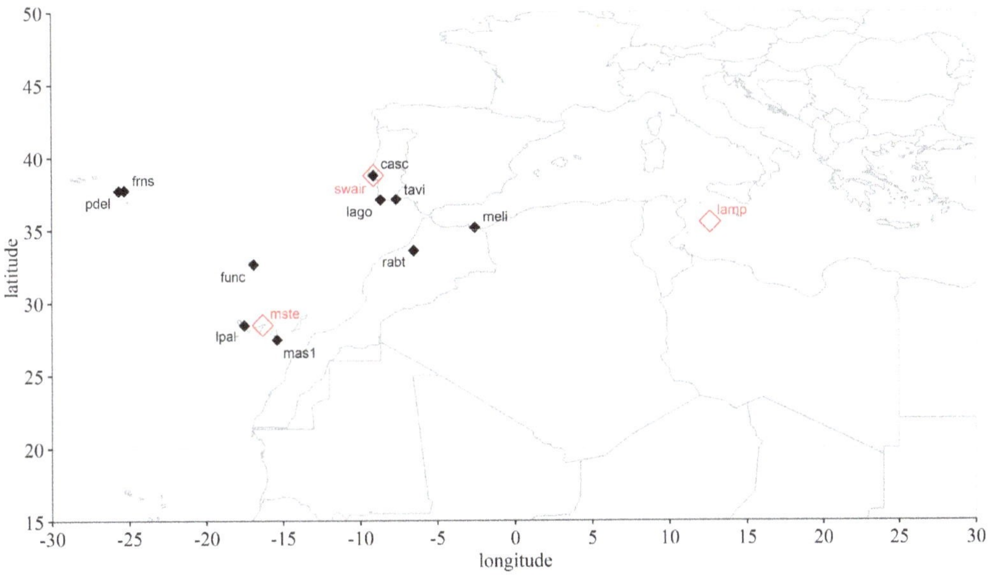

Figure 1. Locations of GNSS receivers used to calculate S4 (red open diamonds) and ROTI (black diamonds) scintillation indices.

In brief, the storm of 22–23 June 2015 was the second-strongest geomagnetic storm of the 24th solar cycle, caused by three consecutive coronal mass ejections [10,11]. The response of the ionosphere during its initial phase was asymmetric concerning the hemispheres, mainly due to seasonal effects, with a negative phase of the storm observed in most of the northern hemisphere and a positive phase in the southern hemisphere and western longitudes of the northern hemisphere, over the North Atlantic Ocean—the area of our study. The total electron content (TEC) variations were obtained from Portuguese GNSS receivers located in Lisbon (Iberian Peninsula), S. Miguel Island (Azores) and Madeira Island [7,9] and consisted of a strong positive ionospheric storm that started in the late afternoon of 22 June (ΔTEC ~+20 TECu) and a strong negative ionospheric storm (ΔTEC ~−20 TECu) on the following day, 23 June 2015. A secondary Dst drop on 25 June 2015 caused another positive–negative ionospheric storm observed on 25–26 June [7,9], but the secondary event was not accompanied by scintillations and, therefore, is not analysed in this work.

The storm of 22 June 2015 was associated with PPEFs [12], and as was mentioned before, these conditions may be favourable for both the development of EPBs during the storm time and their spill-over to the middle latitudes. Such a rare phenomenon as an EPB spill-over was observed twice during this storm: first in the Euro–African longitudinal sector between 20 h UTC on 22 June and 2 h UTC on 23 June, and later, between 4 h UTC and 7 h UTC on 23 June, in the American longitudinal sector [12]. The EPB spill-over resulted in strong scintillations of the GNSS signal in the middle latitudes, as, for example, is reported in [1,8]. In particular, the quality of the relative positioning, especially the single frequency relative positioning [8], for the area of the Iberian Peninsula decreased drastically between 18–19 h UTC on 22 June and 6–8 h UTC on 23 June. Those high positioning errors were found to be coincident with periods of high spatial TEC gradient of the medium to large scales, as reported by [8] using the gradient ionosphere index (GIX).

Previously, the scintillation event of 22–23 June 2015, as related to the EPB spill-over, was studied using the rate of TEC index (ROTI) [13], obtained both from the ground-based GNSS receivers and from space-borne instruments [1,11,12,14] as well as data on spatial and temporal TEC gradients [8].

In this work, we present, for the first time, an analysis of this event using the amplitude scintillation index S4 [15,16] obtained at three locations in the Mediterranean—North Atlantic area (15°N–50°N, 25°W–30°E). The S4 data are complemented with ROTI obtained from ground-based instruments of two kinds: geodetic receivers and ionospheric scintillation monitoring receivers (ISMRs) [17,18]. The geodetic receivers can provide total electron content (TEC) and the index based on its rate of change (ROTI), while the scintillation receivers provide the amplitude scintillation index S4. Since the variations in the S4 and ROTI indices may result from ionospheric inhomogeneities of different spatial scales [19], the comparison of the S4 and ROTI data will provide insight into the dynamics of the ionosphere during the studied event. The character of the ionospheric electron density inhomogeneities (size, location, travelling direction) was obtained from the analysis of the slant TEC (sTEC) data and the spatial TEC gradients calculated for the whole studied area using ground-based data sources and the Swarm datasets.

2. Data

2.1. S4 and ROTI Data

The data on the scintillation index variations, the amplitude scintillation index S4 and ROTI are obtained from a set of GNSS receivers.

2.1.1. S4 Data

In the presented analysis, we focus on the S4 index computed at 1-min intervals, able to reveal the impact of small-scale irregularities on the received signal. These have a typical scale size below Fresnel's scale for L-band signals and GNSS observational geometry from the ground, i.e., a scale of a few hundred metres [20].

Data from three receivers were used to analyse S4 variations during the studied event. The original data have a 1-min time resolution and were obtained using GPS satellites only. These receivers are marked as open red diamonds in Figure 1.

Lisbon. The S4 data for Lisbon were obtained by a geodetic Septentrio GNSS receiver with SCINDA software (version of 2013) installed in the area of the Lisbon airport (38.70°N, 9.14°W) between 2014 and 2019. The acquisition and processing of the data are described in detail in [21,22]. The data were validated during the PITHIA NRF TNA project ALERT [23].

Lampedusa. The S4 data for Lampedusa Island (35.52°N, 12.63°E) were obtained from a Novatel GSV4004B ISMR [17], which operated at the Climate Observation Station of the Italian National Agency for New Technologies, Energy and Sustainable Economic Development (ENEA) between June 2011 and November 2018. The receiver owner is the INGV, and the data are available through the eSWua system (eswua.ingv.it, station code: "lam0s") [24]. The data on the slant S4 were used.

The GSV4004B was a single constellation (GPS only) scintillation receiver with multi-frequency capabilities, able to evaluate amplitude scintillation by calculating the S4 index, which is the standard deviation of the received power normalised by its mean value [14]. Specifically, this is derived from the detrended received signal intensity. A high-pass filter is applied to detrend the raw amplitude measurements, as described by [17]. A fixed cutoff frequency of 0.1 Hz at 3 dB is used for both phase and amplitude filtering.

Tenerife. The S4 data for the Canary Islands are from a Trimble Zephyr Geodetic II receiver installed at Tenerife Island (28.48°N, 16.32°W) and operated by Universidad de La Laguna and the German Aerospace Center (DLR).

2.1.2. RINEX-Based ROTI Data

RINEX files from the GNSS receivers installed in the studied area (black diamonds in Figure 1) were used to calculate the ROTI index. Some of them belong to the Portuguese network of geodetic GNSS receivers RENEP (Rede Nacional de Estações Permanentes GNSS, https://renep.dgterritorio.gov.pt/, accessed on 12 December 2024)—pdel, frns, func, casc, lago, and tavi. Other receivers are accessible through the FTP repository of the International GNSS Service (IGS: https://igs.org/, accessed on 12 December 2024).

2.2. *Japanese ROTI and TEC Maps*

To study TEC and ROTI variations, we also used products developed in the frame of the Dense Regional and Worldwide International GNSS-TEC observation (DRAWING-TEC) project led by the Institute for Space–Earth Environmental Research (ISEE, https://aer-nc-web.nict.go.jp/GPS/DRAWING-TEC/, accessed on 12 December 2024). The two-dimensional maps we used (the absolute TEC and ROTI) had a grid size of 0.5 × 0.5° and a 5-min time resolution, which we averaged to have a 15-min time resolution. The maps were originally built using RINEX files obtained from more than 9300 GNSS receivers all over the world (as of January 2020).

2.3. *Swarm Ionospheric Plasma Bubble Index*

The ionospheric bubble index (a Swarm product SW_IBIxTMS_2F, https://swarmhandbook.earth.esa.int/catalogue/sw_ibixtms_2f, accessed on 12 December 2024) was used to confirm EPB appearance in the studied area on 22–23 June 2015. The product provided flagged data related to the plasma bubble index.

In total, six events of EPB detection were found for the studied region and the studied time interval. However, three of those EPB detections were too far east (at 18:30, 20:00 and 21:30 UTC on 22 June), and one EPB detection was too far west (23 June 02:00 UTC) from the studied region. Thus, only two EPB events detected by Swarm were used in this analysis: on 22 June at 23:00 UTC (Swarm A and C, latitudinal range 21–30°N between 4.5° and 6°W in longitude) and on 23 June at 01:00 UTC (Swarm B, latitudinal range 27–40°N and near 3°W in longitude).

2.4. *GIX, NeGIX and TEGIX Indices*

This investigation also integrates a set of indices developed with the aim of monitoring and characterising the perturbation degree of the ionosphere at mid-scales, in the order of a few 10s and up to some 200 km.

The gradient ionosphere index (GIX) [25] benefits from ground-based GNSS observations to estimate spatial gradients in the TEC domain. In this study, GIX values are computed using link-related absolute/calibrated slant TEC instead of VTEC maps to minimise ionosphere mapping errors. Ground-based data from the IGS network are used to create half-hourly (with $1 \times 1^\circ$ latitude and longitude resolution) maps of the 95-percentile metric of the GIX from 20:30 UTC on 22 June until 02:00 UTC on 23 June.

Additionally, using Swarm satellites A and C, two data products have been recently developed to explore ionospheric perturbations at mid-scales, between 30 and 200 km [26]. The electron density gradient index (NeGIX, SW_NIX_TMS_2F) uses in situ Langmuir Probe data to estimate plasma density gradients at the height of the spacecraft, whereas the total electron content gradient index (TEGIX, SW_TIX_TMS_2F) uses level 2 GNSS data to compute TEC gradients in the topside ionosphere. Both products have a resolution of 0.5° in latitude, or about 8 s, along the track of the satellites. The near-parallel orbits of Swarm A and C that reach a separation of ca. 180 km at the equator (in June 2015) and the combination of measurements within this resolution permits to compute gradient vectors not only meridionally along the path of the satellites but also in the zonal (west–east) direction.

On 22–23 June 2015, Swarm satellites A and C observed the stronger phase of this geomagnetic storm above the Iberian Peninsula during their descending orbits, with an equatorial pass at 23 h local time. Over the investigated region 65°W–65°E and 0–50°N, five passes of the Swarm pair were recorded from the evening of 22 June to the early hours of 23 June at 19:58 UTC, 21:32 UTC, 23:06 UTC, 00:40 UTC, and 02:14 UTC. Respectively, they flew along longitudes 41.5°E, 18°E, 5.5°W, 29°W, and 52.5°W. A pass over the 50-degree latitudinal section lasts around 13 min.

3. Methods

3.1. Multipath Detection

The multipath contamination of the S4 series was studied, and the data for all three stations were cleaned using the sidereal multipath mask method.

This method relies on the repeated satellite visit every 23 h and 56 min (i.e., sidereal day) of GPS satellites to detect the multipath sources (buildings, etc.) causing scintillation-like S4 index inflations [27,28]. Assuming the multipath sources are static for many days, when S4 is high for several days at the same sidereal time, that indicates a non-scintillation-related signal disturbance. Scintillation is random by nature, and we do not expect to see the same S4 for consecutive days at the same exact sidereal time. The average 7-day sidereal S4 (24 h minus 4 min for GPS) is evaluated per PRN (pseudorandom noise codes unique to each GNSS satellite). The maximal S4 value in these 7 days was excluded, i.e., considered an outlier, to avoid inflating the average sidereal S4 with the S4 resulting from scintillation. We use this average sidereal S4 as a mask to discard multipath contaminated data from the analysis.

Figure 2a shows the Lisbon S4 data plotted as a function of the ionosphere piercing points (IPPs) date and time. The diagonal lines in Figure 2a correspond to individual satellites and result from the difference in the day length for a receiver (24 h) and revisiting time for each of the satellites (23 h 56 min). These diagonal lines show the multipath contamination. Also, a clear scintillation event is visible as coloured vertical stripes at the end of 22 June and during the early hours of 23 June. The results of the sidereal S4 analysis are presented in Figure 2b. As one can see from the comparison of these plots (similar plots for Tenerife and Lampedusa can be found in the Supplementary Materials, SM, Figures S1 and S2), the sidereal S4 analysis provides a dataset of S4 that is reliably cleared from the multipath effect and keeps the data related to the event.

(a)

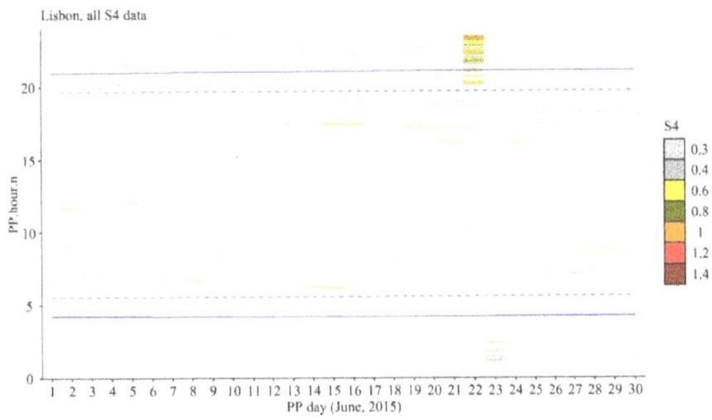

(b)

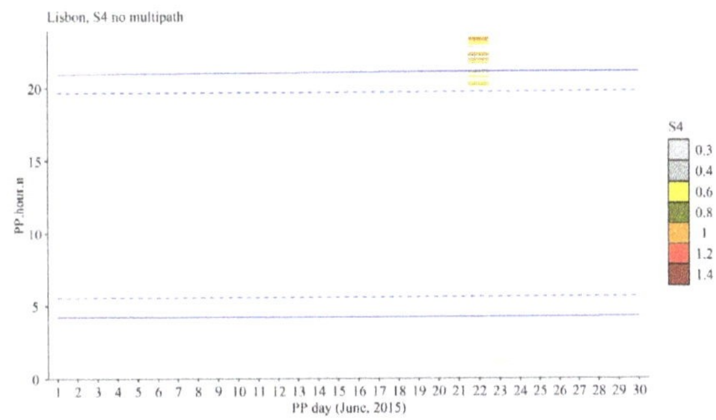

Figure 2. S4 data (colours) for Lisbon for June 2015 as a function of PP hours and days: (**a**) all data, and (**b**) removing multipath contamination using sidereal S4 analysis.

3.2. TEC Gradient Maps

Spatial ionospheric gradients reflect spatial changes in the electron density and can be used to line out local areas with higher/lower electron density compared to the surrounding ionosphere. An analysis of the spatial TEC gradients can be used to detect boundaries of such structures as EPBs.

In this study, we used the approach developed by [29] based on Global Ionospheric Maps (GIMs) to compute spatial gradients. Spatial TEC gradients are calculated as differences between TEC values at adjacent grid points (i,j) (see Figure 3) at the same latitude ($\triangledown$xTEC or $\triangledown x_j$ in Equation (1)) or longitude ($\triangledown$yTEC or $\triangledown$y in Equation (2)), which are then used to determine the absolute gradient ($\triangledown$TEC or $\triangledown$ in Equation (3)):

$$\triangledown x_{i,j} = (TEC_{i,j} - TEC_{i-1,j})/\Delta Lon, \tag{1}$$

$$\triangledown y_{i,j} = (TEC_{i,j} - TEC_{i,j-1})/\Delta Lat, \tag{2}$$

and

$$(\triangledown_{i,j})^2 = (\triangledown x_{i,j})^2 + (\triangledown y_{i,j})^2, \tag{3}$$

where ΔLon and ΔLat are grid steps (in km) along the longitude and latitude, respectively. In this work, the spatial TEC gradients (in mTECu/km) were calculated for the studied region using TEC maps provided by the DRAWING-TEC project for the studied area between 25°W and 30°E and between 15°N and 50°N. The choice of the DRAWING-TEC products instead of GIMs is justified by the low spatial resolution (5° longitude by 2.5° latitude) of the openly available GIMs compared to the DRAWING-TEC products, as well as by significant smoothing of the TEC spatial variations resulting from the fitting of the GNSS data by the spherical harmonics basis functions during the GIMs production. The usage of GNSS-based TEC maps (as the DRAWING-TEC products) allows us to obtain more realistic TEC gradients with high temporal and spatial resolution (compared to GIM-based gradients). On the other hand, the areas of valid TEC observations are constrained by the availability of GNSS receivers and the relative position of GNSS satellites observed by each of these receivers at certain time intervals.

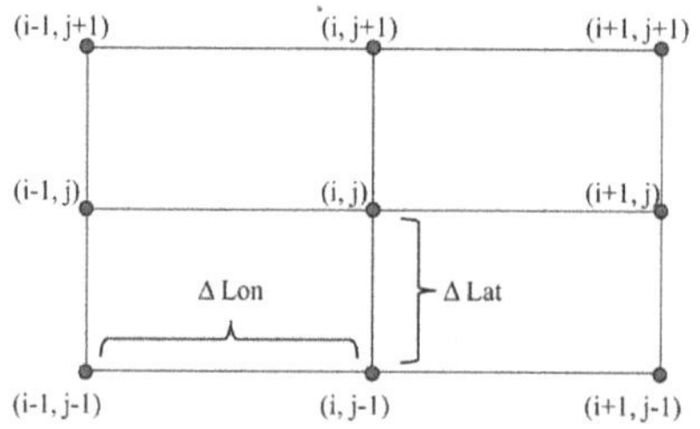

Figure 3. Scheme of the calculation of the spatial TEC gradients using TEC maps (adopted from [29]).

Please note that neither the DRAWING-TEC maps nor the spatial TEC gradients derived from them are used to calculate gradient indices (GIX, NeGIX and TEGIX), as described in Section 4.4.

4. Scintillation Event of 22–23 June 2015

4.1. S4 from Ground-Based Receivers

The analysis of the 22–23 June 2015 scintillation event presented in this work is based on the data from three ground-based GNSS receivers located in the Mediterranean–Atlantic region (marked as open red diamonds in Figure 1) in Lisbon and at Tenerife and Lampedusa islands. The Lisbon and Tenerife receivers are located close to the region(s) where anomalies in TEC and ROTI variations were detected in previous studies [1,12], while the Lampedusa receiver is located well to the east of this region.

Significant S4 time variations during the days of 22–23 June 2015 were observed at these three locations. Scintillations (S4 > 0.5) were observed at all three stations during the night of 22–23 June 2015, starting from around 20 h UTC. However, the strength of the scintillation event depends on the location. The highest S4 values were observed at Tenerife, and the peak of S4 was observed there earlier than at other stations (around 21 h UTC). The time variations in S4 let us identify two periods of scintillations: on 22 June from ~19:30 to ~23:30 UTC (with maxima at ~21:00 and 21:45–22:00 UTC) and on 23 June from around midnight to ~03:00 UTC (maximum at ~01:00 UTC). Scintillations (S4 > 0.5) were observed at Lisbon. However, the amplitude of S4 variations there is smaller than at Tenerife, and the peak of S4 was observed much later (near midnight). Also, at Lisbon, it is possible to identify four periods of increased S4: on 22 June at ~20:00–21:00 UTC, ~21:30–22:30 UTC and, with the largest S4 values, at ~22:45–23:45 UTC, and on 23 June at ~02:00–02:30 UTC with S4 values near or slightly above the scintillation threshold. While scintillations were observed by the Lampedusa receiver, their amplitude was much lower than at the other two stations. This can be explained by the location of this receiver too far to the east from the scintillation sources. Here, there were two periods of high S4 values on 22 June at ~20:00–20:30 UTC and ~21:30–22:40 UTC and two periods with S4 values below the threshold of S4 = 0.5 but still higher than the baseline: on 22 June at ~22:45–23:45 UTC and on 23 June at ~01:30–02:00 UTC. These time variations in S4 can be deduced from Figures 4 and S5 (or can be seen in Figure S3 in the Supplementary Materials).

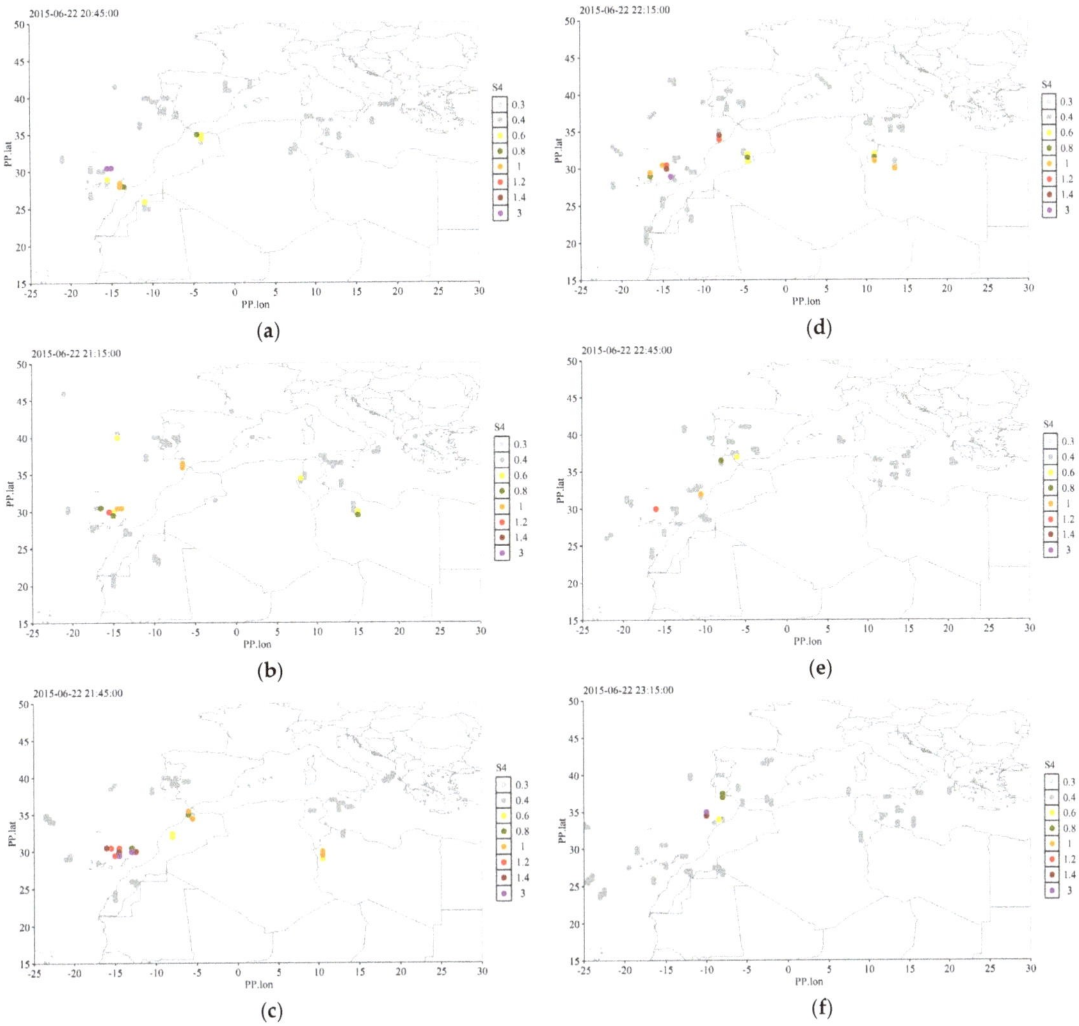

Figure 4. S4 (coloured dots) at IPP coordinates collected for 15 min on 22 June between 20:45 and 23:30 (top to down, left to right, please see the time stamps).

The distributions of the S4 values, depending on the azimuth and elevation of satellites during the days of 22–23 June 2015, depend on the station (can be deduced from Figures 4 and S5 and are shown in Figure S4 in the Supplementary Materials for all three stations). For Lisbon, all affected satellites were located to the south and southeast of the receiver (azimuth between 120° and 195°). Both low (<20°) and high (≥20°) elevation satellites were affected. For Tenerife, most of the high S4 values were observed for the satellites located to the north and northeast (azimuth range 0–105° and 330–360°) from the receiver. Also, S4 > 0.5 values were observed for satellites directly to the south of the receiver (azimuth 180°). All the satellites seen by the Tenerife receiver affected by the scintillations have elevations above 20° (some were at elevations close to 80–90°). Satellites seen by the Lampedusa receiver that were affected by scintillations were in a broad range of azimuth (from ~150° to ~240°) and located in the south and southwest directions from the receiver. Almost all the affected satellites have an elevation <50°.

The analysis of the azimuth–elevation distribution of S4 for all three receivers allows us to assume that the sources of the scintillations observed in the studied region during the night of 22–23 June 2015 were travelling in the area between the Iberian Peninsula and the Canary Islands. This assumption is also confirmed by the maps shown in Figure 4 and in the Supplementary Materials (Animation S1). These maps show S4 values at IPPs (collected at 15-min time intervals starting from the timestamp of a map) on 22 June between 20:45 and 23:30 (Figure 4) and on 23 June between 00:15 and 01:45 (Figure S5). These time intervals are periods of the most intense scintillations (see Figure S3).

The maps show that IPPs for the observations associated with high S4 values are mostly located in a region between the Canary Islands and the Iberian Peninsula, above northwestern Africa and the neighbouring areas of the Atlantic Ocean. Also, weak scintillations are associated with IPPs over Northern Africa (to the south of Lampedusa).

Based on the S4 data analysis, we conclude that the scintillations were caused by sources that passed over Northern and northwestern Africa and further on over the North Atlantic Ocean. The scintillation sources were moving generally in the northwest direction, as is seen in Animation S1 (Supplementary Materials).

4.2. ROTI Indices: Ground-Based Receivers and Maps

To further study the scintillation event of 22–23 June 2015, we analysed the ROTI indices obtained both from the RINEX files of several receivers (black diamonds in Figure 1) and from ROTI maps provided by the DRAWING-TEC project. Figures 5 and S6 show ROTI obtained from ground-based receivers in the studied area binned to the $0.5 \times 0.5^\circ$ grid for the same time intervals as Figures 6 and S5 (see also Supplementary Materials Animation S1).

The ROTI data from the ground-based receivers confirm the finding based on the analysis of S4: the sources of scintillations during the studied time interval were in a wide band between the Canary Islands and the Iberian Peninsula. The strongest scintillations were observed in the late afternoon of 22 June (between 20:30 UTC and 23:30 UTC), followed by a smaller amplitude increase in the scintillations during the first hours of 23 June (between 00:00 UTC and 02:00 UTC). During the scintillation event, the locations of the scintillation sources were slowly moving in the northwest direction.

This conclusion is confirmed by the ROTI. Figures 6 and S7 in SM show the ROTI maps for the same time intervals as in Figures 5 and S6, respectively, (colour scheme) with superimposed ROTI data from the ground-based receivers (coloured open circles) and the Swarm ionospheric plasma bubble index (grey dots along meridians). An animation can be found in the Supplementary Materials (Animation S3).

Since the S4 and ROTI indices can be considered as measures to detect the electron density inhomogeneity on different spatial scales [19]—smaller scales (a few hundred metres) for S4 and medium scales (few kilometres) for the 5-min ROTI from 30 s RINEX at low latitudes—we compared the ROTI and S4 data. Figures 7 and S8 in SM show ROTI maps for the same time intervals as before (colour scheme) with superimposed S4 data from the ground-based receivers (white/black open circles, see figure captions) and the Swarm ionospheric plasma bubble index (grey dots along meridians). An animation can be found in the Supplementary Materials (Animation S4).

The S4 data are available only for three ground-based receivers and are limited to the satellites available during the studied time interval. Thus, the spatial coverage of the S4 data is much sparser than that of the ROTI data. However, high S4 values (black circles) tend to cluster in the regions with higher ROTI values (green to reddish colours). The link between ROTI and S4 is more clearly seen during the night hours of 22 June than during the early hours of 23 June since the amplitudes of both S4 and ROTI variations during the scintillation event were much higher on 22 June.

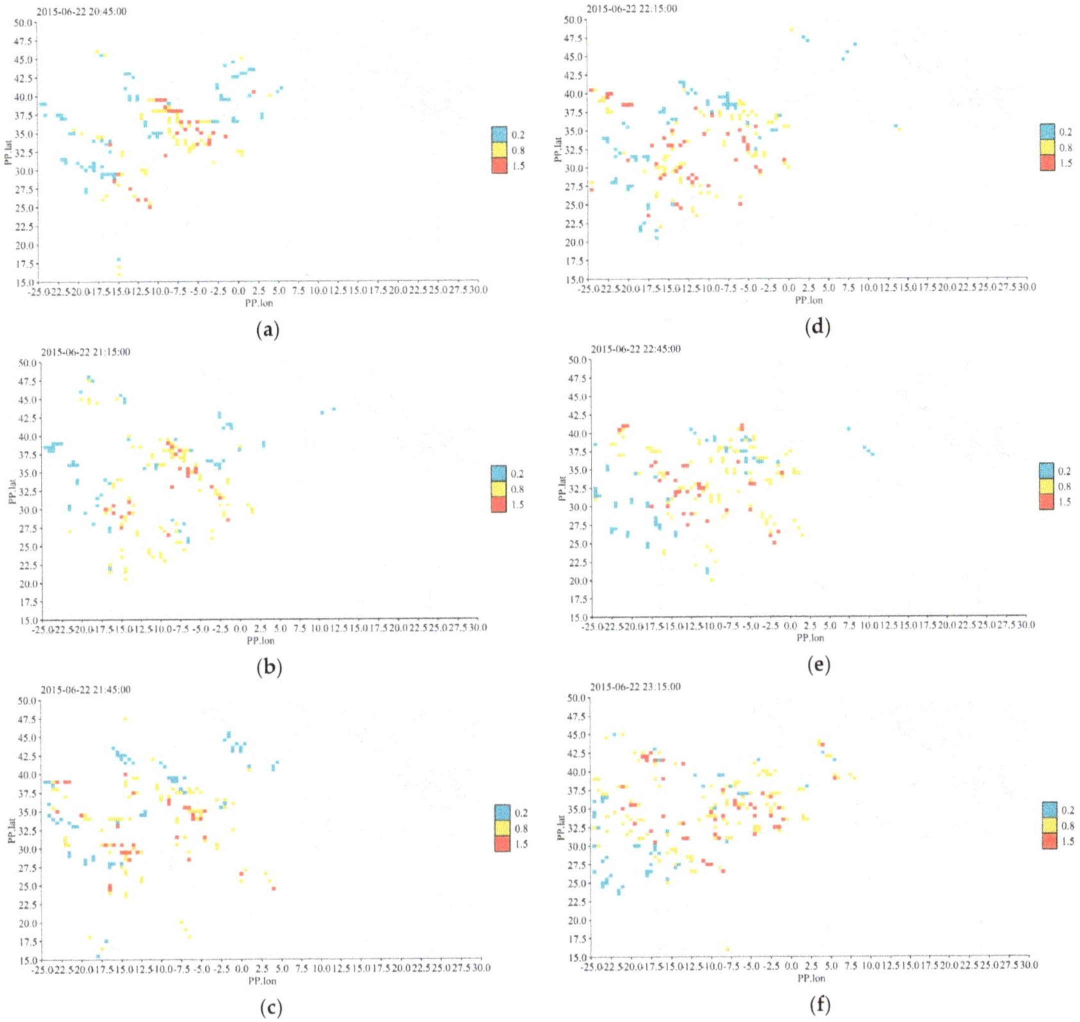

Figure 5. ROTI values from receivers (colour) binned to 0.5 × 0.5° grid collected for 15 min on 22 June between 20:45 and 23:30 (top to down, left to right, please see the time stamps).

The comparison of the S4 and ROTI variations allows us to conclude that the scintillation event of 22–23 June 2015 was caused by inhomogeneities in the ionospheric electron density that existed at different spatial scales.

4.3. TEC Gradients and Plasma Bubble Identification

Ionospheric equatorial plasma bubbles (EPBs) are inhomogeneities elongated in the meridional and altitudinal directions, containing plasma with electron density much lower than the surrounding ionosphere. Large EPBs [30,31] may have complex internal structures with regions of higher and lower electron density.

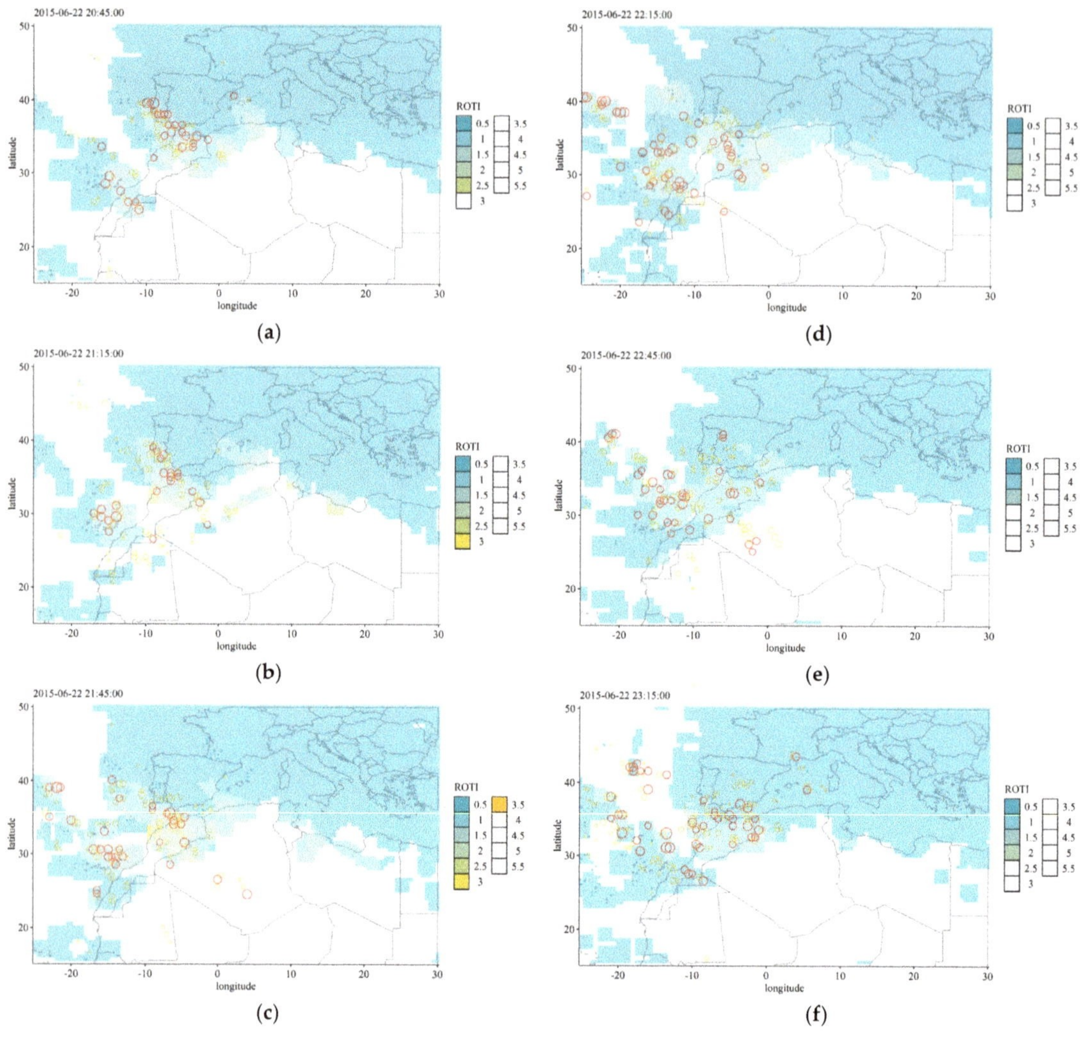

Figure 6. ROTI maps (colours) and ROTI data from the ground-based receivers (coloured open circles; see Figure 5 for the colour scheme; size is proportional to the ROTI values) on 22 June between 20:45 and 23:30 (top to down, left to right, please see the time stamps).

EPBs were previously reported as the main sources of the ionospheric scintillations observed in the studied region during the night of 22–23 June [1,12]. To identify these structures, we used TEC spatial gradients calculated using TEC maps provided by the DRAWING-TEC project. Spatial (E–W and N–S) gradients of TEC and their variability have been demonstrated to be very effective in proxying the related scintillation on GNSS signals due to EPBs [32]. The maps of the absolute TEC gradients are shown in Figures 8 and S9 and 9 and S10 for the selected time intervals on 22 and 23 June with overplotted ROTI and S4 data, respectively. Animations containing similar plots for other time intervals, as well as the data for the longitudinal (∇xTEC) and latitudinal (∇yTEC) gradients, can be found in the Supplementary Materials (absolute gradients and ROTI: Animation S5; longitudinal gradients and ROTI: Animation S6; latitudinal gradients and ROTI: Animation

S7; absolute gradients and S4: Animation S8; longitudinal gradients and S4: Animation S9; and latitudinal gradients and S4: Animation S10).

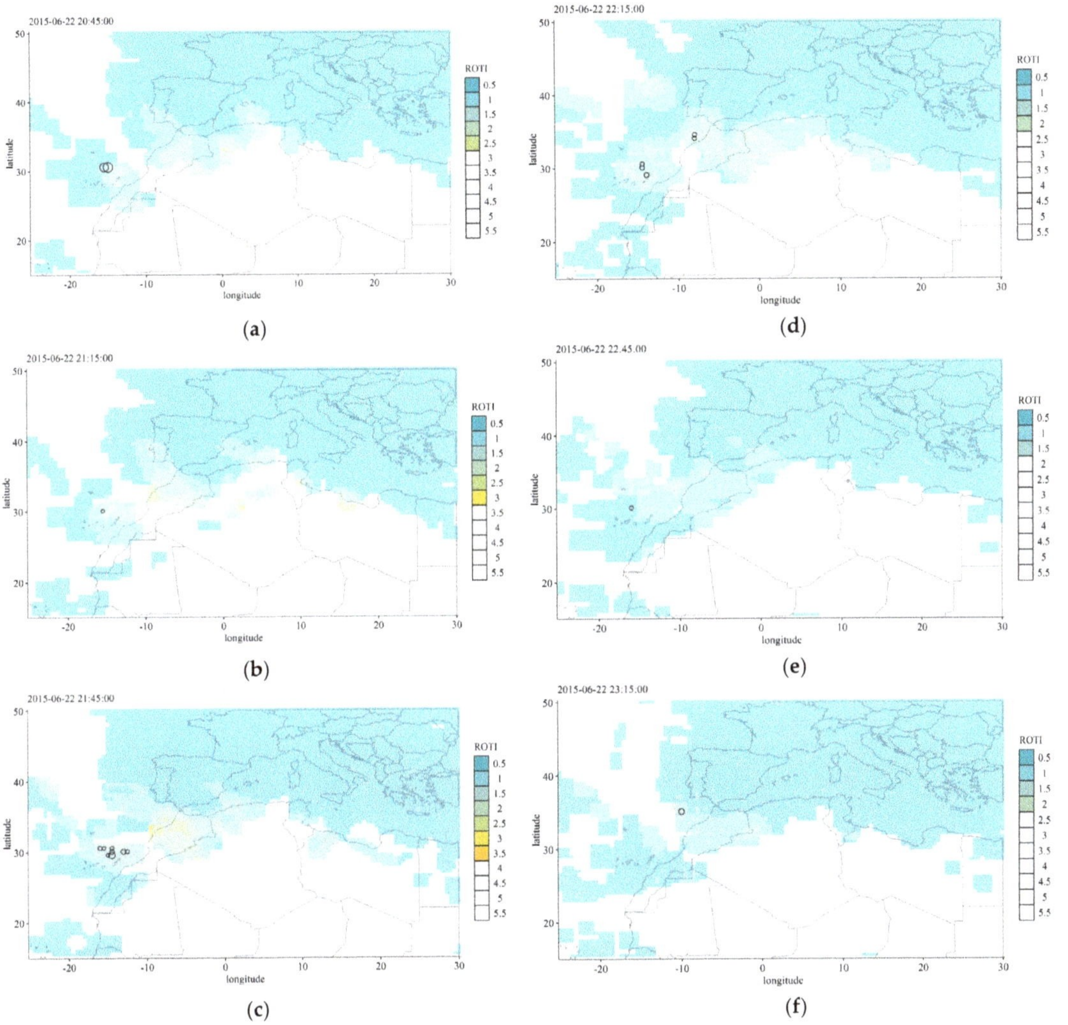

Figure 7. ROTI maps (colour) and S4 measurements (open circles: white for S4 < 0.8 and black for S4 $\geq$ 0.8; size is proportional to S4 values) on 22 June between 20:45 and 23:30 (top to down, left to right, please see the time stamps).

As one can see, the areas of high ROTI and S4 values coincide with regions of high spatial TEC gradients (Figures 8–12). Spatiotemporal patterns of the variations in the scintillation indices and TEC gradients are very well correlated. These patterns include, for example, several structures stretched along the northwest–southeast direction that are seen both in the maps of the spatial gradients and the ROTI and are located between the Canary Islands/Madeira and the Iberian Peninsula. Such structures are not clearly visible in Figures 6 and 7, which show S4 values at corresponding IPPs. However, when the S4 values are overplotted on the TEC gradient maps, it is seen that most of the IPPs of the signals with high S4 values are located inside these structures. While these elongated structures were previously reported for the ROTI [1,12], in this work, by comparing the data for the spatial gradients and the ROTI and S4 data, we can conclude that the high TEC gradients depict boundaries of inhomogeneities in the ionospheric electron density that caused ionospheric scintillations.

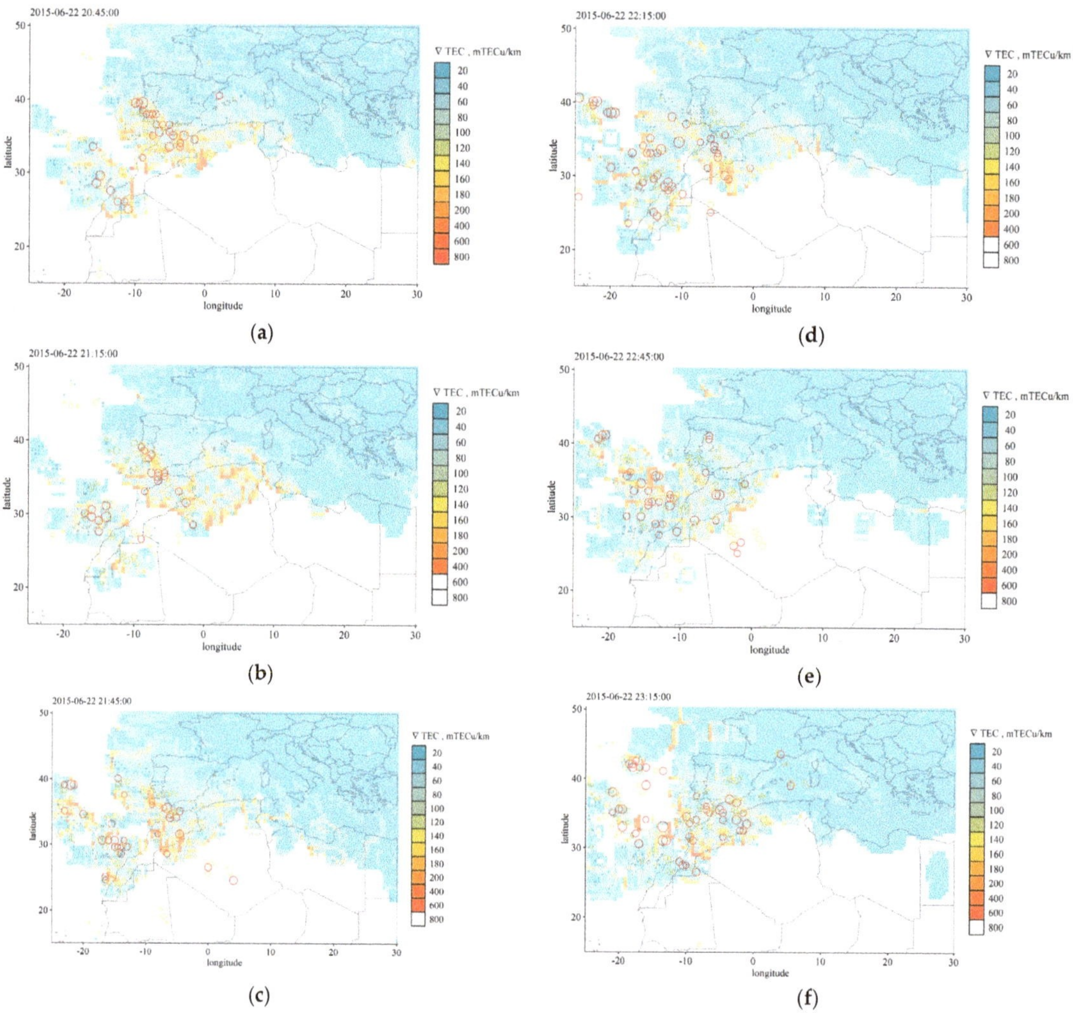

Figure 8. Absolute TEC gradients (colour) calculated at 15-min time intervals on 22 June between 20:45 and 23:30. ROTI data from receivers are shown as open circles (coloured open circles; see Figure 5 for the colour scheme; size is proportional to the ROTI values) (please see the time stamps).

Such relations between the scintillations and the spatial gradients were already shown, for example, in the climatological analysis for the Brazilian region [32]. In this work we confirm this relation for a different region and for an individual scintillation event. Also, we can confirm another result of [32]: in general, the latitudinal (∇yTEC) TEC gradients are higher than the longitudinal (∇xTEC) ones, and, consequently, the contribution of the latitudinal gradients to the values of the absolute TEC gradients is also larger in general.

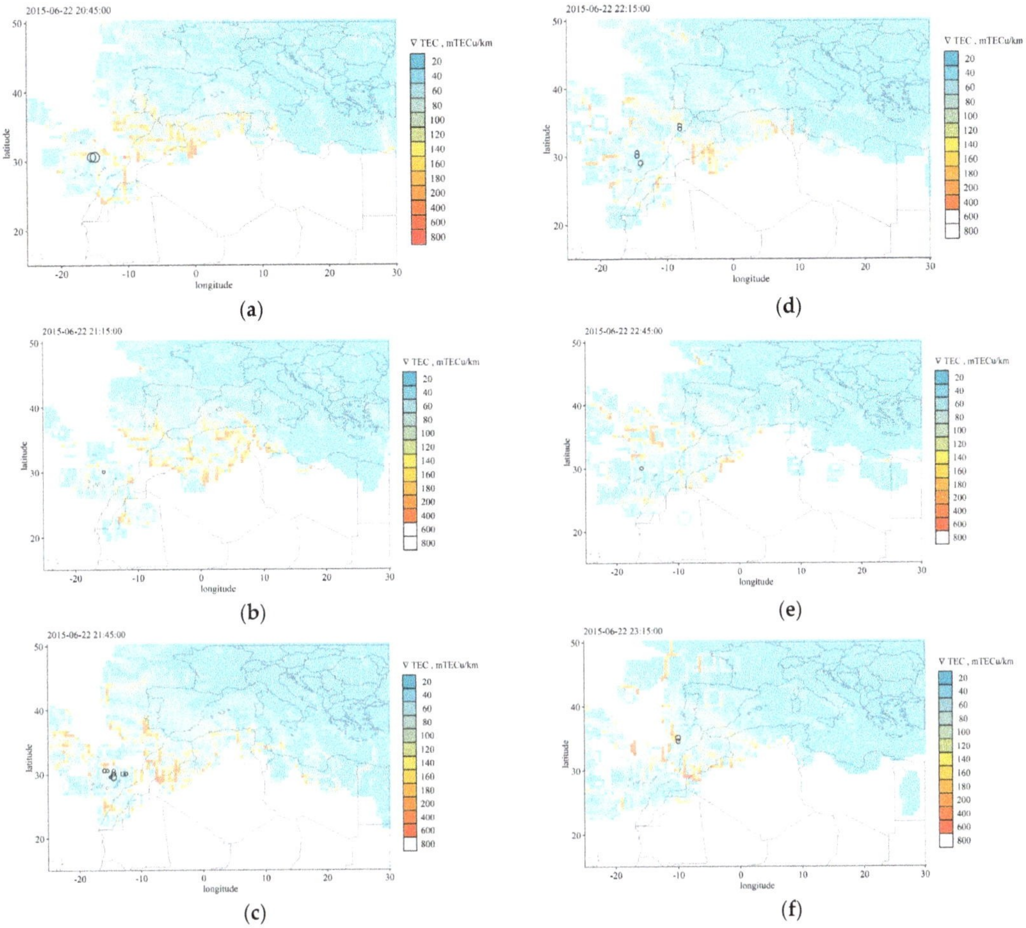

Figure 9. TEC gradients (colour) calculated at 15-min time intervals on 22 June between 20:45 and 23:30. S4 measurements are shown as open circles (see Figure 12 for the colour scheme; size is proportional to S4 values), (top to down, left to right, please see the time stamps).

Figures 7d and S8d show EPBs identified by the Swarm mission on 23 June 2015, 01:00 UTC (vertical lines of grey dots depict Swarm PBI = 1). They are shown again in Figure 10 together with similar plots for 22 June 2015, 23:00 UTC. The plasma bubble identified by the Swarm product at 23:00 UTC on 22 June as containing EPB (vertical lines of grey dots) coincides with areas of high absolute (see Figure 10a,b) and longitudinal (see Supplementary Materials Animations S9 and S10) TEC gradients, but not with the areas of high ROTI values on the ROTI maps (see Figure 10c). On 23 June at 01:00 UTC, the areas identified by the Swarm product as containing EPBs are located between the areas of high TEC gradients (Figure 10d,e)/high ROTI values on the ROTI maps (Figure 10d,f). One of the possible explanations for no relations between the spatial TEC gradients and EPB identified by Swarm at this time is that the structures identified by the Swarm product have a smaller size than the TEC map resolution (0.5 × 0.5°).

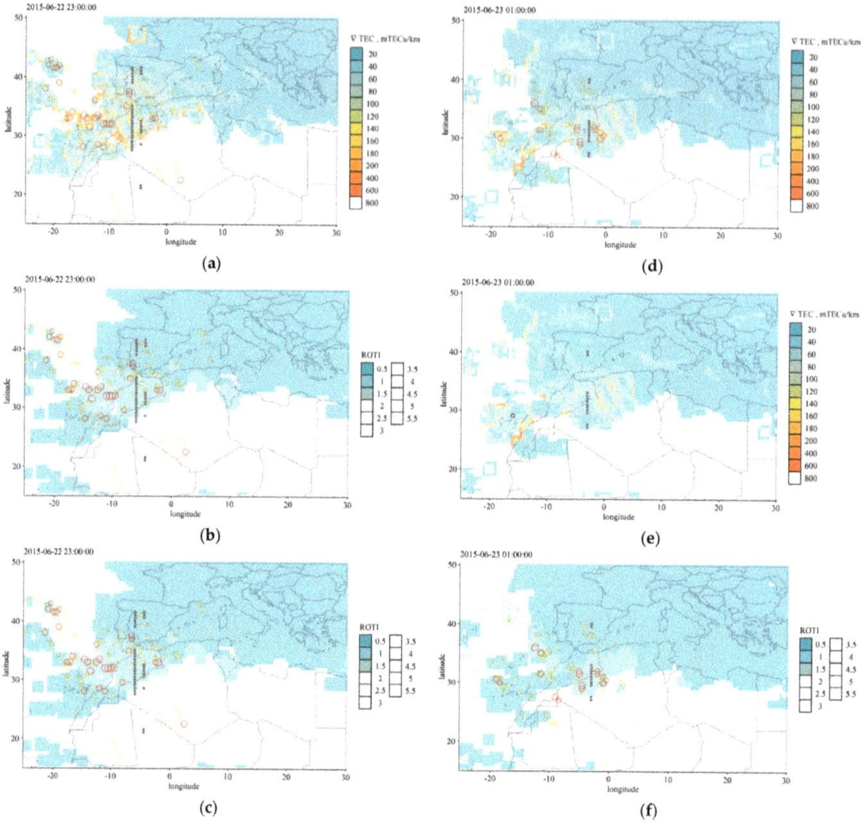

Figure 10. Maps of the absolute TEC gradients ((**a**,**b**,**d**,**e**), colours) and ROTI ((**c**,**f**), colours) together with scintillation indices obtained from ground-based receivers: ROTI ((**a**,**c**,**d**,**f**), coloured circles; see Figure 8 for the colour scheme; size is proportional to the ROTI values) and S4 ((**b**,**e**), white and black circles; see Figure 12 for the colour scheme; size is proportional to S4 values), and the Swarm PBI detections (vertical lines of grey dots) for 22 June 23:00 and 23 June 1:00 UTC.

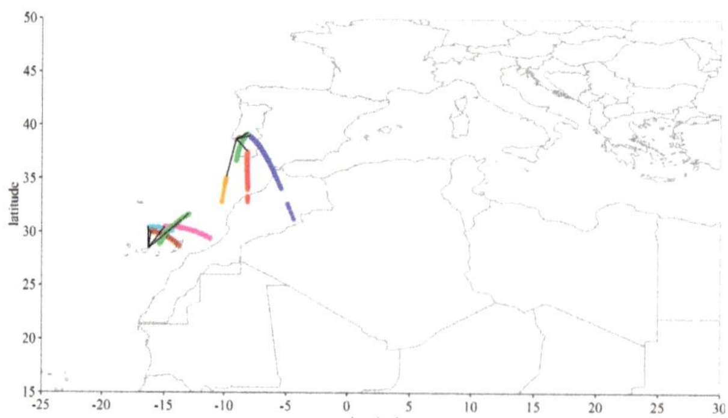

Figure 11. IPPs of several GPS satellites (see text for details)—coloured asterisks, observed at Lisbon and Tenerife during the strong scintillations (slightly different time intervals for different satellites). Black lines connect IPP trajectories to corresponding GNSS receivers and give a simplistic visualisation of corresponding LOSs.

There are other ways to confirm relations between ionospheric scintillations and depletions of the electron density, revealing the presence of EPBs, for example, by the analysis of slant TEC (sTEC) variations obtained for satellites experiencing signal scintillations [14]. To perform this analysis, we selected four satellites for both the Lisbon and Tenerife receivers, with the highest S4 values observed during the night of 22–23 June. These are satellites with PRN G09, G19, G23 and G32 for Lisbon and G01, G04, G11 and G32 for Tenerife. The sTEC values began to rise after the geomagnetic storm commencement in the late afternoon of 22 June 2015, following the positive ionospheric storm development (as described, e.g., in [7,9–11]) (corresponding plots with sTEC and S4 variations can be found in Figures S11 and S12). However, when the S4 values started to grow around 20–22 h UTC (depending on the receiver and PRN), the sTEC values dropped significantly (by 30–50 TECu as measured at Tenerife). The times of the scintillations' onsets and the sTEC decreases are well correlated for each of the satellites. According to [14], this S4-sTEC dynamics can be interpreted as an effect of a() EPB(s) crossing the line of sight (LOS) of a receiver–satellite pair. Figure 11 shows the IPPs of these satellites at the time intervals of the strongest scintillations for each of the satellites. As one can see, the directions of LOS (and locations of corresponding sTEC decreases related to EPBs) coincide very well with areas of high spatial TEC gradients (see Figures 9 and 10), allowing us to conclude that these high spatial gradients result from EPBs crossing the studied area.

4.4. GIX, NeGIX and TEGIX Indices Analysis

To cross-validate the findings of the analysis of the S4, ROTI and ∇TEC data presented above, we used the GIXy component (GNSS-based gradient ionosphere index) and NeGIX and TEGIX (Swarm data-based indices) data. Figure 12 shows half-hourly maps of the 95 percentile of the GIXy (gradient in geographic north–south direction) component from 20:30 UTC on 22 June until 2:00 UTC on 23 June. These maps clearly show clear signatures of the dynamic evolution of TEC gradients in agreement with the other aforementioned results (see Figures 9 and 12). Regions indicated by purple circles on 23 June during 00:00–2:00 UTC show the presence of positive (i.e., higher TEC for the northern IPPs) and negative (i.e., higher TEC for the southern IPPs) GIXs, indicating signatures of high dynamical regions such as IPBs. The provision of a sign in the GIX provides a unique opportunity to identify such regions. The results agree with the Swarm PBI detections (vertical lines of grey dots) in Figure 10 for 22 June 23:00 and 23 June 1:00 UTC. The GIXx (gradient in west–east direction) component does not show significant changes on gradient maps and, therefore, is not included. It must be mentioned that the observed GIX-related TEC gradients have a mid-to-large-scale character in agreement with [8]. Since S4 and ROTI data indicate small- to mid-scale irregularities, it can be concluded that the ionosphere over the south part of the Iberian Peninsula was perturbed at all spatial scales on 22 June 2015. This conclusion is confirmed by gradient estimates from Swarm data on 22 June 2015.

The NeGIX and TEGIX indices are calculated using the Swarm observations. Figures 13 (top) and 14 display the resulting electron density gradients for the orbits of Swarm A and C that navigate westwards from the evening of 22 June to the early hours of 23 June 2015. As previously, the data of only two passes of Swarm on 22 June were used for the analysis: at 21:32 UTC (along the 18°E meridian) and at 23:06 UTC (along the 5.5°W meridian). As one can see from Figure 13, these orbital passes are the ones with the strongest Ne and TEC gradients during the studied time interval. The location of these identified ionospheric perturbations in the region of the Strait of Gibraltar and Northern Africa correlates very well with other observational data presented above. As seen from Figure 14, the most extreme 95-percentile values of the electron density gradients for these two orbits are seen between 20°N and 40°N and surpassed 5000 $e/cm^3{\cdot}km$ for the zonal and meridional components of NeGIX. This value is 10× greater than for the other three orbits of Swarm A and C.

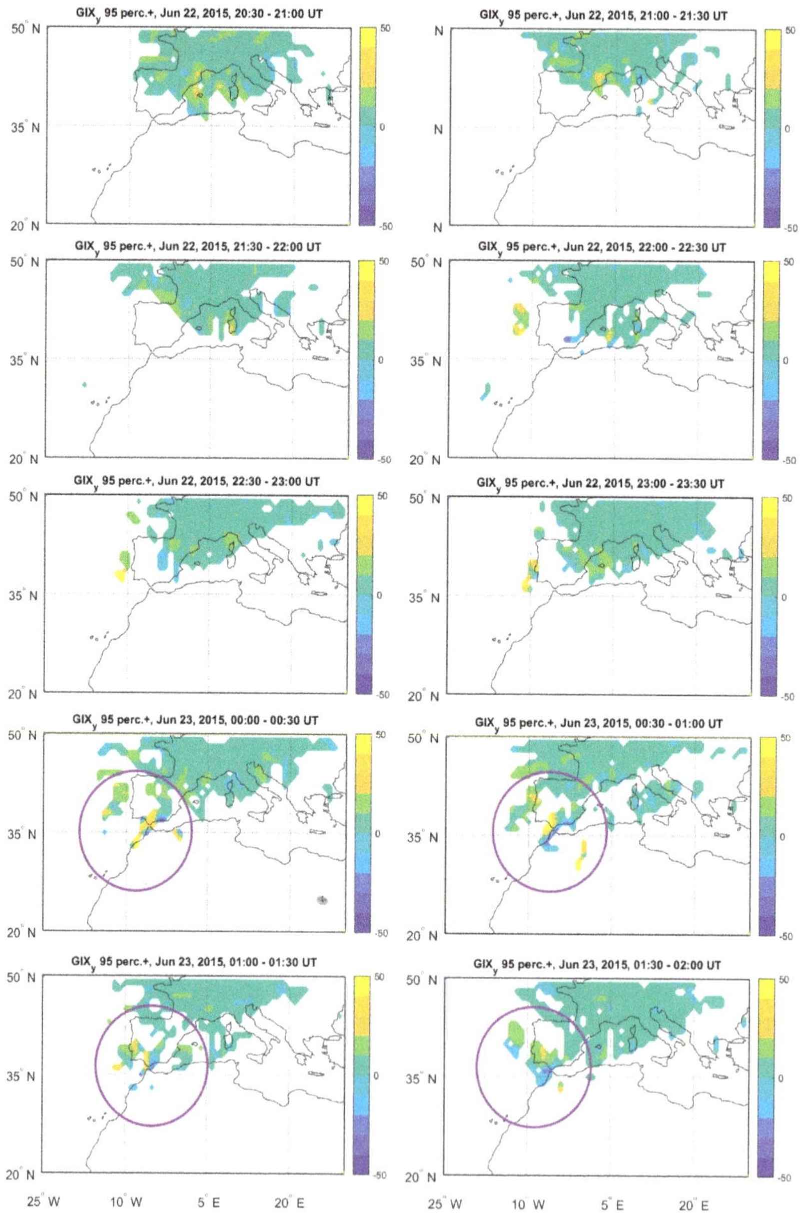

Figure 12. Maps of 95 percentile of the GIXy component (i.e., −ve means higher gradient in south–north direction) in mTECU/km averaged at every 30-min interval from 20:30 UTC on 22 June until 2:00 UTC on 23 June. The dipole length range (50–1000 km) is rather large to obtain sufficient data coverage for mapping. Regions indicated by circles show the existence of +ve and −ve gradients, indicating regions of high gradients.

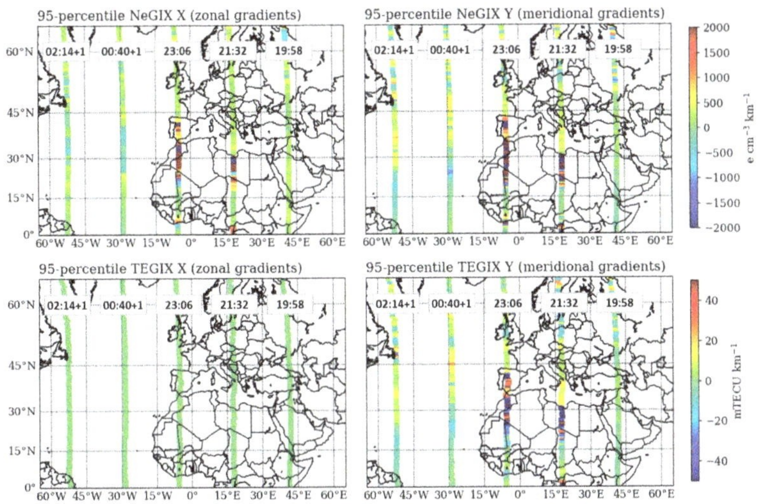

Figure 13. Resulting maps for the 95-percentile values of the X and Y components of NeGIX (**top**) and TEGIX (**bottom**). The indices are determined using data from Swarm A and C, which orbit from the right pass in the evening of 22 June towards the left in the early hours of 23 June. The paths at 21:32 and 23:06 UTC show the strongest gradients for this event and are reviewed in more detail below.

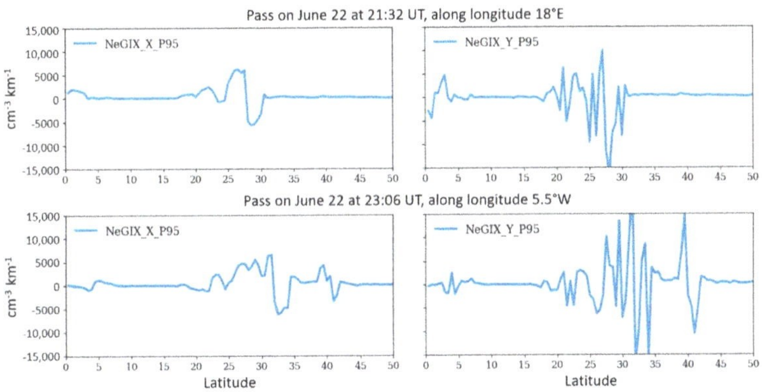

Figure 14. The 95-percentile values for the X (**left**) and Y (**right**) components of NeGIX as a function of latitude. As seen from the maps above, the two orbits of Swarm A and C with the strongest gradient values on 22 June are traced from top to bottom.

Also, the meridional component of TEGIX on the right panels of Figure 15 exposes strong 95-percentile values of TEC gradients over the same latitudinal range, reaching absolute values of more than 100 mTECU/km. For comparison, with red dashed lines, the X and Y components of the spatial gradients extracted from the ∇TEC maps (GNSS data) are overplotted as a function of latitude. The ∇TEC results are limited to latitudes above 30 N. Nevertheless, this comparative analysis with the values of TEGIX for the orbit at 21:32 UTC shows consistent results in magnitude. For the pass at 23:06 UTC, although strong TEC variability in the region is detected by the two methods, a strong difference in the amplitude of gradients is observed. This can be explained by the natural definition of these approaches. While the estimation of ∇TEC gradients is a simplified approach to comparing TEC values at adjacent grid points of TEC maps, the richer sample of Swarm A and C TEC measurements and their combination to form gradient vectors permits the deduction of a 95 percentile. In this context, Swarm permits the overcoming of limitations of data availability and resolution, as generally experienced with ∇TEC maps and ground-based GNSS techniques, especially in the studied region.

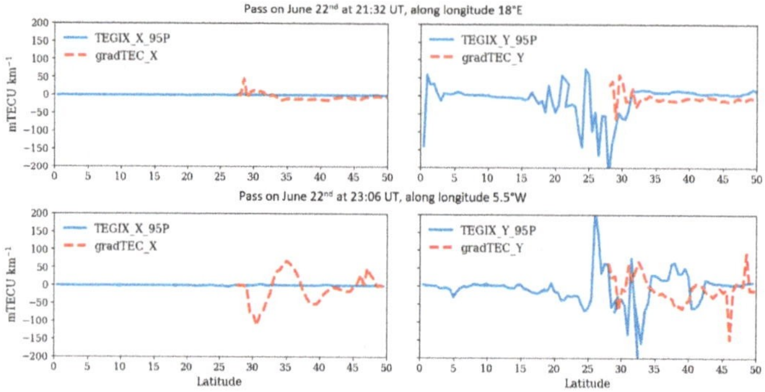

Figure 15. Comparison between the 95-percentile values for the X and Y components of TEGIX (blue lines) and the spatial gradients for similar meridians extracted from ∇TEC maps (red dashed lines) as a function of latitude. The two orbits of Swarm A and C with the strongest gradient values are displayed at 21:32 UTC (**top**) and 23:06 UTC (**bottom**).

Still, the contrast between ground-based and space-borne instruments serves to confirm the location of ionospheric inhomogeneities and the dynamics of the studied parameters.

5. Discussion and Conclusions

Monitoring ionospheric irregularities over the Western Mediterranean is of paramount importance due to the potential for small-scale irregularities to spill over into mid-latitudes, especially during disturbed geospace conditions. In the case event of the June 2015 storm, we contributed to the understanding of the conditions that lead to the penetration into these latitudes and pinpoint how their monitoring and characterisation are crucial, given the growing reliance on GNSS services and technologies, which are highly susceptible to disruptions caused by those kinds of threatening ionospheric irregularities. In fact, the continuous monitoring and analysis of ionospheric irregularities in this region (which is located in the south and southwest parts of the area covered by EGNOS, a satellite-based augmentation system developed by the European Space Agency and EUROCONTROL to enhance the reliability and accuracy of the positioning data) are essential to mitigate these risks and ensure the integrity of critical GNSS-dependent technologies.

The dynamics of the ionosphere in the low–middle latitudes of the Euro–African meridional sector during the geomagnetic storm of 22–23 June 2015 was different from what is normally expected for this region. Large variations in scintillation indices (S4 and ROTI) as well as large spatial TEC gradients were observed there between 19 h UTC on 22 June and 3 h UTC on 23 June. These ionospheric disturbances were caused by a rare event of a spill-over of EPBs from low latitudes triggered by the storm dynamics. Our finding using, for the first time, ground-based measurements of S4 amplitude scintillation indices are in agreement with previous analyses based solely on ROTI and TEC measurements (e.g., [8,11,12,14]). On the other hand, our study allowed us to access the dynamics of the ionospheric inhomogeneities on different spatial scales by comparing ROTI and S4 data and by comparing scintillation indices' variations with changes in the spatial TEC gradients. During the studied event, EPBs were seen as elongated (NW–SE) areas of lower electron density (lower TEC) travelling northwestward through the studied region. High values of S4 and ROTI obtained from individual GNSS receivers were located, as a rule, inside those areas. The spatial and temporal variations in different parameters, such as scintillation indices, TEC gradients and TEC and electron density indices, obtained from the ground-based and space-borne instruments, are consistent and clearly show that spill-over EPBs are responsible for specific ionospheric conditions that were observed during this storm in this meridional sector.

The unusual character of the ionospheric response to the storm of June 2015 was already noted in [8] when compared to the geomagnetic storm of March 2015. During the storm of March 2015, the strength of the ionospheric disturbance (measured by the GIX and SIDX indices and ROTI) gradually decreased from high (60–75°N) to low–middle (30–45°N) latitudes. This observation indicates a growing dissipation of travelling ionospheric disturbances generated by the solar energy input at high latitudes during the storm and subsequently moving equatorward. On the contrary, as is shown in [8], the highest values of the GIX were observed both at high and low–middle latitudes, and even the ROTI values obtained for the low–middle latitudes were higher than for the middle (45–60°N) latitudes. The results of a study presented in this paper allow assuming that the difference in the latitudinal response of the ionosphere to these two storms of 2015 results from the EPB spill-over event. The rapid changes in spatial gradients and positioning errors shown in [8] can also be explained by the effect of passing spilled-over EPBs. It is interesting to note that the GIX indicates medium- to large-scale (here 50–500 km) perturbations of TEC. Thus, a broad spectrum of spatial and temporal scales has been observed in the Western Mediterranean area on 22 June 2015. The related physical processes probably originate in lower latitudes.

In conclusion, this work contributes to a deeper understanding of the longitudinal variations in the North African sector, particularly in the western region, where EPB occurrences are more intense. These variations, whose origin is not completely understood, play a key role in determining the severity and frequency of EPB spill-overs, and their study is willing to contribute to future more accurate forecasting and mitigation strategies. Lastly, this work can be of support towards the planning of enhanced monitoring infrastructure and research efforts in the Western Mediterranean to address the challenges posed by ionospheric irregularities and ensure the continued reliability of GNSS services in the region.

Supplementary Materials: The following supporting information can be downloaded at: https://www.mdpi.com/article/10.3390/rs17030535/s1, Figures S1 and S2: S4 multipath analysis for Lampedusa and Tenerife, respectively (same as Figure 2); Figures S3 and S4: Time variations and azimuth–elevation distribution of S4, respectively, at Lisbon (a), Tenerife (b) and Lampedusa (c), S03: Vertical lines mark the midnight, horizontal lines mark S4 = 0.5, time is in UTC, S04: The horizontal lines mark S4 = 0.5; Figures S5 and S6: Same as Figures 4 and 5, respectively, but for 23 June between 00:15 and 1:45; Figures S7, S8, S9 and S10: Same as Figures 6–9, respectively, but for June 23 between 00:15 and 1:45. Swarm ionospheric plasma bubble index data are shown as grey dots along meridians in (d); Figure S11: S4 (a,c,e,g) and sTEC (b,d,f,h) values obtained at Lisbon for PRNs G09 (a,b), G19 (c,d), G23 (e,f) and G32 (g,h) between 18:00 UTC and 6:00 UTC of the following day for 21–23 June 2015. Colours: 21 June—black, 22 June—red, 23 June—blue, Figure S12: Same as S11 but for Tenerife and PRNs G01 (a,b), G04 (c,d), G11 (e,f) and G32 (g,h). Animation S1: Timelapse of S4 IPP maps between June 22 19:30 and June 23 02:30; Animation S2: Timelapse of ROTI IPP maps between June 22 19:30 and June 23 02:30; Animation S3: Timelapse of ROTI maps between June 22 19:30 and June 23 02:30; Animation S4: Timelapse of ROTI and S4 IPP maps between 22 June 19:30 and 23 June 2:30; Animations S5–S7: Timelapse of absolute, x and y, respectively, TEC gradient maps and ROTI between 22 June 19:30 and 23 June 2:30; Animations S8–S10: Timelapse of absolute, x and y, respectively, TEC gradient maps and S4 between 22 June 19:30 and 23 June 02:30.

Author Contributions: This work has the following contributions: conceptualization—A.M.; methodology—A.M., L.S. and T.B.; software—A.M., R.I., E.P. and D.E.; validation—A.M., L.S. and

T.B.; formal analysis—A.M.; investigation—A.M., L.S., T.B., R.I., E.P., M.M.H., J.A.C. and D.E.; resources—A.M., T.B. and L.S.; data curation—A.M., R.I. and E.P.; writing—original draft preparation—A.M.; writing—review and editing—A.M., L.S., T.B., R.I., E.P., M.M.H., J.A.C. and N.J.; visualization—A.M., J.A.C. and M.M.H.; supervision—A.M.; project administration—A.M.; funding acquisition—A.M. and L.S. All authors have read and agreed to the published version of the manuscript.

Funding: IA is supported by Fundação para a Ciência e a Tecnologia (FCT, Portugal) through the research grants UIDB/04434/2020 (https://doi.org/10.54499/UIDB/04434/2020), UIDP/04434/2020 (https://doi.org/10.54499/UIDP/04434/2020) and UID/04434/2025 . This study is a contribution to the "PROSE: P3-SWE-XXXVII—SWE PRODUCTS FOR SOUTHERN EUROPE—PHASE 1" project funded by ESA. The work was supported through the "ALERT: Assessment of the ionospheric scintillation over Portugal" project as a part of the 3rd PITHIA-NRF TNA. The PITHIA-NRF project has received funding from European Union's Horizon 2020 research and innovation program under grant agreement no. 101007599. A.M. acknowledges the research infrastructure and the access provider INGV of the PITHIA-NRF project (https://www.pithia-nrf.eu/). RI's current research fellowship is funded by the Swarm Space Weather Variability, Irregularities, and Predictive Capabilities for the Dynamic Ionosphere (Swarm-VIP-Dynamic) project, which has been funded by the European Space Agency, contract 4000143413/23/I-EB within the "Esa Solid Magnetic Science Cluster—Research Opportunities: 4dionosphere—Expro+".

Data Availability Statement: The SCINDA data are available: for [21] at Barlyaeva, T.; Barata, T.; Morozova, A. Data from: Datasets of ionospheric parameters provided by SCINDA GNSS receiver from Lisbon airport area. Mendeley Data 2020, V1. http://dx.doi.org/10.22632/kkytn5d8yc.1, accessed on 12 December 2024; for [22] at Morozova, A.; Barlyaeva, T.; Barata, T. Data from: Datasets of ionospheric parameters (TEC, SI, positioning errors) from Lisbon airport area for 2014–2019. Mendeley Data 2022, V2. http://dx.doi.org/10.22632/3z6mjk39jv.2, accessed on 12 December 2024. The RENEP RINEX 2.11 files are available through https://renep.dgterritorio.gov.pt/ (accessed on 18 November 2024). Rinex data from the mas1, lpal, rabt and meli receiver are accessible through the FTP repository of the BKG's GNSS Data Center (https://igs.bkg.bund.de/), part of the International GNSS Service (IGS). Data for the Novatel ISMR receiver operating in Lampedusa (code "lam0s") are accessible through the eSWua repository (https://doi.org/10.13127/eswua/gnss, accessed on 12 December 2024). The GNSS data for the Tenerife station are by courtesy of the Ionosphere Monitoring and Prediction Center at DLR, Germany (https://impc.dlr.de/). TEC and ROTI maps are from the DRAWING-TEC project by ISEE (https://aer-nc-web.nict.go.jp/GPS/DRAWING-TEC/), accessed on 12 December 2024 and available at https://stdb2.isee.nagoya-u.ac.jp/GPS/GPS-TEC/, accessed on 12 December 2024 and https://stdb2.isee.nagoya-u.ac.jp/GPS/GPS-TEC/GLOBAL/RMAP/index.html, accessed on 12 December 2024. Global GNSS-TEC data processing has been supported by JSPS KAKENHI Grant Number 16H06286. GNSS RINEX files for the GNSS-TEC processing are provided from many organizations listed by the webpage. (http://stdb2.isee.nagoya-u.ac.jp/GPS/GPS-TEC/gnss_provider_list.html, accessed on 12 December 2024). Ionospheric plasma bubble index is from the SW_IBIxTMS_2F Swarm product (please visit https://swarmhandbook.earth.esa.int/catalogue/sw_ibixtms_2f, accessed on 12 December 2024 for description) and available at ftp://swarm-diss.eo.esa.int/Level2daily/Latest_baselines/IBI/TMS, accessed on 12 December 2024.

Acknowledgments: Authors are grateful to Martin Kriegel (DLR) for providing Tenerife data. The authors would like to acknowledge the Direção Geral do Território (DGT) and Helena Ribeiro personally for making ReNEP data available (RENEP). The authors thank Giorgio di Sarra, Scientific director of the activities at the ENEA Climate Observation Station, and Damiano Sferlazzo for their support with the ISMR station in Lampedusa.

Conflicts of Interest: The authors declare no conflicts of interest. The funders had no role in the design of this study; in the collection, analyses, or interpretation of the data; in the writing of the manuscript; or in the decision to publish the results.

References

1. Cherniak, I.; Zakharenkova, I. First observations of super plasma bubbles in Europe. *Geophys. Res. Lett.* **2016**, *43*, 11–137. [CrossRef]
2. Pica, E.; Spogli, L.; Cesaroni, C.; Alfonsi, L.; Haralambous, H.; Vallianatos, F.; De Franceschi, G.; Romano, V.; Marcocci, C. Assessing the ionospheric scintillations occurrence on L-band in the southern Mediterranean sector. *Adv. Space Res.* **2024**, *75*, 837–855. [CrossRef]
3. Wei, Y.; Zhao, B.; Li, G.; Wan, W. Electric field penetration into Earth's ionosphere: A brief review for 2000–2013. *Sci. Bull.* **2015**, *60*, 748–761. [CrossRef]
4. Li, G.; Ning, B.; Otsuka, Y.; Abdu, M.A.; Abadi, P.; Liu, Z.; Spogli, L.; Wan, W. Challenges to equatorial plasma bubble and ionospheric scintillation short-term forecasting and future aspects in east and southeast Asia. *Surv. Geophys.* **2021**, *42*, 201–238. [CrossRef]
5. Spogli, L.; Alberti, T.; Bagiacchi, P.; Cafarella, L.; Cesaroni, C.; Cianchini, G.; Coco, I.; Di Mauro, D.; Ghidoni, R.; Giannattasio, F.; et al. The effects of the May 2024 Mother's Day superstorm over the Mediterranean sector: From data to public communication. *Ann. Geophys.* **2024**, *67*, PA218. [CrossRef]
6. Rodrigues, F.S.; Socola, J.G.; Moraes, A.O.; Martinis, C.; Hickey, D.A. On the properties of and ionospheric conditions associated with a mid-latitude scintillation event observed over southern United States. *Space Weather* **2021**, *19*, 2021SW002744. [CrossRef]
7. Morozova, A.L.; Barlyaeva, T.V.; Barata, T. Variations of TEC over Iberian Peninsula in 2015 due to geomagnetic storms and solar flares. *Space Weather* **2020**, *18*, 2020SW002516. [CrossRef]
8. Nykiel, G.; Cahuasquí, J.A.; Hoque, M.M.; Jakowski, N. Relationship between GIX, SIDX, and ROTI ionospheric indices and GNSS precise positioning results under geomagnetic storms. *GPS Solut.* **2024**, *28*, 69. [CrossRef]
9. Barata, T.; Pereira, J.; Hernández-Pajares, M.; Barlyaeva, T.; Morozova, A. Ionosphere over Eastern North Atlantic Midlatitudinal Zone during Geomagnetic Storms. *Atmosphere* **2023**, *14*, 949. [CrossRef]
10. Astafyeva, E.; Zakharenkova, I.; Alken, P. Prompt penetration electric fields and the extreme topside ionospheric response to the June 22–23, 2015 geomagnetic storm as seen by the Swarm constellation. *Earth Planets Space* **2016**, *68*, 152. [CrossRef]
11. Ngwira, C.M.; Habarulema, J.B.; Astafyeva, E.; Yizengaw, E.; Jonah, O.F.; Crowley, G.; Gisler, A.; Coffey, V. Dynamic response of ionospheric plasma density to the geomagnetic storm of 22–23 June 2015. *J. Geophys. Res. Space Phys.* **2019**, *124*, 7123–7139. [CrossRef]
12. Cherniak, I.; Zakharenkova, I.; Sokolovsky, S. Multi-instrumental observation of storm-induced ionospheric plasma bubbles at equatorial and middle latitudes. *J. Geophys. Res. Space Phys.* **2019**, *124*, 1491–1508. [CrossRef]
13. Pi, X.; Mannucci, A.J.; Lindqwister, U.J.; Ho, C.M. Monitoring of global ionospheric irregularities using the worldwide GPS network. *Geophys. Res. Lett.* **1997**, *24*, 2283–2286. [CrossRef]
14. Kashcheyev, A.; Migoya-Orué, Y.; Amory-Mazaudier, C.; Fleury, R.; Nava, B.; Alazo-Cuartas, K.; Radicella, S.M. Multivariable comprehensive analysis of two great geomagnetic storms of 2015. *J. Geophys. Res. Space Phys.* **2018**, *123*, 5000–5018. [CrossRef]
15. Yeh, K.C.; Liu, C.H. Radio wave scintillations in the ionosphere. *Proc. IEEE* **1982**, *70*, 324–360. [CrossRef]
16. Fremouw, E.J.; Leadabrand, R.L.; Livingston, R.C.; Cousins, M.D.; Rino, C.L.; Fair, B.C.; Long, R.A. Early results from the DNA Wideband satellite experiment—Complex-signal scintillation. *Radio Sci.* **1978**, *13*, 167–187. [CrossRef]
17. Bougard, B.; Sleewaegen, J.M.; Spogli, L.; Veettil, S.V.; Monico, J.F. CIGALA: Challenging the solar maximum in Brazil with PolaRxS. In Proceedings of the 24th International Technical Meeting of the Satellite Division of the Institute of Navigation (Ion GNSS 2011), Portland, OR, USA, 20–23 September 2011; pp. 2572–2579.
18. Van Dierendonck, A.J.; Klobuchar, J.; Hua, Q. Ionospheric scintillation monitoring using commercial single frequency C/A code receivers. *Proc. ION GPS* **1993**, *93*, 1333–1342.
19. Mrak, S.; Coster, A.J.; Groves, K.; Nikoukar, R. Ground-based infrastructure for monitoring and characterizing intermediate-scale ionospheric irregularities at mid-latitudes. *Front. Astron. Space Sci.* **2023**, *10*, 1091340. [CrossRef]
20. Wernik, A.W.; Secan, J.A.; Fremouw, E.J. Ionospheric irregularities and scintillation. *Adv. Space Res.* **2003**, *31*, 971–981. [CrossRef]
21. Barlyaeva, T.; Barata, T.; Morozova, A. Datasets of ionospheric parameters provided by SCINDA GNSS receiver from Lisbon airport area. *Data Brief* **2020**, *31*, 105966. [CrossRef]
22. Morozova, A.; Barlyaeva, T.; Barata, T. Updating datasets of ionospheric parameters provided by SCINDA GNSS receiver from Lisbon airport area with full data sets for 2014–2019. *Data Brief* **2023**, *47*, 109026. [CrossRef] [PubMed]
23. Morozova, A. Ionosphere scintillations. In *Space Weather Influence on the Ionosphere and Its Potential Threats to GNSS Services: Focus on the Portuguese Territories*; Morozova, A., Barlyaeva, T., Eds.; Materials of the on-line workshop "Space weather influence on the ionosphere and its potential threats to GNSS services: Focus on the Portuguese territories"; Instituto de Astrofísica e Ciências do Espaço: Lisbon, Portugal, 2023; ISBN 978-989-33-5493-3/978-989-33-5492-6. [CrossRef]
24. Cesaroni, C.; De Franceschi, G.; Marcocci, C.; Pica, E.; Romano, V.; Spogli, L. *Electronic Space Weather Upper Atmosphere Database (eSWua)—GNSS Scintillation Data*; Version 1.0.; Upper atmosphere physics and radiopropagation Working Group, Istituto Nazionale di Geofisica e Vulcanologia: Roma, Italy, 2020. [CrossRef]

25. Jakowski, N.; Hoque, M.M. Estimation of spatial gradients and temporal variations of the total electron content using ground-based GNSS measurements. *Space Weather* **2019**, *22*, 339–356. [CrossRef]
26. Cahuasquí, J.A.; Hoque, M.M.; Jakowski, N.; Vasylyev, D.; Buchert, S.; Nykiel, G.; Kriegel, M.; David, P.; Tagargouste, Y.; Berdermann, J. NeGIX and TEGIX: Two new indices to characterize the topside ionosphere with Swarm (No. EGU24-12608). *Copernic. Meet.* **2024**. [CrossRef]
27. McCaffrey, A.M.; Jayachandran, P.T. Spectral characteristics of auroral region scintillation using 100 Hz sampling. *GPS Solut.* **2022**, *21*, 1883–1894. [CrossRef]
28. Imam, R.; Alfonsi, L.; Spogli, L.; Cesaroni, C.; Mehr, I.E.; Minetto, A.; Dovis, F. Scintillation Climatology from a Software Defined Radio Receiver over Antarctica. *Ann. Geophys.* **2024**, *67*, PA108. [CrossRef]
29. Liu, Q.; Hernández-Pajares, M.; Yang, H.; Monte-Moreno, E.; García-Rigo, A.; Lyu, H.; Olivares-Pulido, G.; Orús-Pérez, R. A New Way of Estimating the Spatial and Temporal Components of the Vertical Total Electron Content Gradient Based on UPC-IonSAT Global Ionosphere Maps. *Space Weather* **2022**, *20*, e2021SW002926. [CrossRef]
30. Yokoyama, T. A review on the numerical simulation of equatorial plasma bubbles toward scintillation evaluation and forecasting. *Prog. Earth Planet. Sci.* **2022**, *4*, 37. [CrossRef]
31. Rino, C.; Yokoyama, T.; Carrano, C. A three-dimensional stochastic structure model derived from high-resolution isolated equatorial plasma bubble simulations. *Earth Planets Space* **2023**, *75*, 64. [CrossRef]
32. Cesaroni, C.; Spogli, L.; Alfonsi, L.; De Franceschi, G.; Ciraolo, L.; Monico, J.F.G.; Scotto, C.; Romano, V.; Aquino, M.; Bougard, B. L-band scintillations and calibrated total electron content gradients over Brazil during the last solar maximum. *J. Space Weather. Space Clim.* **2015**, *5*, A36. [CrossRef]

remote sensing

Article

Technical Possibilities and Limitations of the DPS-4D Type of Digisonde in Individual Meteor Detections

Csilla Szárnya [1,2,*], Zbyšek Mošna [3], Antal Igaz [4], Daniel Kouba [3], Tobias G. W. Verhulst [5], Petra Koucká Knížová [3], Kateřina Podolská [3] and Veronika Barta [1]

1 HUN-REN Institute of Earth Physics and Space Science, 9400 Sopron, Hungary
2 Doctoral School of Earth Sciences, ELTE Eötvös Loránd University, 1171 Budapest, Hungary
3 Department of Ionosphere and Aeronomy, Institute of Atmospheric Physics, Czech Academy of Sciences, 14100 Prague, Czech Republic
4 HUN-REN Research Centre for Astronomy and Earth Sciences, Konkoly Thege Miklós Astronomical Institute, 1121 Budapest, Hungary
5 Royal Meteorological Institute of Belgium, Solar Terrestrial Centre of Excellence, 1180 Brussels, Belgium
* Correspondence: szarnya.csilla@epss.hu

Abstract: During the peak days of the 2019 Leonids and Geminids (16–19 November and 10–16 December), two ionograms/minute and one Skymap/minute campaign measurements were carried out at the Sopron (47.63°N, 16.72°E) and Průhonice (50.00°N, 14.60°E) Digisonde stations. The stations used frequencies between 1 and 17 MHz for the ionograms, and the Skymaps were made at 2.5 MHz. A temporary optical camera was also installed at Sopron with a lower brightness limit of +1 visual magnitude. The manual scaling of ionograms for November and December 2019 to study the behavior of the regular sporadic E layer was also completed. Although the distributions of the stations were similar, there were interesting differences despite the relative proximity of the stations. The optical measurements detected 88 meteors. A total of 376 meteor-induced traces were found on the Digisonde ionograms at a most probable amplitude (MPA) threshold of 4 dB and of these, 40 cases could be linked to reflections on the Skymaps, too. Of the 88 optical detections, 31 could be identified on the ionograms. The success of detections depends on the sensitivity of the instruments and the noise-filtering. Geometrically, meteors above 80 km and with an altitude angle of 40° or higher can be detected using the Digisondes.

Keywords: Digisonde; ionosphere; sporadic E layer (Es); Leonids; Geminids; meteors; meteor plasma trails; GIRO network

1. Introduction

According to the definitions of the International Astronomical Union (IAU) in 2017, meteors are the light and associated physical phenomena (heat, shock, ionization, etc.), which result from the high-speed entry of a solid object from space into a gaseous atmosphere. The meteor phenomenon can be caused by a meteoroid, a comet, an asteroid or any solid matter with the appropriate combination of velocity, mass and mean-free-path in a planet's atmosphere. A meteoroid's size, by agreement, is roughly between 30 μm and 1 m. Meteor showers are groups of meteors produced by meteoroids of the same meteoroid stream.

The Leonid and Geminid meteor showers are widely known meteoroid swarms. The Leonids are associated with comet 55P/Temple-Tuttle [1], while the source of the Geminids is the asteroid-originated object 3200 Phaethon [2]. The Leonid meteor shower arrives between 6 November and 30 November, and the number of incoming meteors usually peaks on 17 November. The Geminid meteor shower takes place between 4 December and 17 December, usually peaking on 14 December.

The peak of the two main meteor showers overlaps with less-known, minor meteoroid swarms. The Leonids' peak is simultaneously present with the Southern- and Northern-Taurids (active: 23 September–8 December; and 13 October–2 December, respectively) and the α-Monocerotids (active: 15 November–23 November). The σ-Hydrids (active: 22 November–4 January), the Monocerotids (active: 5 December–20 December), the Coma-Berenicids (active: 12 December–23 December) and the Antihelion Source (active: 10 December–20 September) swarms happen at the same time as the Geminids' peak [3].

The Southern-Taurids originated from comet Encke and the Northern-Taurids are associated with asteroid 2004 TG_{10}, which, based on its orbital data, is probably a larger chunk of the comet Encke [4]. The origin of the α-Monocerotids and the Coma-Berenicids is unknown. The α-Monocerotids' origin is presumably a long-period comet [5], while the Coma-Berenicids' is probably a moderate-period comet. The Monocerotids are associated with a long-period comet C/1917 F1 (Mellish) [6]. The σ-Hydrids are suggested to originate from long-period comet C/2023 P1 (Nishimura) based on its orbit parameters, but this is not proven yet [7]. The Antihelion Source meteoroid stream is thought to originate from the Earth passing through the debris of comets and asteroids under Jupiter's influence. The radiant point is in the direction opposite to the Sun in the sky, hence the name, antihelion [8].

Some meteors cannot be linked to a specific parental body based on their orbital data, and these are called sporadic meteors. At most, they can be classified as being of cometary or asteroidal origin. These can come from the remains of ancient, disintegrated comets, or from known meteor swarms that have left orbit.

When meteors enter or meteoric dust and metallic material is deposited in the Earth's lower atmosphere, thin and faint layers of ionization can form via vertical ion convergence. This trapping mechanism is the combined effect of the vertical shear of the zonal winds, the meridional winds and the Lorentz force controlling the movement of ions through the local magnetic field. This ion convergence also creates the sporadic E layer (Es) phenomenon [9–11]. At mid-latitude the wind shear theory is also the currently accepted theory for the formation of sporadic E layers.

Under nighttime conditions, the meteoric smoke absorbs free electrons, significantly changing the usual balance between electrons and negative/positive ions [12]. Atomic oxygen also destroys negative ions, but this does not affect the electrons attached to meteoric smoke particles [13].

It should also be noted that Jacobi et al. [14] found a significant increase in sporadic E activity after the Geminids, with an average delay of 2.5 days. This is not the case for the Leonids, although a distinct increase of sporadic E activity in association with the Leonids event 1996 has been reported in low latitudes by Chandra et al. [15].

According to Stuart (Table 10-2) [16], the brighter a meteor, the higher the electron line density (electrons per meter of trail length). An expected value of roughly 10^{15} el/m is associated with a visual magnitude of 2.5, while −2.5 is associated with 10^{17} el/m. It should be noted that the Stuart's [16] figures are based on the Sugar's [17] tables, who based their figures on Manning and Eshleman's [18] tables.

McKinley [19] noted that for bright meteors (below 0 magnitude), the enduring ionization's height is greater than the corresponding height of the maximum light emitted. For 0 magnitude, these heights are roughly equal, while for fainter meteors (above 0 magnitude) the maximum of the visible light is above the radio-echo's height.

Meteor trails are usually classified as underdense or overdense. A common approximation is to consider the meteor trail as an infinitely long right circular cylinder of electrons, through which an incident electromagnetic wave passes without being significantly modified. Stuart [16] defines underdense trails in this way. If this assumption is not valid and the incident wave is significantly modified, the trail is overdense. This arises if ionization occurs in the wake of a meteor to such an extent that it can be approximated as a metallic reflecting surface that expands and spreads out over time, causing the electron line density to decrease to the point where the underdense approximation is true. This is also discussed in Maruyama et al. [20]. The article states that in their campaign, the ionosondes that they use can only detect events with lifetimes of more than 15 s, which are considered overdense traces [21]. Kozlovsky et al. [22] give a more tangible definition. A meteor echo is underdense if the electron line density is below 2.4×10^{14} el/m. Above this quantity, it is an overdense meteor echo, and close to it is a transitional echo.

It is worth noting that polarization effects can also occur. For traces with a dielectric constant close to -1, a plasma resonance phenomenon can occur if the polarization of the incident electric field is such that it is perpendicular to the axis of the trail. In this case, the reflection coefficient can double or the reflected energy can quadruple [16]. The magnetic field may also play a role. Based on previous observations [23], if the meteor's trail and the magnetic field lines align, this can stabilize the trail and increase its lifetime. Oppenheim and Dimant [24] found in 3D simulations that if the difference in the field alignment is 6° or below, it can significantly enhance the lifetime of the meteoric trails.

The initial mass and the velocity of the meteor strongly influence the observable characteristics of meteor trails. For a given mass, the higher the velocity, the higher the altitude of the meteor trails. And at a given speed, the higher the mass, the lower the altitude of the trails [16]. Theories indicate that the length distribution of the meteor trails is independent of the electron line density that is designated as the criterion for terminating the trail and is also independent of the sensitivity of the radar receiver and the power output of the radar transmitter [25].

Leonid meteors have an average speed of 71 km/s (~40.7 degree/s for 100 km height) when entering the Earth's atmosphere [26], and their average size is 10 mm [27]. The chemical composition is mainly Mg, Fe, Ca and Na, although it is perhaps worth noting that the Na content of the smaller meteors is ~30% lower [28,29] which is the same for the Perseids. However, Leonid meteors easily disintegrate in the atmosphere and the Na in them evaporate faster because of that, unlike in the Perseids [28].

Geminid meteors have an average speed of 35 km/s (~20.1 degree/s for 100 km height) when arriving in the atmosphere [30], and their average size is 1 mm [31]. The chemical composition is mainly Mg, Ca, Fe, Na and Ni [32–34]. It is worth noting that the Ni abundance is higher than in the other meteoroid showers, while the Na abundance is lower [34].

Fragmentation of the incoming meteor leads to the dispersion of smaller pieces over a larger area than the parent body [19]. Statistical analysis shows that the percentage of fragmentation [35] is less pronounced in the asteroid-originated Geminids than in the comet-originated Leonids [36].

Optical observations [37] and meteor radars [38,39] are the most often used meteor detecting methods, but during major meteor falls, meteor echo traces were often observed on ionograms, which were classified into separate categories [40]. However, this classification led to controversy, which was finally resolved by Maruyama et al. [20].

Maruyama et al. [20] analyzed data from a campaign session made during the Leonids of 2001, which was a peak number producing event [41,42]—three 10C-type ionosondes and one DPS4 Digisonde took campaign measurements over Japan from 03:00 (UT, 12:00 JST) on 17 November 2001 to 09:00 (UT, 18:00 JST) on 21 November 2001. The ionosonde stations were at Wakkanai (45.39°N, 141.69°E), Kokubunji/Tokyo (35.71°N, 139.49°E) and Yamagawa (31.20°N, 130.62°E) and they operated with a 1 ionogram/minute setting. At Ogimi/Okinawa (26.68°N, 128.16°E) the Digisonde was taking 11 ionogram/every 15 min interval (the number of its findings was multiplied by 15/11 for convenience). The frequency sweep to create the ionograms (see Section 2.3) went from 1 MHz to 30 MHz. The automatic gain control (AGC) was turned off; also no post processing or noise reductions were used on the data. Between 16:00 UT and 24:00 UT on 18 November, a significant number of meteor-induced echoes were observed, which corresponded to the predicted peak period of the Leonid meteor shower of 2001. Maruyama et al. [20] concluded that these meteor-induced echoes were backscattering (Fresnel scattering) from the region of increased electron line density along the meteor's path. Since the measurements were made using ionosondes, because of the settings and the capabilities of the instruments, only traces longer than 15 s were recorded according to the article, and these were categorized as overdense echoes [21].

At Sodankylä Geophysical Observatory (67.37°N, 26.63°E), in a year of measurements of their frequency-modulated continuous-wave chirp sounder, 28 overdense cases around 90 km in height with 1 min time resolution ionograms were selected, which could be combined with optical measurements and a colocated all-sky interferometric meteor radar. Most of the events originated from the Geminids of 2015. The article itself [22] proves that their proprietary ionosonde can also detect the ionization caused by meteors. In addition, it was found in the analysis of overdense cases that the decay time of the meteor's trail depends on the initial electron line density: a less dense trail decays slower, while a denser trail decays faster. They suggest that this might be the result of two competing mechanisms: the presence of meteoric dust may decrease the rate of diffusion and recombination, which can lead to an increased rate of decay for the denser trails.

The DPS-4D Digisonde is an advanced type of ionosonde, with added features and enhanced sounding ability [43] (see Section 2.3). In case studies [44,45], it was proven that these types of ionosondes can also detect individual meteor-induced traces, but the exact mapping of what these types of devices can do has not yet been conducted so far. The aim of this study is to explore the potential of meteor observation with the DPS-4D Digisonde, the technical details to look out for and the features that can be useful.

Following the Introduction, the data collection and processing methods—the camera and processing software used for the optical images, Metrec (Section 2.1), the data produced by the processing (Section 2.2) and the results from the processing of the Digisonde measurements (Section 2.3), the ionogram and the Skymap data (Sections 2.4.1–2.4.3)—will be presented. After the data summary, a detailed analysis of the measurements will be presented—what was found (Section 3.1.1), what was not found (Section 3.1.2) and the other features that were observed (Sections 3.2.1–3.2.3). The observations are followed by a discussion (Section 4) and finally conclusions (Section 5) are drawn.

2. Methods and Data

2.1. Meteor Camera and Optical Processing

Optical data were acquired using a Watec 902H2 Ultimate camera. It was equipped with a Computar HG2610AFCSHSP objective (2.6 mm focal length, 30 mm effective lens aperture and toward the zenith 122° × 97° field of view). This temporary station at Széchenyi István Geophysical Observatory (47.632°N, 16.718°E) was operated using the Metrec automatic meteor detection software [46]. In this work, the lower limit in brightness is roughly +1 magnitude. According to the software description, the magnitude estimation's accuracy is ±0.5 magnitude. The process also includes position accuracy estimation.

The Metrec software determines the right ascension (degree) and the hour angle (h) in the equatorial coordinate system II using a reference file of bright stationary stars (281 references were used for this processing). Furthermore, it identifies the parental meteor shower based on the meteor's orbit, calculates the velocity (degrees/s) and the duration of the event, and gives a rough estimate of the length of the meteor's light trail based on the number of oversaturated pixels and the direction (degree).

It should be noted that this software is not suitable for high-precision orbit determination. However, it can tell where the meteor was in relation to the Digisonde and whether the instrument had any chance of detecting the object at all. So, for this study, it was decided to be precise enough.

2.2. Optical Data

During the campaign measurements (Leonid campaign: 16–19 November 2019 and Geminid campaign: 10–16 December 2019), most of the nights were overcast at Sopron, and thus, only the pictures from the night of 18 November and dawn of 19 November 2019 from the Leonid campaign, part of 13 December's night and the night of 14 December—dawn of 15 December 2019 from the Geminid campaign could be used (Table 1).

The Metrec software took all occurring meteor showers into consideration when processing the picture. During the Leonid campaign, there were four active meteor showers: the Leonids (LEO), the Southern- and Northern-Taurids (STA, NTA) and the α-Monocerotids (AMO). During the Geminid campaign, there were five: the Geminids (GEM), the Coma-Berenicids (COM), the σ-Hydrids (HYD), the Monocerotids (MON) and the Antihelion Source (ANT).

The Metrec software detected 9 meteors during the Leonid campaign and 79 meteors during the Geminid campaign. According to the identification of the software, from the 9 meteors of the Leonid campaign, only 3 were from the Leonid meteor shower. One was from the Southern-Taurids, while the rest were detected as sporadic meteors (SPO). From the 79 meteors of the Geminid campaign, 21 meteors were from the Geminids, 5 from the Coma-Berenicids, 4 from the sigma-Hydrids, 6 from the Monocerotids and 6 from the Antihelion Source, and 37 were identified as sporadic meteors.

There was also an extra optical detection, thanks to the European Fireball Network. On 17 November 2019, at 4:15 a.m. (UT), a bright Leonid fireball passed not far from Průhonice's Digisonde, and its signal was recorded on the ionograms and on the Skymaps. A separate case study was conducted of this [45]. Since the fireball is a separate phenomenon, although it is included in the tables and descriptions, it is omitted from the analyses and is not shown in the histograms.

Table 1. The summary of the successful optical detections at Sopron for the campaign sessions.

Campaign	Number of Optical Detections
Leonid campaign of 2019 Night: 2019-11-18. (from: 17:05 UT) Dawn: 2019-11-19. (until 3:52 UT)	LEO: 3 NTA: 0 STA: 1 AMO: 0 SPO: 5
Geminid campaign of 2019 Nights: 2019-12-13. (22:40–23:22 UT) 2019-12-14. (from: 17:21 UT) Dawn: 2019-12-15 (until 4:48 UT)	GEM: 21 COM: 5 HYD: 4 MON: 6 ANT: 6 SPO: 37

The software determines the duration of the event, the parental meteor shower (given the time frame, it also lists the possible showers), the brightness (accuracy: ±0.5 magnitude), the right ascension and declination (equatorial coordinate system II) of the beginning and the end point, the number of pixels saturated by the meteor, the direction of the meteor, the velocity (degree/s), the position accuracy and the signal-to-noise ratio. From the right ascension and declination, the azimuth and altitude coordinates were also calculated in the horizontal coordinate system.

2.3. Digisonde Technique

As a consequence of Snell's law, the electromagnetic wave that enters the environment with the changing electron concentration will refract or reflect. All electromagnetic waves with a frequency smaller than the plasma frequency entering the ionosphere at vertical incidence will be reflected back to the Earth (Equation (1)) [47]. The highest frequency for which the electromagnetic wave will return is the critical frequency of the layer.

The equation of the plasma frequency is:

$$f_p = \frac{\omega_{pe}}{2\pi}, \ \omega_{pe} = \sqrt{\frac{n_e e^2}{m^* \varepsilon_0}} \quad (1)$$

where f_p is the electron's plasma frequency (Hz), ω_{pe} is the electron's plasma angular frequency (rad/s), n_e is the electron density (number of electrons/m^3), e is the electron's charge (1.609×10^{-19} C), m^* is the effective mass of the electron (9.11×10^{-31} kg) and ε_0 is the vacuum permittivity (8.85×10^{-12} F/m).

When the sounding signal's frequency reaches the plasma frequency, the stratification, electron content and virtual height of the different ionospheric layers can be derived from the measurements using inversion models. Typically, ionosondes use frequencies between 1 and 20 MHz, which makes the observation of the E and F layers possible. The Digisonde makes a pulse transmission with a peak pulse power of 300 W. Because of the pulsed nature of the signals, the average power will be lower. It should be noted that the precise power transmitted in a given direction is difficult to determine because of the complicated antenna pattern [48]. (The transmitter antenna has different gains for different elevation angles, as well as for different azimuths.)

The D region is undetectable for such frequencies due to its plasma frequency usually being below the minimum used sounding frequency. The upper sounding limit can be adjusted (for instance for DPS instruments) through the year according to the expected maximum value of the observed critical frequencies. The advantage of the sounding program adjustment could be the shortening of the sounding time.

The result of the measurements taken by an ionosonde—the virtual height frequency values—are called ionograms. The delays caused by the local atmospheric conditions are not taken into consideration in the making of the ionograms, and the conditions of a wave propagating in the vacuum are assumed at first. The difference in the delay in the time of flight is small for the E layer, but for the F layer, this difference is significant, and inversion is necessary for the layer's true height.

Deriving the real height from the virtual height is not necessarily trivial. At an 80 km altitude, the difference between the virtual and real height in the vertical direction is relatively small (~2–3 km), less than the vertical resolution of the Digisonde (~5 km). However, in the case of the reflections from the oblique direction (Figure 1a), both the virtual height and the critical frequency suffer a distortion and will appear at higher values compared to the values in the vertical direction. The difference between the virtual and real height derived from the oblique direction can be in the order of ~10 km, even at the height of 80 km.

At Průhonice (50.00°N, 14.60°E, station code: PQ052, geomagnetic coordinates for 2019: 49.33°N, 98.35°E) and at Sopron (47.63°N, 16.72°E, station code: SO148, geomagnetic coordinates for 2019: 46.70°N, 99.56°E), a Digisonde DPS-4D (hereinafter Digisonde) type of ionosonde station was installed. It has been providing measurements since January 2004 in Průhonice and since June 2018 in Sopron [49] as a part of the GIRO Network [50]. Through the addition of four receiver antennas surrounding the central transmitter [43,51,52], the Digisonde is capable of not just the vertical sounding, but also the oblique sounding. This kind of arrangement makes the determination of a reflection's direction possible, based on the phase difference of the reflected signal as observed by the four receiver antennas. On the ionograms, these are color-coded: warm colors for southern and western directions, and cold colors for northern and eastern directions (Figure 1a).

The regular ionograms are automatically scaled. This usually provides the necessary parameters (e.g., critical frequencies of the E (foE) and F (foF) layers) for the setup of the drift measurements. If there are movements in the plasma, the Doppler shift effect will be registered for the reflected sounding waves. Digisondes can estimate the speed and direction of plasma motions, by emitting signals of the same frequencies, and measuring the Doppler shift and the direction (made possible by the receiver antenna arrangements) of the reflected signals. The final product of the measurements is the so-called Skymap. The Skymap displays the measurements of the three velocity vector components, namely vertical (v_z), north (v_n) and east (v_e) directions and these are represented graphically in color-coded east–west and north–south planes (Figure 1b).

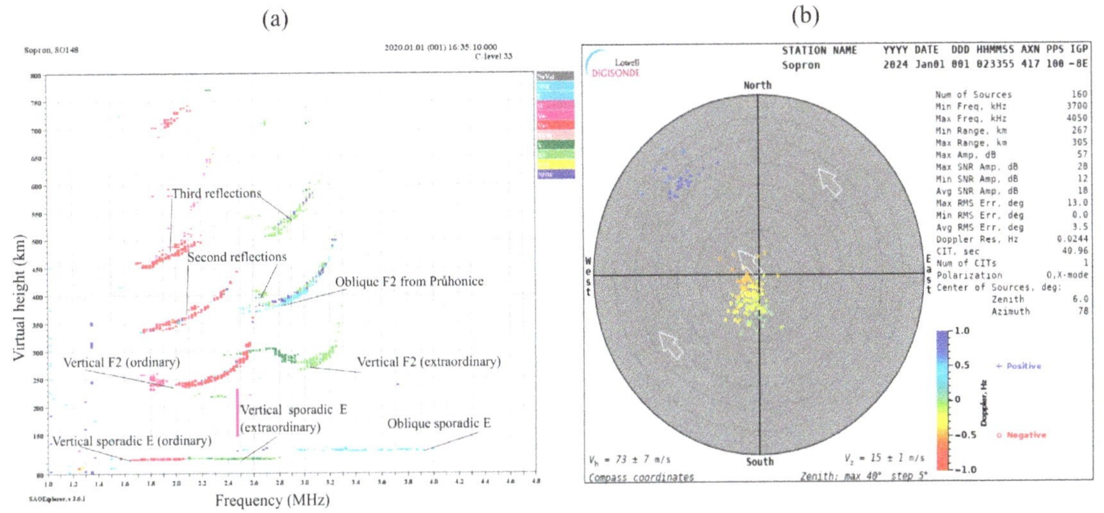

Figure 1. An example of an ionogram and a Skymap. (**a**) The ionogram (taken in Sopron) displays the virtual height at which the reflections occurred at the given frequency. The parameters of the ionospheric layers can be derived from it. (**b**) The Skymap displays the Doppler shifts from which the resulting plasma velocity vector can be computed. The measurements are made at a specific frequency, which also determines the height as a function of ionospheric parameters.

There were synchronized campaign measurements at both stations for the Leonids and Geminids in 2018 and 2019. Synchronization was important because the two stations are very close to each other in ionospheric terms, so local and regional effects could be easily separated by simultaneous measurements (Supplementary Figure S1). The campaigns of 2018 were used as a test. Further, the results were used in the settings for the campaigns of 2019 (Table 2).

In all the cases, the campaigns started after sunset (roughly around 16:30 UT) and ended at sunrise (roughly around 6:30 UT). In 2018, the stations measured one ionogram/minute, but this setting was upgraded for the campaign of 2019 to two ionograms/minute. The drift measurements were not made in high time resolution in the 2018 campaign, but for the Leonids and Geminids of 2019, there was one Skymap/minute in Sopron (in Průhonice only during the Leonid campaign).

Meteor-induced traces were manually searched on the ionograms based on the following criteria: (1) The virtual height is between 80 km and ~160 km. (2) The reflections are well identified and clustered in a line. (3) They show no similarity to the regular sporadic E layer based on their lifetime and evolution. (4) In questionable situations, synchronized measurements from the other station were used to check the ionogram of the other station (Supplementary Figure S1), which effectively filtered out possible false hits caused by regular sporadic E.

Identifying the reflections from the meteor trails on Skymaps was more demanding, as usually 1–5 reflections/meteor were found, and not always. The height and direction had to match those seen on the ionograms or, where appropriate, optical sightings. Most nights, the Skymaps were empty, which simplified the identifications. However, there were also cases where other plasma flows were present, and reflections were observed in the right direction and altitude range. These were recorded as possible hits.

Table 2. The campaign settings and measurement times of 2019 at the different stations.

Campaign	Leonid Campaign		Geminid Campaign	
2019 (UT)	Sopron	Průhonice	Sopron	Průhonice
Digisonde 2 ionogram/minute campaign dates: night-dawn from-till	11.16–11.17 16:30–6:30	11.16–11.17 16:30–6:30	12.10–12.11 19:40–20:00 21:00–06:00	-
			12.11–12.12 16:30–05:30	-
	11.17–11.18 16:30–6:30	11.17–11.18 16:30–6:30	12.12–12.13 17:30–05:30	-
			12.13–12.14 16:30–05:30	12.13–12.14 16:00–08:00
	11.18–11.19 16:30–6:30	-	12.14–12.15 16:30–05:30	12.14 18:00–20:00 12.15 0:00–03:00
			12.15–12.16 16:30–05:30	-
Skymap	1/minute	1/minute (until 6:30 17 November)	1/minute	-
Optical data	11.18–11.19 17:05–3:52 **9 detections**	-	12.13 22:40–23:22 **3 detections** 12.14–12.15 17:21–4:48 **76 detections**	-

2.4. Digisonde Records

2.4.1. Background Sporadic E Activity

On ionograms, the faint meteor-induced traces and regular sporadic E layer can occur in the same height range, at altitudes of ~80–130 km. Although sporadic E is essentially considered a summer phenomenon, it can in fact occur at any time [11]. During the campaigns, there were also several occasions when the background sporadic E activity made it difficult to identify the meteoric traces.

To make it easier to dissect the phenomena, for November and December 2019, the ionograms for the month were manually scaled, except for the days of the campaign measurements. In Průhonice, the ionograms were obtained every 5 min during November and every 15 min during December of 2019, while in Sopron, they were taken every 5 min in both months. A total of 10,406 ionograms were processed manually for Průhonice and 14,069 for Sopron (Figure 2, Supplementary Figure S2, ref. [53]). Of the processed ionograms, the phenomenon occurred in 20.2% over Průhonice and 19.1% over Sopron.

In Sopron, 65.61% of the detected sporadic E occurred in November and 34.39% in December. In Průhonice, 79.07% of the detected sporadic E occurred in November and 20.93% in December.

The distributions of foEs over the two stations and for the two months are very similar. The main difference is that at Průhonice, a second peak in foEs occurs at higher frequencies in both months, which is not present in Sopron's distributions. It is also observed that while the median for Průhonice barely changes for the two months, Sopron shows a decrease of 0.3 MHz for December.

The distributions of virtual height (h'Es) for November are also very similar for the two stations, but a ~5 km shift in the values can be observed, which is within the vertical resolution of the Digisonde. Whether this is because of an instrumental setting or a physical phenomenon is still under investigation. However, the distributions of December show a significant divergence for the stations. In the case of Průhonice, the result is a flat, sloping curve, while in Sopron, a sharp rise is followed by a sharp fall, and the distribution has a significantly higher skew and kurtosis. The medians show a difference of 10 km which is significant for two stations in such close proximity.

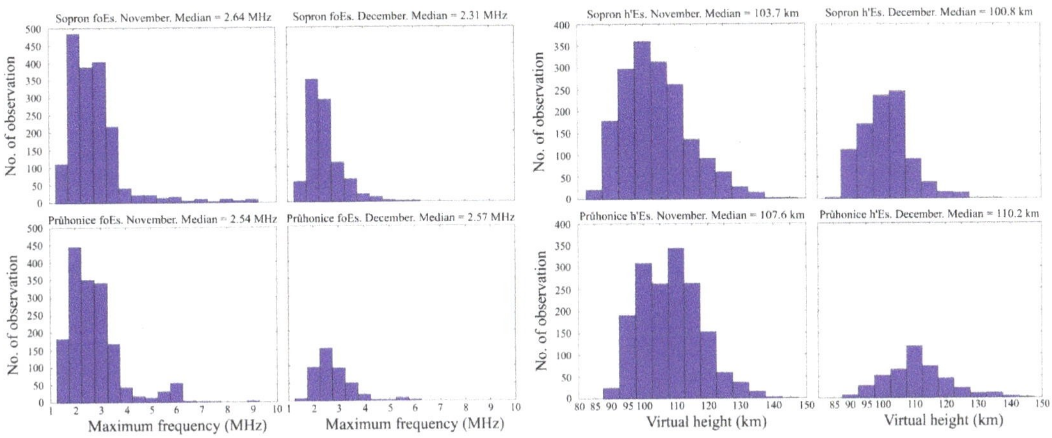

Figure 2. The distribution of the sporadic E layer critical frequency (foEs) and virtual height (h'Es) for Průhonice and Sopron stations, regarding November and December of 2019. For histograms showing the 2 months together, see Supplementary Figure S2.

2.4.2. Ionograms

The ionograms (see Section 2.3) for the 2 ionogram/minute campaign were manually reviewed for both stations. The campaign settings were not quite the same at the two stations—there were a lower number of performed high cadence measurements in Průhonice for various reasons (Table 2). For the times when there were no available 2 ionograms/minute measurements in Průhonice, the regular ionograms (taken every 5 or 15 min) were checked.

When the distinctive, meteor-induced traces appeared on the ionograms (Figure 3), the time of the first and last sighting of the trace was recorded. From these, the lifetime of the trace can be calculated. In rare cases, the trace disappeared for one ionogram, but reappeared on the following ionogram. Such picture jumps were flagged. The maximum and minimum frequencies reached during the trace lifetime were also recorded. The virtual height was read from the ionogram of the first appearance (vertical resolution ~5 km). The direction of reflections is also indicated [53].

According to the manuals [48], ionograms should be scaled at the 6 dB most probable amplitude (MPA) threshold noise reduction (see Supplementary Figure S3). But at the 6 dB MPA, the meteor-induced traces can be jagged, faint and often not even fully visible, probably due to the signal processing techniques. Therefore, the ionograms were reviewed at 4 dB, and the data were read at this threshold. A value of 2 dB would have been better, but due to the noise, most of the recorded ionograms could not be manually analyzed at that MPA threshold. The traces have been classified according to their appearance on different MPAs [53].

Two phenomena were commented on next to the ionograms. Sometimes, a meteor-induced trace split in two on the ionogram, and a distance of at least 5 km appears between the now split traces. This is recorded as a split in the trace. The other comment was added when the meteor-induced trace overlapped with a regular sporadic E, making the finding somewhat unreliable [53].

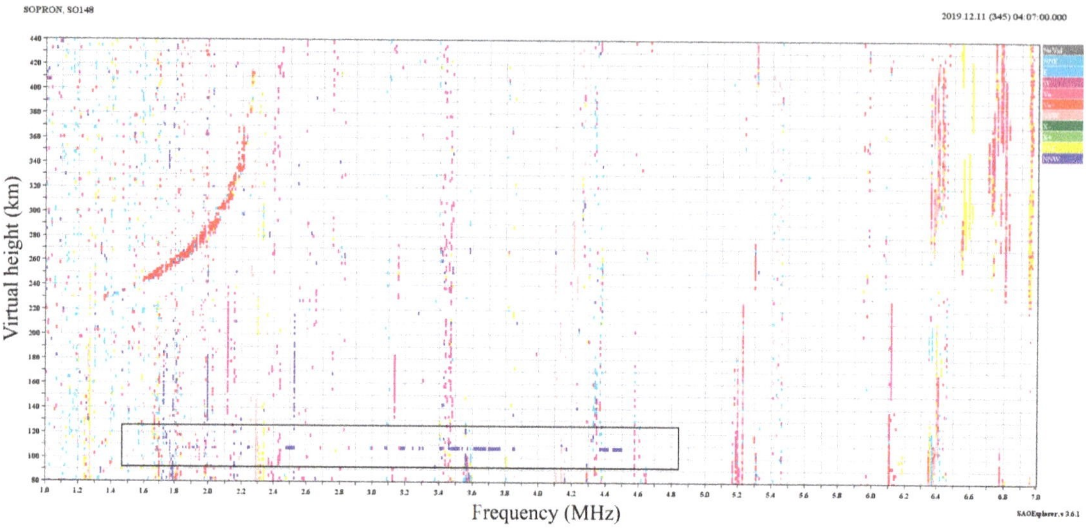

Figure 3. Example of a meteor-induced trace on an ionogram. The NNW (dark blue reflections in the black bracket) reflections are at 106 km virtual height. The maximum frequency of the trace is 4.5 MHz. The image was taken at 4 dB MPA. Even at that, the trace is jagged, and parts of it are missing.

The detected meteor-induced traces in the ionograms and the resulting Skymaps are summarized in Table 3. During the Leonid campaign, at the 4 dB MPA threshold, Sopron registered 59 cases which could be identified as a meteor-induced trace, while Průhonice registered 41 with 4 additional findings in the regular ionograms (Table 3). The Geminid campaign yielded 141 meteor-induced traces for Sopron and 91 for Průhonice with 40 additional findings in regular ionograms (Table 3).

Table 3. The detected meteor-induced traces on the ionograms and linked to them, the Skymap findings. The numbers in brackets are the amounts found on the regular ionograms for the ionograms when Průhonice did not have 2 ionogram/minute measurements. For the Skymaps, the bracketed, question marked numbers are the possible findings—there were reflections in the right direction and height on the Skymaps, but they overlapped with other plasma flows, making the detections uncertain [53].

Detection	Sopron	Průhonice
Leonid campaign on ionogram	59	41 (+4)
Leonid campaign on Skymap	12	6 (+4?)
Geminid campaign on ionogram	141	91 (+40)
Geminid campaign on Skymap	22 (+4?)	-

2.4.3. Skymap

The drift measurements (see Section 2.3) were performed at 2.5 MHz. The Sopron station took one Skymap/minute during both campaigns, while Průhonice station recorded them only during the Leonid campaign, and even then only until 06:30 of 17 November 2019.

Since this method is focused on plasma movements, only 1 Skymap was taken every minute, and the measurement was carried out at a specific frequency, so it is not surprising that few meteoric reflections were captured. However, regarding the fact that a fireball that lasted for 20 min while slowly dissipating and the Skymaps showed only 2–5 reflections [45], the number of successful detections with drift measurements is surprisingly high.

In the case of the Leonid campaign, for 59 identified cases on the ionograms, there were 12 events on the Skymaps at Sopron (see Table 3). When Průhonice was taking the drift measurements, for the 41 findings on the ionograms, there were 6 episodes on the Skymaps (see Table 3), followed by 4 other possible detections (the reflections' direction and height range supports the assumption that they could have been from meteoric trails, but they were overlapping with other plasma flows).

In the case of the Geminid campaign (only Sopron took drift measurements) for the 141 findings on the ionograms, there were 22 identified records on the Skymaps with 4 additional possibilities or suspicious detections (see Table 3).

3. Results and Observations

3.1. Results of the Comparison between the Different Data

3.1.1. Optically Detected Meteors Present in the Digisonde Data

Of the 88 optically detected meteors, 31 were found by the Digisonde measurements [53]. In addition, another case has been identified as a fireball event [45] by the European Fireball Network [54].

During the Leonid campaign, there was one fireball detection near Průhonice and nine optical meteor records in Sopron. The fireball and eight from nine optical detections could be linked to the meteor-induced traces found on the ionograms. The effect of the fireball and one meteor could be found on the Skymaps, too. The meteors detected optically and also on the ionograms were as follows: one Southern-Taurid, four sporadic, three Leonid. The fireball was identified as a Leonid, too (see Supplementary Table S1).

During the Geminid campaign, from 79 optical detections 23 were found on the ionogram: 21 of them in Sopron's measurements, 2 of them in Průhonice's. Four of the detections could be linked to reflections on the Skymaps, too (Figures 4 and 5). The meteors detected via the two different methods were as follows: one Antihelion Source, one σ-Hydrid, one Monocerotid, seven Geminid and thirteen sporadic [53]. Around the optical detections, the lowest noise reduction settings were checked too, and some of these findings were acquired at the 2 dB MPA threshold, though it should be noted that most of our manual review was with the 4 dB MPA threshold setting (see Section 2.4.2 and Supplementary Table S1).

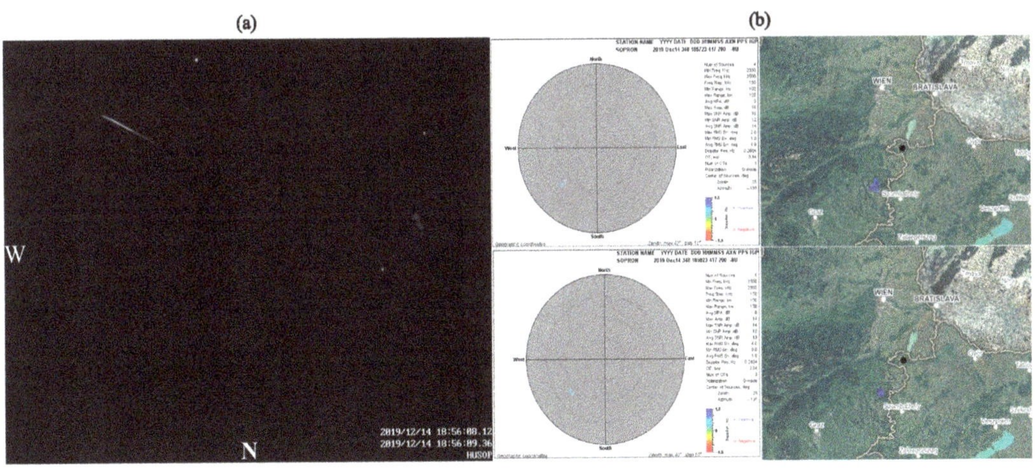

Figure 4. (**a**) A sporadic meteor optically detected at Sopron in SSW direction (0.5 magnitude, the image is upside down, and the contrast was heightened). This meteor caused a meteor-induced trace to appear on the ionograms (see Figure 5) which lasted for 2 min. Reflections tied to the meteoric trace also appeared on Skymaps (**b**). A map projection of the reflected points (with blue dots) is included to show the locations; Sopron station is marked with a black dot.

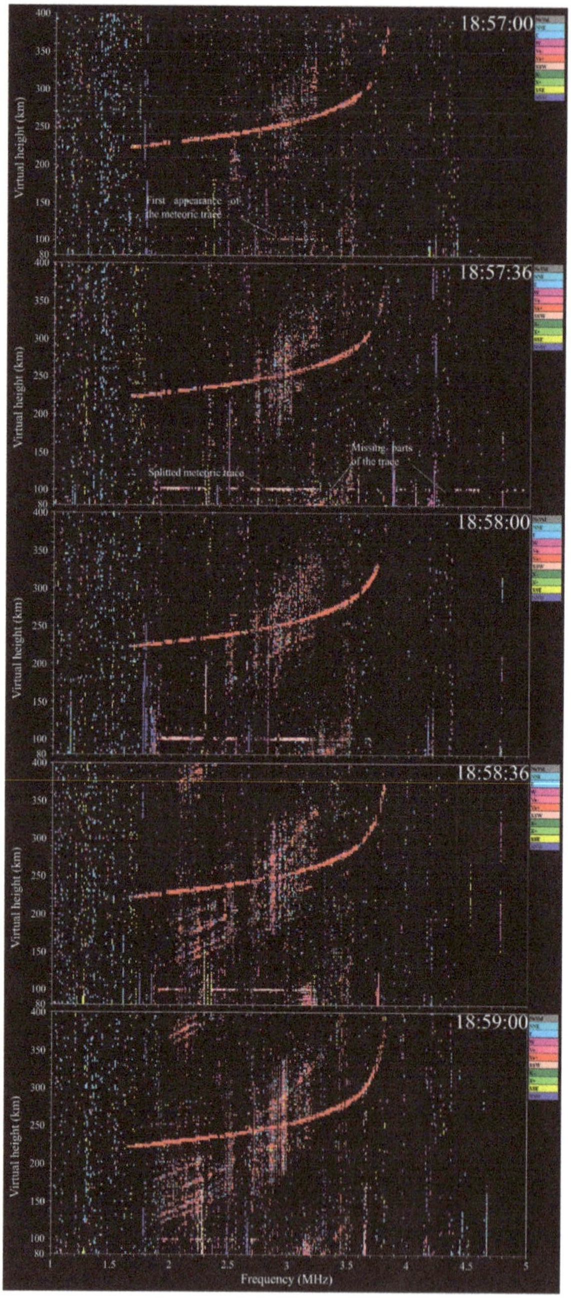

Figure 5. The meteor in Figure 4 caused a trace to appear on the ionogram at 18:57:00 (UT), 50 s after the meteor's passing. The ionograms display reflections taken at 4 dB MPA threshold. The meteor-induced trace first appeared as a faint sporadic E-like layer. At the next frame, the trace splits into two layers, probably due to windshear (see Supplementary Figure S4). Parts of the trace are missing, probably due to the noise filtering and the MPA settings. The trace's last appearance was at 18:59:00 (UT).

3.1.2. Optically Detected Meteors Not Present in the Digisonde Data

In total, 57 of optical detections could not be linked with ionogram records [53]. It should be noted that the optical camera's field of view was wider (122° × 97° field of view) and covered a larger area than the Digisonde can cover (field of view is ~45° × 45°) [48]. Particularly during the Geminids, it was common for sightings to occur at the very edge of the optical recordings, which was beyond the range of the Digisonde.

Of the 57 undetected cases, there were only six cases where Sopron's Digisonde should have detected the meteor, but did not. Also, in nine other cases, there is a slight probability that Průhonice could have detected the meteor, but regular measurements were being taken at the time (one ionogram/15 or 5 min). In the other cases, there are circumstances that clarify or explain why no trace did appear on the ionograms (see Supplementary Table S2).

Such possible explanations are as follows: (1) the meteor was out of the range covered by the Digisonde, (2) it was at such a low altitude angle (horizontal coordinate system), that it was not detectable by the Digisonde, (3) regular sporadic E could cover up the traces, (4) the signal processing could filter out the faint traces (the received signals with an amplitude below the estimated MPA were not detected because of the Digisonde settings at Sopron, detailed in Section 2.4.2; also see Figure 5), and (5) additional circumstances for not detecting meteors with the Digisonde can be that its plasma trail was below 80 km in height (see Supplementary Table S2).

Also, there were cases where there were signals in the noise that seemed to be a good candidate (some points in line at the same height) but could not be clearly categorized as a meteor-induced trace (see Supplementary Table S2).

3.2. Observations on the Digisonde Data Regarding Meteors

3.2.1. Necessary Conditions for the Successful Detection of Meteors Using the Digisonde

The transmission antennas used for the ionosonde are optimized for vertical transmissions, but significant power is emitted even at zenith angles close to 90°. Thus, meteor traces close to the horizon are still illuminated by the ionosonde's transmissions. However, the angle of arrival detection of received echoes used here works by applying digital beamforming with seven beams, namely one vertical and six at different azimuths and centered around a zenith angle of 30°, as shown in Figure 6 [48]. Thus, reflections with zenith angles much larger than 45° are harder to detect. All the detected meteors at Sopron had at least an ~40° altitude angle, which supports our original assumptions. This is also consistent with what was observed by Sodankylä [22].

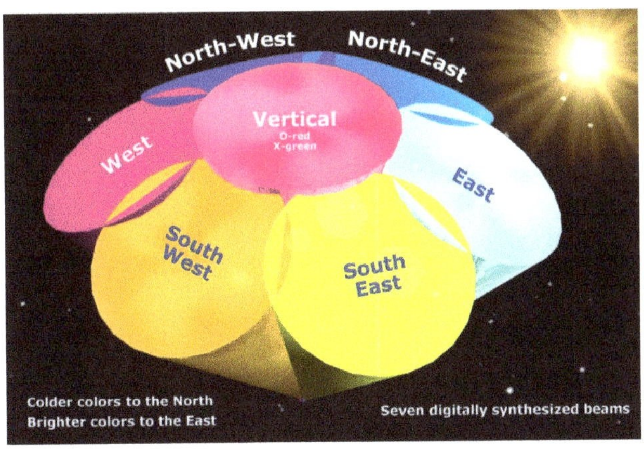

Figure 6. Schematic depiction of the vertical and oblique synthesized beams for angle of arrival indication in the ionograms. Origin of figure: Reinisch [48], Figures 1–23, p. 55.

The acquired raw data were subjected to a noise filtering method—this is the setting of the most probable amplitude (MPA) threshold (see Supplementary Figure S3). This method filters out reflections based on the estimation of how likely it is that the reflection is a real one (Figures 3 and 5). At the 6 dB MPA threshold—the standard for manual scaling—only a fraction of the findings were present. For this reason, for the manual review of the data, the threshold was set at 4dB MPA. The setting of 2 dB MPA would have probably yielded even more, but that was too noisy to work with, while 2 dB data were checked in certain cases (around the time of optically detected meteors), but other than that, it was not reviewed methodically.

The presence of a regular sporadic E layer can also prevent the detection of meteors. For example, in Sopron station's observation on 13 December 2019 (one day before the peak of the Geminid meteor shower) between 1:42:00 and 2:52:36 (UT), there were no detected meteor-induced traces on the ionograms, while a strong regular sporadic E layer was present.

3.2.2. Observations of the Digisonde Detections Regarding Meteors

In the case of the detected fireball [45], it was discussed that a huge part of the trace (between 5 and 14 MHz) consisted of W and SSW reflections, when physically the fireball was located in the E and NNE directions relative to the Digisonde of Průhonice. The explanation lies in the effect of the geometrical parameters of the antenna system that can cause 180° phase jumps. In light of this, it was surprising that there were only two cases where part of the signal suffered a phase jump. The direction determination of the Digisonde proved to be almost entirely reliable.

What also proved surprising was the delay in the appearance of the traces compared to the optical observations. This time delay ranged from 12 to 132 s. It is also worth noting that during the nine matches of the Leonid campaign (including the fireball's case), these delay times were mostly in the explainable range, with a maximum of 62 s. However, for matches under the Geminid campaign, these numbers tended to be higher.

It should be noted that the virtual heights of the Leonids found by both measurement methods were surprisingly high: the values ranged from 113 to 137 km, with one exception (which was ~100 km high). In contrast, the Geminids were generally found at lower heights (~80–115 km) on the ionograms.

The number of detections may also depend largely on the specifications of the instruments. Although both stations use the DPS-4D type of Digisonde, Průhonice detected significantly more meteor-induced traces during the same time periods. For example, during the period 16:30–24:00 on 13 December 2019, Průhonice recorded 27 meteor trails, while Sopron recorded 12. Also, "high-frequency" traces (>5 MHz), which were very rare in Sopron, were often detected in Průhonice's measurements (Figure 7, ref. [55]).

3.2.3. Other Observations

Compared to the optical observations, there was much more trace detection on the ionograms (altogether 376 [53], Table 3) taking into account the whole periods of the two meteor showers. The distribution of the maximum frequency was remarkably different for the stations. The distribution of the meteor-induced traces' virtual heights was different during the two campaign sessions (but similar for the two stations at one given campaign). It was also distinct from the average behavior of sporadic E (Figure 8). Although the shape of the height distribution during the Leonid campaign was reminiscent of the regular ones, the height distribution of the Geminid campaign with their double peaks was reminiscent of the height distributions of the long-enduring echoes of McKinley (Figures 5–13) [19].

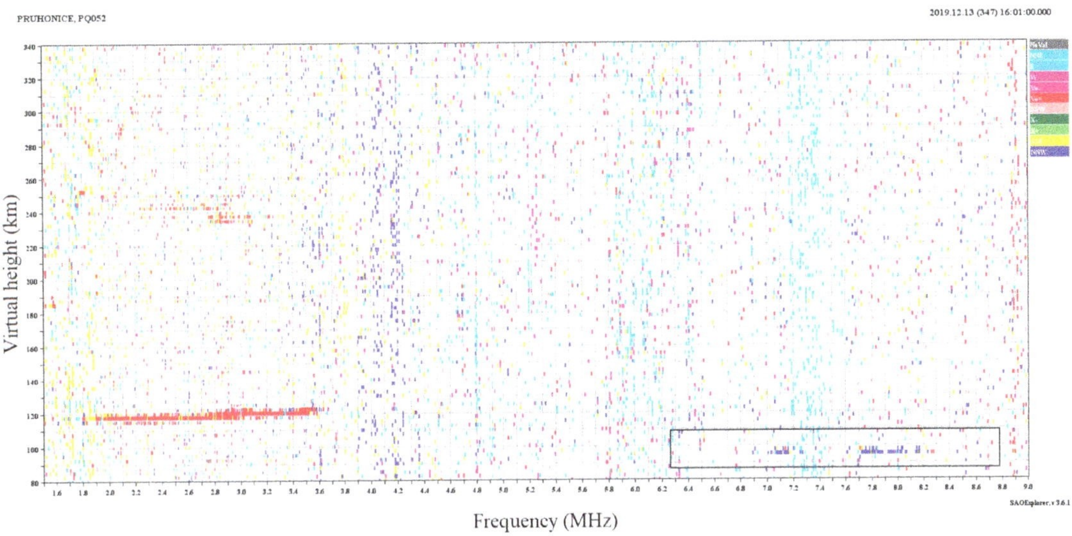

Figure 7. Ionogram taken on 2019.12.13. 16:01:00 (UT), displayed for 4 dB MPA, showing a strange but sometimes occurring meteor-induced trace in the black bracket: NNW reflections (dark blue) at 95 km virtual height between 6.9 and 8.2 MHz. These kinds of traces—where only a small part of the trace is visible and even that is at high frequency—are rare in Sopron, but far more common in Průhonice.

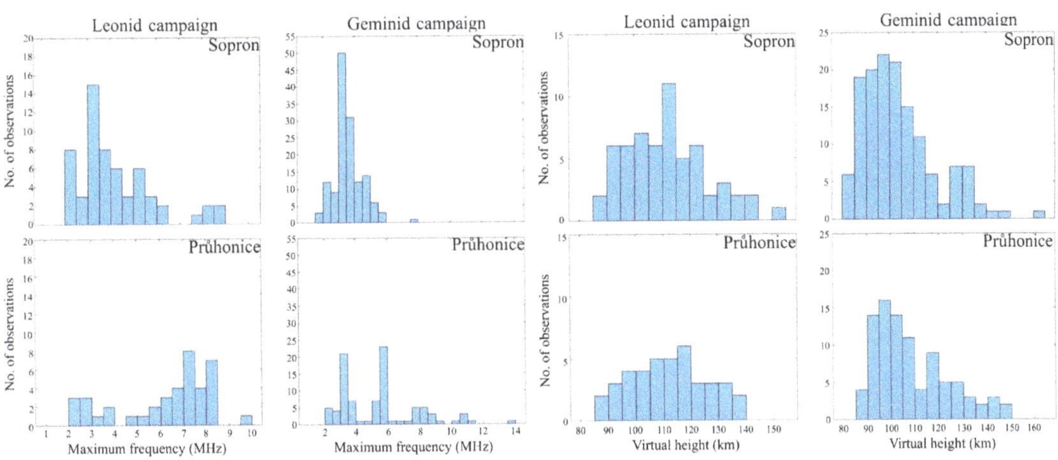

Figure 8. The distribution of the meteor-induced trace heights is very similar in the two stations for one campaign, but highly different for the two campaign sessions. But the distribution of the maximum frequency does not even resemble the regular sporadic E distributions, and there is a huge difference between the two stations and the two campaigns. The fireball event was excluded from the histograms.

It should be noted that for the reflections from the oblique direction, both the height and the frequency are distorted, and higher values are obtained than what would be the case in vertical reflections. This can reach the values of ~10 km at height (Figure 9).

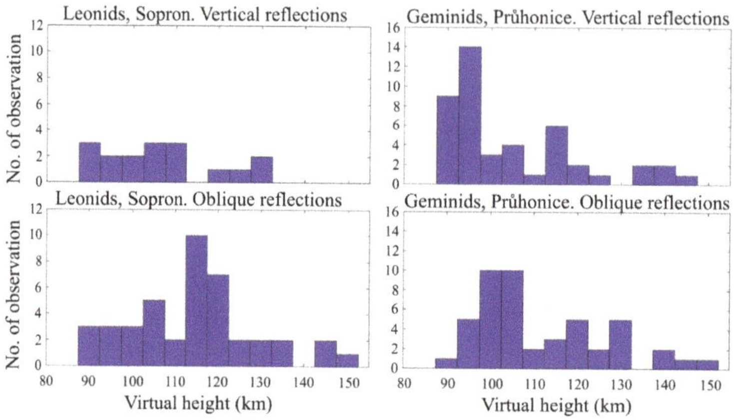

Figure 9. Virtual height distributions of meteor-induced traces of Sopron's Leonids campaign and Průhonice's Geminids campaign. The vertical and oblique tracks are separated and below each other. A difference of about 10 km between the vertical and oblique directions can be observed, especially for the Geminids.

The most surprising finding was that the maximum frequency reached by the meteor-induced traces and the brightness showed anticorrelation—the brighter a meteor was, the lower the maximum frequencies that it reached. The original expectations were the opposite. (Technically, the correlation coefficient is positive—see Figure 10a—but in the case of brightness, the lower the number, the brighter the meteor is. For example, a +1 magnitude meteor is less bright than a −0.5 magnitude meteor. So, in this case, the positive correlation meant the dimmer the meteor, the higher the frequencies that its trace reaches.) It should be noted that this cannot be considered statistically conclusive on the basis of 31 cases (25 if only the 4 dB MPA cases are taken into account; see Figure 10c,d). In contrast, velocity (degree/s) was slightly correlated with the maximum frequency (Figure 10b)—the faster a meteor, the higher the maximum frequency that it tended to reach (Figure 10b,d). This correlation is far better pronounced when the data points acquired at the 2 dB MPA threshold are discarded and the data points used were all acquired at the 4 dB MPA threshold. In that case, the correlation is significant.

The meteor trail length does not show any correlation with any of the other measured parameters.

The lifetime of the trace does not seem to depend on the calculated or observed physical properties of the meteors (duration of the event, brightness, etc.). However, in the case where the trace split in two—presumably due to the wind shear caused by mesospheric winds (see Supplementary Figure S4)—the duration is generally longer.

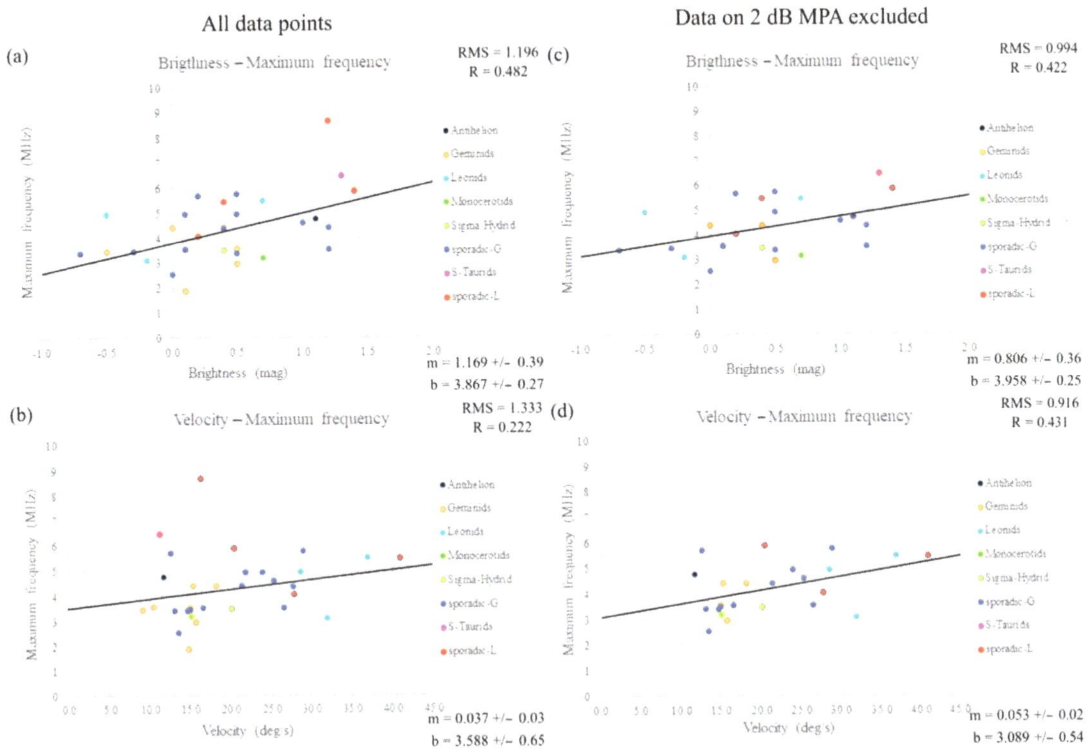

Figure 10. Matches between optical and Digisonde measurements (excluding the fireball event) and their corresponding meteor brightness/meteor velocity—maximum frequency values. In the left column (**a**,**b**) all data points are plotted, while in the right column (**c**,**d**) matches found at the 2 dB MPA threshold were excluded. For sporadic meteors, the L and G markings indicate the meteor shower period during which they were identified as sporadic. The standard deviation (*RMS*) of the linear fit ($m \cdot x + b$) and the linear (Pearson) correlation coefficient (R) are shown in the figures. (**a**) For the brightness, there is a significant, positive correlation with the maximum frequency reached by the traces. Since the magnitude of the brightness is lower the brighter a meteor is (−0.5 magnitude is brighter than +1 magnitude), this means that the physical quantities are anticorrelated. The brighter a meteor, the lower the frequency reached by its trace. (**b**) Velocity also shows a slight (non-significant) positive correlation with maximum frequency. (**c**) Excluding data at the 2 dB noise filtering threshold slightly changes the fitted values of brightness and maximum frequency, though the errors are a bit smaller, but overall, this does not change the correlation (still significant and medium positive). (**d**) Using only data at the 4 dB MPA threshold, however, changes the correlation of meteor velocity and maximum frequency dramatically. It is now significant instead of non-significant. (The results of the statistical calculations are summarized in Supplementary Table S3).

4. Discussion

Using Sopron's Digisonde, 29 optically detected meteors were found in the measurements. At Průhonice station, two of these meteors and one fireball were found (confirmed by the European Fireball Network). Of the remaining 57 optical detections, the meteors with too large zenith angles were not detected by the Digisonde. This is to be expected due to the applied method for the angle of arrival detection (as described in Section 3.2.1) and the fact that these signals would need to travel a long distance through the D-region of the ionosphere and can therefore be expected to experience more attenuation. Of these 57 optically detected meteors, there were a total of six meteors that were not detected by

Sopron's or Průhonice's Digisonde, and no comprehensive explanation can be found to fully clarify the absence of echoes in the ionograms. One of the most probable explanations can be the fact that these meteors have not formed a stable ionization layer existing for a sufficient time so that the Digisonde could detect it. It is also possible that it did form, but the mesospheric winds and/or turbulence ripped it apart or "blew it out" of the Digisonde's range. It is also possible that the meteor had its ionization effect below 80 km, which is below the detection range of the Digisonde.

During the presence of regular sporadic E layers, for a couple of hours, no meteor-induced traces were recorded. Presumably, this is not because there were no incoming meteors—as it can be seen in the ionogram observations comparing the two stations—but because strong, stable and prominent reflections from the regular sporadic E layers dominated the ionograms in the same height region where meteor-induced traces are most likely to occur which prevents the identification of meteor-induced reflections. The traces that do appear tend to reach a higher maximum frequency than the regular sporadic E, or they are well separated from it in height, and usually come from an oblique direction, very rarely from a vertical direction (±15° from zenith).

During the Leonid campaign, Sopron's Digisonde detected 59 and Průhonice detected 45 events, respectively, and during the Geminid campaign, Sopron detected 141 and Průhonice detected 131 events, respectively. A contributing factor to these numbers is that the 2019 Leonids produced anomalously few meteors [56]. The mass of bright meteors (−2.5–+2.5 visual magnitude) ranged from ~1 g to ~10^{-2} g on average [16]. There are also smaller meteors, but these can only be detected via intensified video or meteor radar [57]. If these extra traces found on the ionograms are attributed to small meteors (below ~10^{-2} g)—not detected via the +1 magnitude lower limit camera—then their abundance and ionization (which in most cases produced a more distinct trace and higher frequencies than those also detected optically) can be understood and explained.

The height distribution of the meteor-induced traces in the ionograms are different for the two campaign sessions, but quite similar for the two stations in a given campaign. This could be attributed to the fact that the higher a meteor's velocity, the higher the ionization effect appears [16,19,58]. The Leonids enter the Earth's atmosphere at 71 km/s velocity (~40.7 degree/s for a 100 km height), and the Geminids do so at 35 km/s (~20.1 degree/s for a 100 km height). Also, the shift between the vertical and oblique reflections should be taken into consideration. The difference in distributions may also be caused by the different material quality of the incoming meteoroids. Not only the composition but also the propensity for fragmentation differs between the two swarms. Leonids of cometary origin fragment more easily. Geminids of asteroidal origin, on the other hand, are more compacted in space because of the Sun, and are more difficult to fragment [34,36].

The virtual height values themselves are broadly consistent with those found by Maruyama et al. [20], except one particular type of observation. The study [20] reported that among the events that they identified as meteor echoes, there was one that appeared during the Leonid peak and was detected at all four stations. These echoes occurred at virtual heights of 200–250 km and, according to the reported images, slowly descended. Based on the observed high virtual height and hourly duration, this was probably not a meteor echo, but an intermediate descending layer; nevertheless it occurred well before the sunrise, while generally the descending layers appear at around 6:00 local time [59]. The phenomenon was not described in detail when Maruyama et al. [20] was published [60].

In contrast to the height distribution, the frequency distributions are very different from campaign to campaign and from station to station. Since Průhonice's Digisonde has a quality antenna system resulting in a much higher sensitivity [55], the variation per station is not surprising, nor is the fact that Průhonice recorded higher frequencies. Also, the two stations applied different settings. At the Průhonice station, all the measurements were saved, while at the Sopron station, there were filtering parameters set, which could lead to data loss. This may also have contributed to the fact that within Průhonice's data, higher average frequencies were identified.

Regarding the frequency distributions, another factor to consider is the radar backscatter cross section (σ, Stuart, p. 837, ref. [16]):

$$\sigma = \frac{\mu_0^2 e^2 \lambda_0^4 N_m^2 v_m^2}{4096\pi^6 m_e^2 D^2} e^{-\frac{8\pi^2 r_0^2}{\lambda_0^2}} \quad (2)$$

N_m is the electron line density(el/m), μ_0 is the magnetic permeability of a vacuum (4π × 10^{-7} H/m or 1.26 × 10^{-6} N/A^2), m_e is the mass of an electron (9.11 × 10^{-31} kg), e is the charge of an electron (1.609 × 10^{-19} C), λ_0 is the wavelength of the wave used in the measurements, D is the ambipolar diffusion coefficient, r_0 is the initial meteor trail radius (m), and v_m is the initial speed of the meteor.

Assume a situation in which two meteors from different meteor showers, but with the same physical parameters (N_m, r_0), arrive at the same altitude and face the same ambipolar diffusion coefficient (D). The radar cross section depends quadratically on the speed of the meteor (v_m). This means that Leonids at 71 km/s are four times more detectable than Geminids at 35 km/s (respectively ~40.7 degree/s and ~20.1 degree/s for a 100 km height). Fragmentation may also play a role, affecting the Leonids to a greater extent. Incoming meteors may break up into smaller pieces, which can then scatter over a larger area, making them easier for the Digisonde to detect [19]. This was also reflected somewhat in the data.

Based on the definitions of underdense and overdense meteor echoes of Kozlovsky et al. [22] and the typical signal strength–time plots of the radio signals reflected from meteor trails (see Supplementary Figure S5) [16], the Digisonde can detect overdense meteor trails, and presumable even underdense meteor echoes, too. Although the Digisonde transmits a relatively low power, various signal processing and noise and interference removal techniques are used to allow the detection of faint echoes [48]. Many detections endured for only one ionogram and reached low frequencies (below 3 MHz) [53]. However, in the absence of an independent measurement method (e.g., meteor radar) and a higher time resolution ($\tau = \frac{r_e c^2 a_l}{4\pi^2 f^2 D}$ [22] where τ is the decay time [s], r_e is the classical electron radius ~2.82 × 10^{-15} m, c is the speed of light, D is the diffusion coefficient, a_l is the electron line density [el/m] and f is the frequency [Hz]), which could specify the decay time from which the electron line density could be derived, it was not possible to clearly categorize the meteor-induced traces found on the ionograms, and whether they were truly underdense or not is unknown.

The analysis of the Digisonde measurements associated with the optical findings revealed that there is a slight correlation between the meteor's speed and the maximum frequency reached by the occurring trace. In general, the faster a meteor was, the higher the frequency that the trace reached on the ionograms. This makes sense—the faster a meteor is, the greater the friction and the greater the ionization effect that it can have. For brightness, however, there was a slight anticorrelation with the Digisonde measurements. The brighter a meteor was, the lower the maximum echo frequencies that it reached on the ionograms. This is completely contrary to our preliminary expectations, as the expected electron line density should be higher the brighter that the meteor is [16–18]. It should be noted that 31 cases (25 if the data points acquired at 2 dB MPA are excluded; see Figure 10) do not allow statistically significant and robust conclusions to be drawn; however, this is a very interesting and surprising result. Further research and campaigns would be needed to be able to make statements with certainty.

Another interesting feature is the time-delayed appearance of the traces on the ionograms compared to the detection time of the optical measurements, which ranges from 12 to 132 s. Of this, up to ~40–60 s can be attributed to the measurement settings. The schedule of 1 min measurements was [hh:mm:00]—first ionogram, [hh:mm:20]—drift measurement, [hh:mm:40]—second ionogram. If a meteor arrives at say [hh:mm-1:55] and would reach say 5 MHz, but the Digisonde is currently transmitting and receiving a 6 MHz signal, the meteor's trace will not show up on the [hh:mm:00] ionogram, and can appear only on the [hh:mm:40] ionogram. However, in comparison, it seems that there can be yet another extra

~60 s between the optical detection and Digisonde detection. The reason given for this can be that the Digisonde may have a limited horizontal resolution, and the meteor's effect may need time to diffuse enough to be detected via the instrument (f.e. layer formation). The original incoming meteors can have a length in the kilometer range (for sporadic meteors: the most probable trail length is 15 km, while typical lengths are up to 50 km [17]) but with a diameter of only a few meters [24]. The meteor trail length does not show any correlation with any of the other measured parameters. In the case of the terminating electron line density, this is consistent with Eshleman's theory [25]. Another factor may be the interaction between the angle of arrival and the direction of the magnetic field. In the case of the instrumental part, it is also possible that the meteor trace appears on the ionograms earlier, but only at the 2 dB MPA setting, and is therefore drowned in the noise. It is also curious that for the Leonids (all eight cases identified) this time delay is in the explainable range—the maximum delay is 62 s, which follows the case of the measurement setup described. These numbers are much higher for the Geminids. At this point, the role of the fragmentation—which is more pronounced in the Leonids—should also be mentioned. As a meteoroid breaks up, smaller pieces may scatter over an appreciable cross section larger than that of the parental meteoroid [19]. Whether this is the result of a physical process or an instrumental feature remains to be explained.

The lifetime of the traces on the ionograms is the most interesting result of all, because no correlation could be found with any of the other measured/observed parameters. It was noticed that when a meteor-induced trace splits in two on the ionograms, it tends to have a longer lifetime. These phenomena usually appear in clusters and are presumably related to mesospheric winds. Maruyama et al. [20] and Kozlovsky et al. [22] also describe that a meteor-induced trace can become a sporadic E patch-like formation due to wind shear (Supplementary Figure S4), which is observed for meteor-induced traces with long lifetimes (there, by the way, they refer to traces that persist for ~20–40 min). However, there were also traces with a lifetime of 2–3 min, which did not show any signs of wind shear (splitting). According to Oppenheim and Dimant [24], the evolution of meteor trails can be influenced by the magnetic field in addition to turbulence, which is in agreement with other previous observations [23]. Field alignment is when the direction of the meteor's trace and the magnetic field are nearly the same, and this can increase the lifetime of the meteor's trace. According to 3D simulations [24], near-perfect alignment (deviation below 6°) has a significant effect. A simplified comparison was also attempted in the present study, but the quality of the data is not sufficient for a credible comparison, so no significant conclusion can be drawn from that.

5. Conclusions

In 2019, on the days around the Leonids' and Geminids' peak, campaign measurements were performed using the DPS-4D Digisondes at the Sopron and Průhonice stations. During the sessions, two ionograms and one drift measurement were taken in one minute. The two stations are very close to each other in ionospheric terms. Synchronized measurements were made because in this way it is easy to separate regional and local effects. In addition, a temporary optical monitoring station was installed in Sopron.

Preliminarily, two months of ionograms (excluding the campaign periods) for the two stations were processed manually to study the behavior of the regular sporadic E layer. Although there are small differences (the median of the virtual height shows a 5 km difference, the maximum frequency of Průhonice displays a second peak at ~6 MHz), the distributions were mostly similar in both frequency and virtual height for the stations.

The main outputs of this study can be highlighted as follows:

- DPS-4D represents a convenient tool to detect meteor-induced traces. The settings of the instrument play a key role in the traces' localization.
- A substantial time delay within the optical and Digisonde detection was identified and needs further investigation.

- The height distribution of the meteor-induced traces in the ionograms characterizes the particular campaigns (differs for the Geminis and the Leonids) but is quite similar for the two stations in a given campaign.
- The distributions of the maximum frequencies reached by the meteor-induced traces are very different from campaign to campaign and from station to station, unlike the height distribution. This is probably due to the different sensitivity of the antenna systems, with the handcrafted antennas of Průhonice being much more sensitive [55].
- A meteor's speed and the maximum recorded frequency of the induced trace shows a slight positive correlation.
- The brightness of a meteor and the maximum echo frequencies detected on the ionogram are anticorrelated. This is contrary to the preliminary knowledge and needs further studies.
- The lifetime of the traces on the ionograms does not express any correlation with the other measured/observed parameters.
- A regular sporadic E activity can obscure the meteor-induced traces for the Digisonde measurements.

The analysis of the acquired data showed that it is possible to detect individual meteor-induced traces using the DPS-4D type of Digisonde, too. The meteors that were observed in the optical data but would have a large zenith angle seen by the ionosonde were not detected by the latter because of the applied method for the angle of arrival detection.

In these cases, where the instrument detects meteors, the Digisonde behaves as a meteor radar that operates on 1–17 MHz frequencies, though the upper limit depends on the actual setting and can be adjusted. At least a two ionogram/minute time resolution is needed for a proper study. The MPA threshold and the noise filtering mechanisms can have a huge effect on the detections. The Digisonde recorded more meteors' traces than the optical instrument which had a lower brightness limit of +1 magnitude. The detection of the meteor-induced traces also depends on the sensitivity of the antenna systems. Průhonice's Digisonde—which is modified and much more sensitive—generally detected more traces in a given time period. Because both stations are located close to each other, such a discrepancy should be attributed to the technical/instrumental differences between the two stations rather than the different physical processes above both stations.

The regular sporadic E activity can affect or even prevent successful meteor detections with the Digisondes.

Finding the reflections of a meteoric origin is also possible on Skymaps, but with a much lower probability with the current setting: only one Skymap was made every minute and they were measured/monitored with a constant 2.5 MHz frequency. For a successful detection, it was necessary that the meteor-induced traces were able to reflect the emitted signals at 2.5 MHz very briefly before the time of the measurement.

The distribution of the virtual heights of the meteor-induced traces detected on the ionograms showed that both stations obtained a similar distribution for a given campaign, but different distributions for the Leonids and Geminids campaigns. This is probably due to the different velocities and also probably a consequence of the different material quality of the meteoroid swarms.

Supplementary Materials: The following supporting information can be downloaded at https://www.mdpi.com/article/10.3390/rs16142658/s1, Figure S1: The importance of synchronized measurements can prevent false positive matches caused by sporadic E. Figure S2: Histograms displaying two months of sporadic E's virtual height and critical frequency for the two stations. Figure S3: The effects of the MPA settings. Figure S4: Distortion of a meteor trail that can cause it to appear as a split trace on the ionograms. Figure S5: Time plots of the signal strength of radio reflections from meteor trails. Table S1: Summary of the optically detected meteor present on the Digisonde measurements. Table S2: Summary of the optically detected meteor not present on the Digisonde measurements. Table S3: Summary of the results of the statistical calculations.

Author Contributions: V.B. was the conceptual author of the campaign measurements, is the operator of Sopron's Digisonde and helped with extensive knowledge regarding various ionospheric phenomena. K.P. contributed to the statistical analysis and provided insight into astronomical processing. P.K.K. helped with extensive knowledge regarding various ionospheric phenomena and technical knowledge of measurement. T.G.W.V. contributed technical knowledge of measurement and technical knowledge of the Digisonde to the article. D.K. is the operator of Průhonice's Digisonde and helped in the review of the Skymaps. A.I. installed the temporary optical camera at Sopron, he evaluated the optical data and has extensive knowledge of meteors. Z.M. manually scaled ~24,000 ionograms for the background Es analysis and also helped with extensive knowledge regarding various ionospheric phenomena. C.S. manually reviewed all the ionograms of the campaign measurement described in the manuscript, manually checked the Skymaps when it was needed, compared and analyzed the resulting data and pre-wrote the manuscript. All authors have read and agreed to the published version of the manuscript.

Funding: The work was supported also by the GINOP-2.3.2-15-2016-00003 (titled "Kozmikus hatások és kockázatok") Hungarian national project. This article was made possible by the "Multiinstrumental investigation of the midlatitude ionospheric variability" bilateral project of the Czech Academy of Sciences and Hungarian Academy of Sciences (MTA-19-03 and NKM 2018-28) and the Researcher Mobility Program of HUN-REN (KMP-2023/77). The contribution of VB was partially supported by Bolyai Fellowship (GD, no. BO/00461/21) and by OTKA, Hungarian Scientific Research Fund (Grant No. PD 141967) of the National Research, Development and Innovation Office.

Data Availability Statement: The data are available in a publicly accessible repository: The ionograms, the manual scaling of Es, the optical data and tables containing all the information used for this article are available in the Mendeley Data repository with in-detail descriptions. Szárnya, Csilla; Igaz, Antal; Mošna, Zbyšek; Kouba, Daniel; Barta, Veronika (2024), "Data and evaluation of Digisonde campaign measurements at Průhonice and Sopron stations during the 2019 Leonids and Geminids", Mendeley Data, V1, doi: 10.17632/22n6tvjvv4.1. The software used to evaluate the optical data is freeware and can be found here: https://www.metrec.org/, 18 July 2024. The Digisonde data (ionograms and Skymaps) can be found in the GIRO Network's database: https://giro.uml.edu/didbase/ and https://giro.uml.edu/driftbase/ (accessed on 18 July 2024). The software used to scale and view the ionograms (SAO Explorer) and Skymaps (DriftExplorer) is available at https://ulcar.uml.edu/SAO-X/SAO-X.html (accessed on 18 July 2024) and at https://ulcar.uml.edu/Drift-X.html (accessed on 18 July 2024), respectively. The geomagnetic coordinates were calculated for 2019 using the utilities of https://wdc.kugi.kyoto-u.ac.jp/igrf/gggm/index.html (accessed on 18 July 2024).

Conflicts of Interest: The authors declare no conflicts of interest.

References

1. Yeomans, D.K.; Yau, K.K.; Weissman, P.R. The Impending Appearance of Comet Tempel–Tuttle and the Leonid Meteors. *Icarus* **1996**, *124*, 407–413. [CrossRef]
2. Williams, I.P.; Wu, Z. The Geminid Meteor Stream and Asteroid 3200 Phaethon. *Mon. Not. R Astron. Soc.* **1993**, *262*, 231–248. [CrossRef]
3. McBeath, A.; Arlt, R. Meteor Shower Calendar: October–December 2001. *WGN J. Int. Meteor Organ.* **2001**, *29*, 69–77.
4. Porubčan, V.; Kornoš, L.; Williams, I.P. The Taurid Complex Meteor Showers and Asteroids. *Contrib. Astron. Obs. Skaln. Pleso* **2006**, *36*, 103–117. [CrossRef]
5. Jenniskens, P.; Betlem, H.; de Lignie, M.; Langbroek, M. The Detection of a Dust Trail in the Orbit of an Earth-Threatening Long-Period Comet. *Astrophys. J.* **1997**, *479*, 441. [CrossRef]
6. Neslušan, L.; Hajduková, M. The Meteor-Shower Complex of Comet C/1917 F1 (Mellish). *Astron. Astrophys.* **2014**, *566*, A33. [CrossRef]
7. Greaves, J. The Remarkable Similarity of the Orbit of C/2023 P1 Nishimura and the σ Hydrid Meteor Shower. *eMeteorNews* **2023**, *8*, 281–282.
8. Jenniskens, P.; Nénon, Q.; Gural, P.S.; Albers, J.; Haberman, B.; Johnson, B.; Morales, R.; Grigsby, B.J.; Samuels, D.; Johannink, C. CAMS Newly Detected Meteor Showers and the Sporadic Background. *Icarus* **2016**, *266*, 384–409. [CrossRef]
9. Whitehead, J.D. The Formation of the Sporadic-E Layer in the Temperate Zones. *J. Atmos. Terr. Phys.* **1961**, *20*, 49–58. [CrossRef]
10. Axford, W.I. The Formation and Vertical Movement of Dense Ionized Layers in the Ionosphere Due to Neutral Wind Shears. *J. Geophys. Res.* **1963**, *68*, 769–779. [CrossRef]
11. Haldoupis, C. A Tutorial Review on Sporadic E Layers. In *Aeronomy of the Earth's Atmosphere and Ionosphere*; Springer: Dordrecht, The Netherlands, 2011; pp. 381–394.

12. Friedrich, M.; Rapp, M.; Blix, T.; Hoppe, U.-P.; Torkar, K.; Robertson, S.; Dickson, S.; Lynch, K. Electron Loss and Meteoric Dust in the Mesosphere. *Ann. Geophys.* **2012**, *30*, 1495–1501. [CrossRef]
13. Hoppe, U.P.; Rapp, M. Structure, Composition, and Dynamics of the Middle Atmosphere and Lower Ionosphere during a Major Meteor Shower. *Ann. Geophys.* **2013**, *31*, 1829–1831. [CrossRef]
14. Jacobi, C.; Arras, C.; Wickert, J. Enhanced Sporadic E Occurrence Rates during the Geminid Meteor Showers 2006–2010. *Adv. Radio Sci.* **2013**, *11*, 313–318. [CrossRef]
15. Chandra, H.; Sharma, S.; Devasia, C.V.; Subbarao, K.S.V.; Sridharan, R.; Sastri, J.H.; Rao, J.V.S.V. Sporadic-E Associated with the Leonid Meteor Shower Event of November 1998 over Low and Equatorial Latitudes. *Ann. Geophys.* **2001**, *19*, 59–69. [CrossRef]
16. Stuart, W.D. Ionized Regions. In *Radar Cross Section Handbook*; Kluwer Academic/Plenum Publishers: New York, NY, USA, 1970; Volume 2, pp. 773–839.
17. Sugar, G.R. Radio Propagation by Reflection from Meteor Trails. *Proc. IEEE* **1964**, *52*, 116–136. [CrossRef]
18. Manning, L.A.; Eshleman, V.R. Meteors in the Ionosphere. *Proc. IRE* **1959**, *47*, 186–199. [CrossRef]
19. McKinley, D.W.R. *Meteor Science and Engineering*; McGraw-Hill: New York, NY, USA, 1961.
20. Maruyama, T.; Kato, H.; Nakamura, M. Ionospheric Effects of the Leonid Meteor Shower in November 2001 as Observed by Rapid Run Ionosondes. *J. Geophys. Res.* **2003**, *108*, SIA 4-1–SIA 4-13. [CrossRef]
21. Ceplecha, Z.; Borovička, J.; Elford, W.G.; ReVelle, D.O.; Hawkes, R.L.; Porubčan, V.; Šimek, M. Meteor Phenomena and Bodies. *Space Sci. Rev.* **1998**, *84*, 327–471. [CrossRef]
22. Kozlovsky, A.; Shalimov, S.; Kero, J.; Raita, T.; Lester, M. Multi-Instrumental Observations of Nonunderdense Meteor Trails. *J. Geophys. Res. Space Phys.* **2018**, *123*, 5974–5989. [CrossRef]
23. Heritage, J.L.; Fay, W.J.; Bowen, E.D. Evidence That Meteor Trails Produce a Field-Aligned Scatter Signal at VHF. *J. Geophys. Res.* **1962**, *67*, 953–964. [CrossRef]
24. Oppenheim, M.M.; Dimant, Y.S. First 3-D Simulations of Meteor Plasma Dynamics and Turbulence. *Geophys. Res. Lett.* **2015**, *42*, 681–687. [CrossRef]
25. Eshleman, V.R. The Theoretical Length Distribution of Ionized Meteor Trails. *J. Atmos. Terr. Phys.* **1957**, *10*, 57–72.
26. Beech, M. Large-Body Meteoroids in the Leonid Stream. *Astron. J.* **1998**, *116*, 499–502. [CrossRef]
27. Höffner, J.; von Zahn, U.; McNeil, W.J.; Murad, E. The 1996 Leonid Shower as Studied with a Potassium Lidar: Observations and Inferred Meteoroid Sizes. *J. Geophys. Res. Space Phys.* **1999**, *104*, 2633–2643. [CrossRef]
28. Borovička, J.; Štork, R.; Bocek, J. First Results from Video Spectroscopy of 1998 Leonid Meteors. *Meteorit. Planet. Sci.* **1999**, *34*, 987–994. [CrossRef]
29. Vojáček, V.; Borovička, J.; Koten, P.; Spurný, P.; Štork, R. Catalogue of Representative Meteor Spectra. *Astron. Astrophys.* **2015**, *580*, A67. [CrossRef]
30. Hocking, W.K. Real-Time Meteor Entrance Speed Determinations Made with Interferometric Meteor Radars. *Radio Sci.* **2000**, *35*, 1205–1220. [CrossRef]
31. Jewitt, D.; Li, J. Activity in Geminid Parent (3200) Phaeton. *Astron. J.* **2010**, *140*, 1519–1527. [CrossRef]
32. Borovička, J.J. Video Spectra of Leonids and Other Meteors. In Proceedings of the Meteoroids 2001 Conference, Kiruna, Sweden, 6–10 August 2001; Volume 495, pp. 203–208.
33. Trigo-Rodríguez, J.M.; Llorca, J.; Fabregat, J. Chemical Abundances Determined from Meteor Spectra—II. Evidence for Enlarged Sodium Abundances in Meteoroids. *Mon. Not. R. Astron. Soc.* **2004**, *348*, 802–810. [CrossRef]
34. Kasuga, T.; Watanabe, J.; Ebizuka, N. A 2004 Geminid Meteor Spectrum in the Visible–Ultraviolet Region. *Astron. Astrophys.* **2005**, *438*, L17–L20. [CrossRef]
35. Vojáček, V.; Borovička, J.; Koten, P.; Spurný, P.; Štork, R. Properties of Small Meteoroids Studied by Meteor Video Observations. *Astron. Astrophys.* **2019**, *621*, A68. [CrossRef]
36. Narwa, R.C.; Bathula, P.K.; Yellaiah, G. Statistical Analysis of Meteoroid Fragmentation during the Geminid and Leonid Meteor Showers. *Indian J. Radio Space Phys.* **2021**, *49*, 110–121.
37. Koten, P.; Rendtel, J.; Shrbený, L.; Gural, P.; Borovička, J.; Kozak, P. Meteors and Meteor Showers as Observed by Optical Techniques. In *Meteoroids: Sources of Meteors on Earth and Beyond*; Ryabova, G.O., Asher, D.J., Campbell-Brown, M.J., Eds.; Cambridge University Press: Cambridge, UK, 2019; p. 90.
38. Fukao, S.; Hamazu, K. Observations by Atmospheric Radar. In *Radar for Meteorological and Atmospheric Observations*; Springer: Tokyo, Japan, 2014; pp. 435–485.
39. Chen, J.S.; Wang, C.Y.; Su, C.L.; Chu, Y.H. Meteor Observations Using Radar Imaging Techniques and Norm-Constrained Capon Method. *Planet. Space Sci.* **2020**, *184*, 104884. [CrossRef]
40. Ellyett, C.D.; Goldsbrough, P.F. Relationship of Meteors to Sporadic E, 1. A Sorting of Facts. *J. Geophys. Res.* **1976**, *81*, 6131–6134. [CrossRef]
41. McNaught, R.H.; Asher, D.J. Leonid Dust Trails and Meteor Storms. *WGN J. Int. Meteor Organ.* **1999**, *27*, 85–102.
42. Maslov, M. Leonid Predictions for the Period 2001–2100. *WGN J. Int. Meteor Organ.* **2007**, *35*, 5–12.
43. Reinisch, B.W.; Galkin, I.A.; Khmyrov, G.M.; Kozlov, A.V.; Lisysyan, I.A.; Bibl, K.; Cheney, G.; Kitrosser, D.; Stelmash, S.; Roche, K.; et al. Advancing Digisonde Technology: The DPS4. *AIP Conf. Proc.* **2008**, *974*, 127–143.

44. Kereszturi, Á.; Barta, V.; Bondár, I.; Czanik, C.; Igaz, A.; Mónus, P.; Rezes, D.; Szabados, L.; Pál, B.D. Review of Synergic Meteor Observations: Linking the Results from Cameras, Ionosondes, Infrasound and Seismic Detectors. *Mon. Not. R. Astron. Soc.* **2021**, *506*, 3629–3640. [CrossRef]
45. Szárnya, C.; Chum, J.; Podolská, K.; Kouba, D.; Koucká Knížová, P.; Mošna, Z.; Barta, V. Multi-Instrumental Detection of a Fireball during Leonids of 2019. *Front. Astron. Space Sci.* **2023**, *10*, 1197832. [CrossRef]
46. Molau, S. The Meteor Detection Software MetRec. In Proceedings of the Meteoroids, Stará Lesná, Slovakia, 20–23 August 1998; pp. 131–134.
47. Davies, K. *Ionospheric Radio*; IEE Electromagnetic Wave Series; Peter Peregrinus: London, UK, 1990.
48. Reinisch, B.W. *Digisonde 4D Technical Manual (Version 1.0)*; Lowell Digisonde International: Lowell, MA, USA, 2009.
49. Bór, J.; Sátori, G.; Barta, V.; Szabóné-André, K.; Szendrői, J.; Wesztergom, V.; Bozóki, T.; Buzás, A.; Koronczay, D. Measurements of Atmospheric Electricity in the Széchenyi István Geophysical Observatory, Hungary. *Hist. Geo Space Sci.* **2020**, *11*, 53–70. [CrossRef]
50. Reinisch, B.W.; Galkin, I.A. Global Ionospheric Radio Observatory (GIRO). *Earth Planets Space* **2011**, *63*, 377–381. [CrossRef]
51. Reinisch, B.W. Modern Ionosondes. In *Modern Ionospheric Science*; European Geophysicæal Society: Katlenburg-Lindau, Germany, 1996; pp. 440–458.
52. Reinisch, B.W.; Huang, X.; Galkin, I.A.; Paznukhov, V.; Kozlov, A. Recent Advances in Real-Time Analysis of Ionograms and Ionospheric Drift Measurements with Digisondes. *J. Atmos. Sol. Terr. Phys.* **2005**, *67*, 1054–1062. [CrossRef]
53. Szárnya, C.; Igaz, A.; Mošna, Z.; Kouba, D.; Barta, V. Data and Evaluation of Digisonde Campaign Measurements at Průhonice and Sopron Stations during the 2019 Leonids and Geminids. *Mendeley Data* **2024**, *V1*. [CrossRef]
54. Borovička, J.; Shrbený, L.; Spurný, P. Automation of the Czech Part of the European Fireball Network: Equipment, Methods and First Results. *Proc. Int. Astron. Union* **2006**, *2*, 121–130. [CrossRef]
55. Mošna, Z.; Barta, V.; Berényi, K.A.; Mielich, J.; Verhulst, T.; Kouba, D.; Urbář, J.; Chum, J.; Koucká Knížová, P.; Marew, H.; et al. March and April 2023 ionospheric storms in the period of Solar cycle 25. *Front. Astron. Space Sci.* 2024, *submitted*.
56. Vaubaillon, J.; Colas, F.; Jorda, L. A New Method to Predict Meteor Showers. *Astron. Astrophys.* **2005**, *439*, 761–770. [CrossRef]
57. Jenniskens, P. *Meteor Showers and Their Parent Comets*; Hawkes, R.L., Ed.; Cambridge University Press: Cambridge, UK, 2007; Volume 101.
58. Lukianova, R.; Kozlovsky, A.; Lester, M. Recognition of Meteor Showers From the Heights of Ionization Trails. *J. Geophys. Res. Space Phys.* **2018**, *123*, 7067–7076. [CrossRef]
59. Oikonomou, C.; Haralambous, H.; Leontiou, T.; Tsagouri, I.; Buresova, D.; Mošna, Z. Intermediate Descending Layer and Sporadic E Tidelike Variability Observed over Three Mid-Latitude Ionospheric Stations. *Adv. Space Res.* **2022**, *69*, 96–110. [CrossRef]
60. Haldoupis, C.; Meek, C.; Christakis, N.; Pancheva, D.; Bourdillon, A. Ionogram Height–Time–Intensity Observations of Descending Sporadic E Layers at Mid-Latitude. *J. Atmos. Sol. Terr. Phys.* **2006**, *68*, 539–557. [CrossRef]

remote sensing

Article

Evaluation of MAX-DOAS Profile Retrievals under Different Vertical Resolutions of Aerosol and NO_2 Profiles and Elevation Angles

Xin Tian [1], Mingsheng Chen [1], Pinhua Xie [2,*], Jin Xu [2], Ang Li [2], Bo Ren [3], Tianshu Zhang [2], Guangqiang Fan [2], Zijie Wang [1], Jiangyi Zheng [2] and Wenqing Liu [2]

1 Information Materials and Intelligent Sensing Laboratory of Anhui Province, Institutes of Physical Science and Information Technology, Anhui University, Hefei 230601, China; xtian@ahu.edu.cn (X.T.)
2 Key Laboratory of Environmental Optical and Technology, Anhui Institute of optics and Fine Mechanics, Hefei Institutes of Physical Science, Chinese Academy of Sciences, Hefei 230031, China
3 School of Environmental Science and Optoelectronic Technology, University of Science and Technology of China, Hefei 230026, China
* Correspondence: phxie@aiofm.ac.cn

Abstract: In the Multi-Axis Differential Absorption Spectroscopy (MAX-DOAS) trace gas and aerosol profile inversion algorithm, the vertical resolution and the observation information obtained through a series of continuous observations with multiple elevation angles (EAs) can affect the accuracy of an aerosol profile, thus further affecting the results of the gas profile. Therefore, this study examined the effect of the vertical resolution of an aerosol profile and EAs on the NO_2 profile retrieval by combining simulations and measurements. Aerosol profiles were retrieved from MAX-DOAS observations and co-observed using light detection and ranging (Lidar). Three aerosol profile shapes (Boltzmann, Gaussian, and exponential) with vertical resolutions of 100 and 200 m were used in the atmospheric radiative transfer model. Firstly, the effect of the vertical resolution of the input aerosol profile on the retrieved aerosol profile with a resolution of 200 m was studied. The retrieved aerosol profiles from the two vertical resolution aerosol profiles as input were similar. The aerosol profile retrieved from a 100 m resolution profile as input was slightly overestimated compared to the input value, whereas that from a 200 m resolution input was slightly underestimated. The relative deviation of the aerosol profile retrieved from the 100 m resolution as input was higher than that of the 200 m. MAX-DOAS observations in Hefei city on 4 September 2020 were selected to verify the simulation results. The aerosol profiles retrieved from the oxygen collision complex (O_4) differential slant column density derived from MAX-DOAS observations and Lidar simulation were compared with the input Lidar aerosol profiles. The correlation between the retrieved and input aerosol profiles was high, with a correlation coefficient $R > 0.99$. The aerosol profiles retrieved from the Lidar profile at 100 and 200 m resolutions as input closely matched the Lidar aerosol profiles, consistent with the simulation result. However, aerosol profiles retrieved from MAX-DOAS measurements differed from the Lidar profiles due to the influence of the averaging kernel matrix smoothing, the different location and viewing geometry, and uncertainties associated with the Lidar profiles. Next, NO_2 profiles of different vertical resolutions were used as input profiles to retrieve the NO_2 profiles under a single aerosol profile scenario. The effect of the vertical resolution on the retrieval of NO_2 profiles was found to be less significant compared to aerosol retrievals. Using the Lidar aerosol profile as the *a priori* aerosol information had little effect on NO_2 profile retrieval. Additionally, the retrieved aerosol profiles and aerosol optical depths varied under different EAs. Ten EAs (i.e., 1, 2, 3, 4, 5, 6, 8, 15, 30, and 90°) were found to obtain more information from observations.

Keywords: MAX-DOAS; aerosol; profile; vertical resolution; elevation angle; NO_2

1. Introduction

Most kinds of gaseous pollutants and aerosols in the boundary layer are closely linked to human activities. The risks posed by these pollutants to humans and other organisms primarily occur within this layer. The study of the three-dimensional distribution, evolution, and transport of atmospheric components in the boundary layer is an important part of understanding the formation mechanism of atmospheric pollution. Nonetheless, due to the limitations in detection technology, it has been difficult to carry out in-depth research on the related atmospheric physical and chemical processes. Therefore, an accurate estimation/monitoring of the distribution of and variation in gaseous pollutants and aerosols in the atmospheric boundary layer is crucial for explaining and predicting the transport and diffusion processes of atmospheric pollution components and for improving the monitoring level and forecasting ability of the atmospheric environment. This further provides important data for research work on the atmospheric environment and air-quality forecasting models.

Multi-Axis Differential Optical Absorption Spectroscopy (MAX-DOAS) technique [1–6] is a passive ground-based atmospheric remote-sensing technology that uses the scattered light of the sun as the light source to perform qualitative and quantitative measurements of trace gases. The spatio-temporal distribution information of various trace gases (e.g., NO_2, SO_2, HCHO, HONO, BrO, and CHOCHO) [7–15] and aerosols [14–16] can be obtained simultaneously. The differential slant column densities (dSCDs) of oxygen-dimer (O_4) and trace gases are obtained by multi-elevation angle observations based on MAX-DOAS. The tropospheric vertical column densities (VCDs), surface volume mixing ratios (VMRs), and vertical profiles of aerosol extinction and trace gas mixing ratios are obtained by combining with the atmospheric radiative transfer model (RTM). Today, numerous inversion algorithms within the MAX-DOAS community are regularly used that rely on different mathematical inversion approaches. These algorithms can mainly be divided into two categories: either based on the optimal estimation (OE) method proposed by Rodgers [17] or based on the parameterization approach, usually relying on pre-calculated differential air mass factor (dAMF) look-up tables (LUTs) at multiple wavelengths. The inversion algorithms comprise a two-step inversion procedure. Firstly, O_4 dSCDs are used to retrieve the aerosol extinction profile and aerosol optical depth (AOD). Next, the tropospheric (0–4 km) trace gas profiles, VCDs, and VMRs are retrieved based on the aerosol profiles derived in step 1 and the trace gas dSCDs [18,19]. In order to study the accuracy of the MAX-DOAS inversion algorithm, extensive comparative studies have been carried out based on large-scale comprehensive observation experiments and simulation verification, which mainly included (1) remote-sensing techniques based on synthetic data that tested the performance of the algorithms for the retrieval of the atmospheric state [2,4,20–22]; (2) three-dimensional distribution of trace gases and aerosols obtained through observation data, which were then compared with data from other instruments installed at different heights (airborne, tethered balloons, towers, etc.) [23–26]; (3) intercomparisons of different algorithms [18,19,23]. In order to study the effect of physical characteristics and the spatio-temporal distributions of aerosols on the remote sensing of trace gases, Tian et al. [22] studied the effect of the aerosol extinction and profile shape on the retrieved NO_2 profiles based on synthetic data. The vertical resolution of the profile inversion algorithm used in the previous study was 200 m. In addition, the retrieved aerosol profile in step 1 was also required as the *a priori* information to retrieve the trace gas profile. The aerosol profile had a non-linear influence on the gas profile inversion. Therefore, an accurate inversion of an aerosol profile is the premise for the gas profile retrieval. The accuracy of an aerosol profile is not only related to the value of aerosol extinction and the shape of the profile, but is also closely related to its vertical resolution. At the same time, the retrievals of trace gas and aerosol by MAX-DOAS are based on the information from a set of spectral measurements taken at different elevation angles (EAs). The elevation angles (especially low-elevation angles) also affect the information obtained by MAX-DOAS observations. Therefore, the effect of the vertical resolution of the aerosol profile and EAs of MAX-DOAS observations on gas profile retrieval was examined in this study. A simulation of synthetic data and verification of actual observations were adopted.

The paper is structured as follows. Section 2 outlines the main methodology and parameter settings of this study. The location and system introduction of MAX-DOAS are detailed in Sections 2.1 and 2.2, respectively. Section 2.3 elucidates the retrieval methods of aerosol and trace gas profiles. Section 2.4 delineates the analysis strategy and parameter settings of the radiative transfer model for the simulation. Lidar aerosol profiles pretreatment are introduced in Section 2.5. The effects of the vertical resolution of the aerosol profiles and EAs on the retrieval profile are discussed in Section 3. Finally, Section 4 concludes and summarizes the results.

2. Instrumentation and Methodology

2.1. Site Description

The measurement site was situated on the campus of the Anhui Institute of Optics and Fine Mechanics (31.91°N, 117.16°E, 20 m), Chinese Academy of Sciences (AIOFM, CAS), located on Science Island, a suburban area of Hefei in Anhui Province, China. Figure 1 illustrates the location of the measurement site in relation to the surrounding topography. Additionally, Figure 1 displays relationship between the site's MAX-DOAS and Lidar. Farm areas are in the north, west, and northeast, whereas areas in the southeast are highly urbanized. Furthermore, there is an airport located in the northwest and a power plant in the northeast.

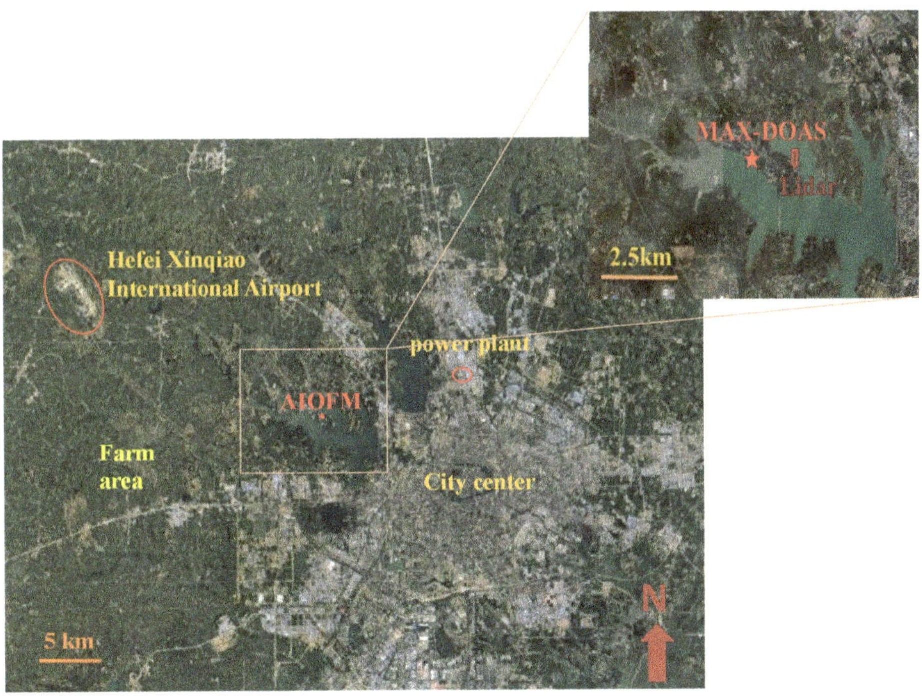

Figure 1. The locations of observation sites of MAX-DOAS and Lidar (AIOFM) and surrounding areas.

2.2. MAX-DOAS and Lidar Setup

The MAX-DOAS systems was positioned on the campus of the Anhui Institute of Optics and Fine Mechanics (AIOFM), specifically on the roof of a 6-storey building. The spectrometer used in the MAX-DOAS measurements was a Princeton Instruments (PI) 160 (IsoPlane-160, Princeton, USA), which was equipped with an area array of charge-coupled device (CCD) detectors, comprising 2048 × 512 pixels (Acton, Princeton, USA). The wavelength range of the spectrograph was from 290 to 410 nm, with a spectral resolution of 0.35 nm. The telescope restricted the field of view angle to less than 0.2°. During the measurements, the sequential elevation angles (EAs) were set at 1°, 2°, 3°, 4°, 5°, 6°, 8°, 15°, 30°, and 90°. Each individual measurement consisted of an average of 100 spectral recording scans, with the integration time automatically adjusted to optimize the signal-to-noise ratio based on the light intensity. The duration of each sequence of observations at 10 EAs was approximately 10 min. The system generally operated between the local time (LT) hours of 6:00 and 18:00. The MAX-DOAS data from an azimuth of 0° (north) were used for analysis in this study. A detailed description of the instrument can be found in Ren et al. [27].

A differential absorption Lidar was operated to retrieve aerosol extinction (AE, 316 nm) and O_3 profile at the Science Island Experimental Field, which was located approximately 1.0 km away from the MAX-DOAS (red rectangle point in Figure 1). The Lidar system employed in this study was developed by the AIOFM [28]. The maximum statistical error associated with the Lidar profile was found to be below 10% [29]. The assumed radar ratio for the measurements was set at 50 sr. Due to the observation blind spot of the Lidar, the aerosol profiles with altitudes of ≥0.3 km were used [27].

2.3. DOAS Analysis and Profile Retrieval

2.3.1. Spectral Analysis

Spectra were analyzed using the QDOAS software (http://uv-vis.aeronomie.be/software/QDOAS/ (accessed on 13 November 2023)) developed by BIRA-IASB. The O_4 and NO_2 dSCDs were retrieved in the spectral fitting window of 338–370 nm, which included several strong absorption bands. The measured Fraunhofer spectrum in the zenith was used as the reference spectrum for each cycle to eliminate the impact of the Fraunhofer reference spectrum (FRS) [1]. The parameter settings and the trace gas absorption cross-sections for the retrieval of NO_2 dSCDs were based on recommendations from the CINDI-2 campaign (2nd Cabauw Intercomparison of Nitrogen Dioxide-measuring Instruments, 2016 [30]), as shown in Table 1. An example of an O_4 and NO_2 fitting result, which was obtained from these 2D MAX-DOAS on 4 September 2020 at a 1° EA and at 7:45 LT (local time), is shown in Figure 2. The typical root mean square error (RMSE) for NO_2 and O_4 was <0.001.

Table 1. Baseline DOAS analysis settings used for the O_4 and NO_2 slant column retrieval, where "√" expresses the cross-section used in the retrieval.

Parameter	Source	Species
		O_4 and NO_2
Fitting spectral range		338–370 nm
NO_2 (220 K, 294 K)	[31], I_0-corrected (10^{17} molec. cm^{-2})	√
O_4 (293 K)	[32]	√
O_3 (223 K, 243 K)	[33], I_0-corrected (10^{20} mole. cm^{-2})	√
HCHO (297 K)	[34]	√
BrO (223 K)	[35]	√
Ring		RING_DOAS_SAO2010 [36]
Polynomial degree		Order 5 (6 coefficients)
Intensity offset		Constant

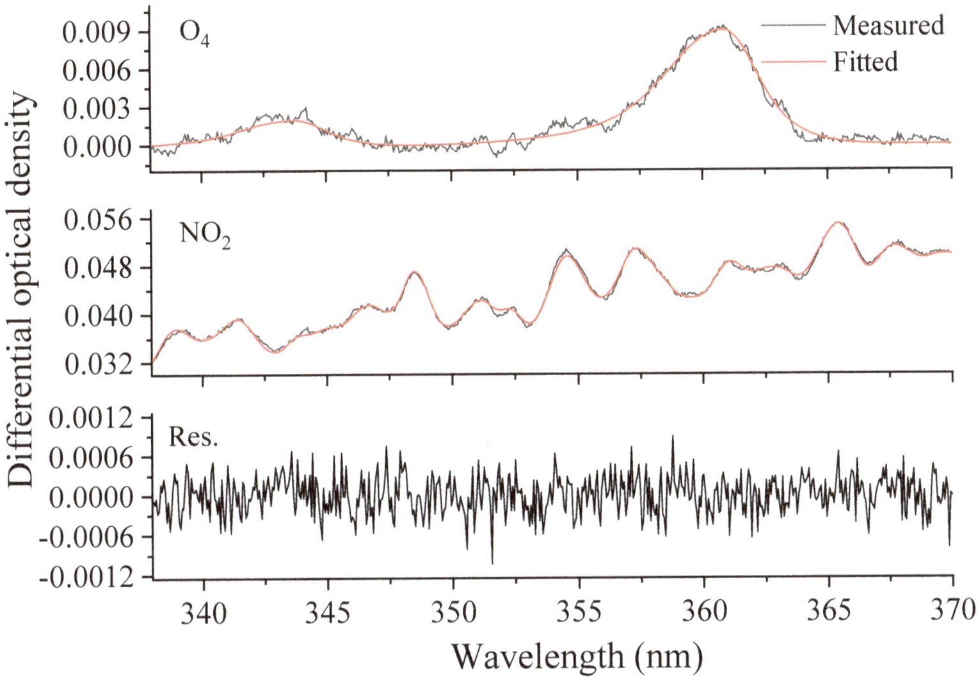

Figure 2. Example of an O_4 and NO_2 slant column fit obtained with the 2D MAX−DOAS on 4 September 2020, at approximately 7:45 LT, at a 1° elevation angle. The red and black curves indicate the fitted absorption structures and the derived absorption structures from the measured spectra, respectively.

2.3.2. Profile Retrievals

The dSCD obtained from the DOAS spectral analysis represents the trace gas concentration along the effective optical path length. Thus, the dSCD must be converted to vertical column density (VCD) by applying the air mass factor (AMF) as follows:

$$\mathrm{VCD} = \frac{\mathrm{SCD_i} - \mathrm{SCD_{i=90^\circ}}}{\mathrm{AMF_i} - \mathrm{AMF_{i=90^\circ}}} = \frac{\mathrm{dSCD_i}}{\mathrm{AMF_i} - \mathrm{AMF_{i=90^\circ}}}, \tag{1}$$

where *i* represents the elevation angle.

VCD is defined as the integrated concentration of trace gas concentration through the atmosphere along a vertical path. AMF is often used to describe the absorption path of a gas in the atmosphere. The atmospheric RTM is able to calculate the AMF.

In this study, the PriAM algorithm, based on the SCIAMACHY radiative transfer model (SCIATRAN), which was used to retrieve trace gas VCDs and profiles, was developed by the Anhui Institute of Optics and Fine Mechanics, Chinese Academy of Sciences (AIOFM, CAS), in cooperation with the Max Planck Institute for Chemistry (MPIC) [24,37,38]. PriAM also consists of a two-step inversion procedure based on the optimal estimation method (OEM) [8,18,22,27,39]. PriAM can retrieve trace gas and aerosol profiles for any arbitrary vertical grid. In this study, vertical layers (altitude grid) with a 200 m resolution at an altitude range below 4.0 km were used. Parameter settings used in the general PriAM retrieval for aerosol and NO_2 profiles are shown in Table 2.

Table 2. Parameter settings used in the general PriAM retrieval for aerosol and NO_2 profiles.

Parameters	
Type of radiative transfer model	Scattered light in a spherical atmosphere
Vertical resolution	200 m resolutions at an altitude range below 4.0 km
A priori profile	Aerosol: Boltzmann shape with an AOD of 0.3 and a height of 1.5 km (See Table S1 for details) NO_2: Boltzmann input profile with a NO_2 VCD of 1.0×10^{16} molec. cm^{-2} (See Table S1 for details)
Wavelength (nm)	360
Single scattering albedo (SSA)	0.9
Asymmetry parameter (AP)	0.72
Surface albedo	0.06
Solar zenith angle (SZA, °)	From observation
Relative azimuth angle (RAA, °)	From observation
Azimuth angle (°)	0 (North)

In the OEM, the impact of the deviation between the retrieved profile and the true profile, caused by the retrieval profile shifting towards the *a priori* profile at a higher altitude [17], is referred to as the "smoothing effect". Information about the vertical resolution and sensitivity is provided by the averaging kernel matrix (AVK). The trace of AVK is equivalent to the degrees of freedom for signal (DOFS), which quantifies the total number of independent pieces of information contained in the measurements compared to the *a priori* one. The maximum value of the AVK for each layer can represent the sensitivity for that specific layer in the retrieval. The envelope, formed by connecting the maximum AVK values of each layer, illustrates the change in sensitivity with height. In this study, the height corresponding to an envelope value decreasing to 10% of the maximum value is termed the upper sensitive height limit (H_m). An altitude range of 0 to H_m is considered the sensitive zone for profile retrieval by MAX-DOAS.

2.4. Analysis Strategy and RTM Parameters

This study employed both simulations and actual measurements to verify the impact of the vertical resolution of the input profile and the elevation angle on the retrieved trace gas profile. The depicted analysis strategy in Figure 3 outlines the approach. Firstly, the simulation focused on assessing the effect of the vertical resolution of the input profile and EAs on the retrieved profile using synthetic data. The vertical resolution of the input profile was set to 100 or 200 m. AOD and NO_2 VCD were set to 0.3 and 1.0×10^{16} molec. cm^{-2}, respectively. O_4 and NO_2 dSCDs were calculated from input profiles by RTM, which were then used in aerosol and NO_2 profile retrievals. The retrieved and input profiles were subsequently compared. It is important to note that aerosols influence NO_2 profile retrieval. When using the two-step inversion algorithm PriAM to retrieve the NO_2 profile, the *a priori* aerosol profile information was obtained from the aerosol profile at a 200 m resolution input. Secondly, the study utilized MAX-DOAS and Lidar observations to validate the simulation results. Part 1 focused on verifying the effect on aerosol profile retrieval: (1) the impact of different EAs on aerosol profile retrieval was investigated; (2) aerosol profiles observed by Lidar with different vertical resolutions were used as input profiles, and the aerosol profiles were then retrieved following the simulation strategy. The retrieved aerosol profile with the Lidar profile as input was compared with the aerosol profile from the MAX-DOAS and Lidar observations. Part 2 addressed the verification of the effect on NO_2 profile retrieval: (1) the influence of different EAs on NO_2 profile retrieval was examined; (2) the retrieved NO_2 profiles were compared under the *a priori* aerosol profiles from the scenarios of the MAX-DOAS and Lidar observations.

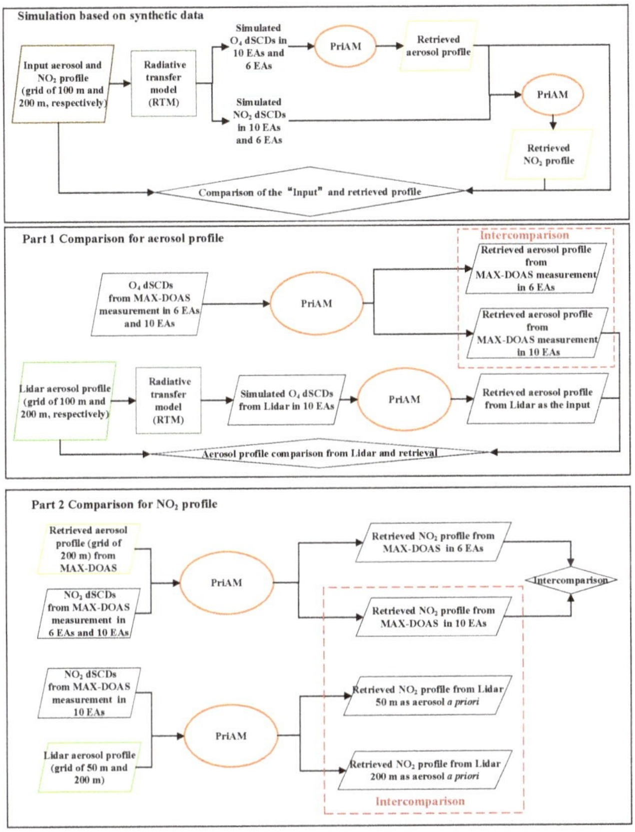

Figure 3. Flow diagram illustrating the strategy employed to analyze the effects of the vertical resolutions of the aerosol profile and EAs of the MAX-DOAS observations on the retrievals of aerosol and NO_2 profile.

In this study, the SCIATRAN version 2.2 [39] was utilized for the forward model calculations, consistent with PriAM. The EAs were configured for four scenarios: 10 EAs for Setting I (i.e., 1, 2, 3, 4, 5, 6, 8, 15, 30, and 90°), aligning with the MAX-DOAS measurement settings; the low EAs were reduced for Setting II, with six EAs (i.e., 2, 4, 8, 15, 30, and 90°); the low EAs were retained for Setting III, with six EAs (i.e., 1, 2, 3, 4, 5, and 90°); and the low EAs were omitted for Setting IV, with six EAs (i.e., 5, 6, 8, 15, 30, and 90°). In the simulation of synthetic data and real atmospheric inversion, the scenario with a single scattering albedo (SSA) = 0.9, asymmetry parameter (AP) = 0.72, and surface albedo = 0.06 was used in the simulation. These values were the long-term averages from the Aerosol Robotic Network (AERONET) at the Hefei site [37,40]. A solar zenith angle (SZA) = 60° and relative azimuth angle (RAA) = 120° were also applied in the simulation of synthetic data. In the real atmospheric inversion, the SZA and RAA were derived from actual measurements. Three profile shapes were employed in the simulation, which are exponential, Boltzmann, and Gaussian shapes. Exponential profiles depict a systematic decrease in concentration with altitude, representing situations where emissions typically occur at the surface. Boltzmann profiles signify situations typical for well-mixed boundary layers. Gaussian profiles illustrate scenarios involving pollutant transport, which typically occurs above the boundary layer. In this study, a Boltzmann profile corresponding approximately to the boundary layer height and a Gaussian profile representing pollutant transport in the boundary layer were designed. The detailed parameters used for the RTM in the simulation are presented in Table S1.

2.5. Pretreatment of Lidar Aerosol Profiles

Before comparing the Lidar aerosol profile with the MAX-DOAS retrieval result, the Lidar aerosol profile needed to be preprocessed to match the vertical resolution of the MAX-DOAS result. Firstly, as the aerosol extinction measurements were acquired by two different systems operating at distinct wavelengths (MAX-DOAS: 360 nm; Lidar: 316 nm), the Lidar's extinction profiles needed to be converted to the MAX-DOAS wavelength using the aerosol Ångström exponent. The conversion was performed as follows:

$$(\frac{\lambda_a}{\lambda_b})^{-\alpha} = \frac{\tau(\lambda_a)}{\tau(\lambda_b)}, \quad (2)$$

where λ and α represent the wavelength and Ångström exponent, respectively. $\tau(\lambda)$ denotes the aerosol optical depth. The Ångström exponent was derived from real-time observation data obtained by the Cimel Electronique sun photometer (CE318) located approximately 7 m away from the MAX-DOAS system. The wavelength range used for calculating the Ångström exponent ranged from 440 to 870 nm. Secondly, the vertical layer of the Lidar aerosol profile was processed into three altitude grids of 50, 100, and 200 m. Lastly, the gas and aerosol profiles were both smoothed in the OEM profile inversion algorithm retrieval. To reduce possible sources of discrepancies in the comparison of an OEM retrieved profile and Lidar profile, the Lidar profile resolution and information content had to be degraded by "smoothing" it with the corresponding MAX-DOAS AVK matrix (**A**) according to the following equation [17]:

$$\tilde{x} = \mathbf{A}x + (1 - \mathbf{A})xa, \quad (3)$$

Here, x_a is the *a priori* profile; $\tilde{x}$ represents the profile smoothed by the MAX-DOAS AVK matrix **A**; x perfectly represents the Lidar profile.

3. Results and Discussion

The MAX-DOAS profile inversion algorithm establishes a functional relationship between the measurement vector and parameters of the atmospheric state based on the RTM. The optimal state between the measurements and prior values for the aerosol and gas was calculated by linear and nonlinear iteration. In order to investigate the effect of the observation information on the MAX-DOAS profile retrieval, the simulation of synthetic data and the verification of actual observations were adopted. Specifically, the number of elevation angles and the vertical resolution of the input profiles were used to represent different observation information obtained by MAX-DOAS.

3.1. Aerosol Results

3.1.1. Effect on the Aerosol Profile Retrieval

Aerosol profiles with the same concentration but varying resolutions serve as representations of the diverse MAX-DOAS observation information. Consequently, these aerosol profiles were used to simulate the effect of MAX-DOAS observation information on aerosol profile retrievals. The input aerosol profiles were configured with vertical resolutions of 100 m and 200 m. Initially, the O_4 dSCDs were simulated using the RTM based on the "assumed input aerosol profiles" with different vertical resolutions, while considering various settings of EAs. Subsequently, the simulated O_4 dSCDs were used as inputs in the PriAM algorithm for retrieving the aerosol profile.

The aerosol profiles retrieved from various scenarios are depicted in Figure 4. In general, the retrieved aerosol profile from all scenarios at an AOD of 0.3 closely approximated the input aerosol profile. However, the retrieved aerosol profiles from a 100 m resolution as input exhibited slight differences compared to those from 200 m as input. Moreover, the retrieved scale height of the aerosol profile in a Boltzmann shape from 100 m resolution as input was slightly higher than that from 200 m as the input profile. The aerosol profiles retrieved using a 200 m resolution as input displayed an underestimation below 1.2 km, particularly for aerosol extinction below 500 m. In contrast, the aerosol profile results obtained using a 100 m resolution as input exhibited an overestimated trend below 1.2 km, primarily for aerosol extinction between 400 m and 1.0 km. The concentration value at the peak of the retrieved Gaussian shape profile from the 100 m resolution input was higher than the input value. Conversely, the concentration value from the 200 m resolution input was underestimated. It is worth noting that, for the retrieved exponential aerosol profile, the value in the first layer was underestimated in both resolutions used as input. In general, the aerosol profile below 2.0 km retrieved from a 100 m resolution input exhibited a slight overestimation compared to the input value. Conversely, the aerosol profile obtained from a 200 m resolution input showed a slight underestimation. However, the relative deviation in the 100 m resolution input was marginally higher than that in a 200 m resolution input. It is worth noting that the vertical resolution of the input profile does not have a significant effect on the shape of the retrieved profile. The retrieved aerosol profiles differed slightly among different EAs, particularly from a 100 m resolution input. When the aerosol profile with a 200 m resolution was used as the input, the difference between the retrieved results and the input value for EAs using Setting IV was minimal. However, the smallest deviation for the 100 m resolution input was observed in EAs utilizing Setting III. This is mainly due to the fact that low elevation angles contain more information with respect to gas profile shapes than high elevation angles [41].

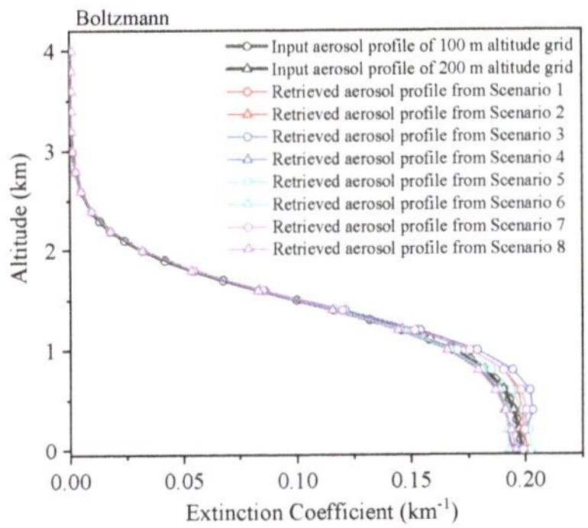

(a)

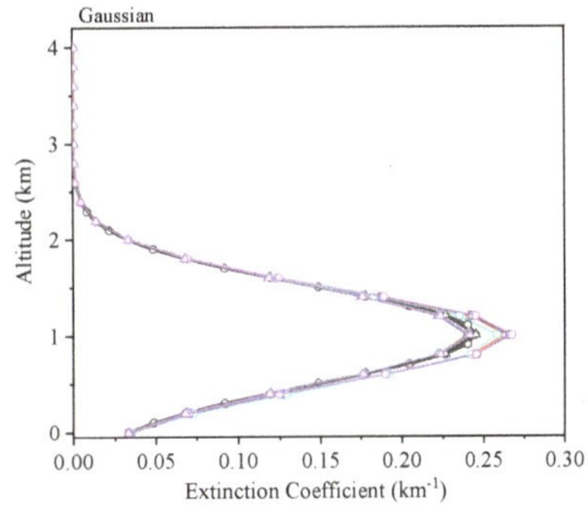

(b)

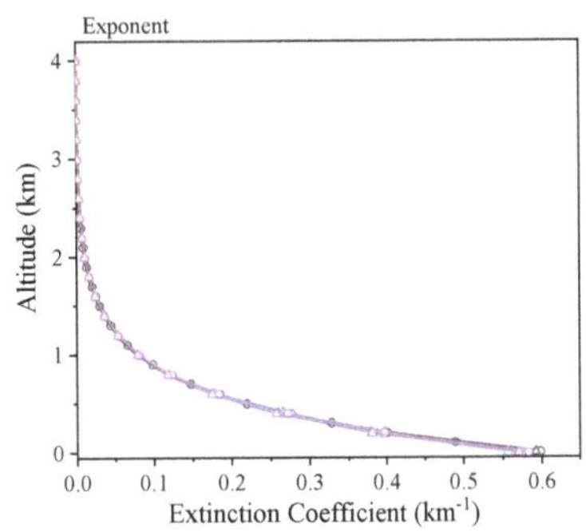

(c)

Figure 4. The comparison of aerosol profiles retrieved by PriAM with the corresponding input aerosol profiles in (**a**) Boltzmann shape, (**b**) Gaussian shape, and (**c**) exponential shape under different scenarios.

The colors and symbols refer to the input and retrieved aerosol profile under the scenarios shown in the top right corner of (**a**). (1) Scenario 1: retrieved from a 100 m resolution input based on Setting I with 10 EAs; (2) Scenario 2: retrieved from a 200 m resolution input based on Setting I with 10 EAs; (3) Scenario 3: retrieved from a 100 m resolution input based on Setting II with 6 EAs; (4) Scenario 4: retrieved from a 200 m resolution input based on Setting II with 6 EAs; (5) Scenario 5: retrieved from a 100 m resolution input based on Setting III with 6 EAs; (6) Scenario 6: retrieved from a 200 m resolution input based on Setting III with 6 EAs; (7) Scenario 7: retrieved from a 100 m resolution input based on Setting IV with 6 EAs; (8) Scenario 8: retrieved from a 200 m resolution input based on Setting IV with 6 EAs. The EAs settings are detailed in Table S1. Circle and triangle symbols represent the retrieved aerosol profile from 100 m and 200 m resolution inputs, respectively. Colors indicate the results for different EAs settings.

The discrepancy between the simulated O_4 dSCDs by the PriAM algorithm and the input O_4 dSCDs (simulated using the RTM from the input aerosol profile) can serve as a metric to evaluate how PriAM effectively utilizes the measured aerosol information (Figure 5). Table 3 presents the correlation coefficient (R), slopes, and intercepts of the linear regressions for the eight scenarios. The O_4 dSCD simulated by PriAM closely approximated the input value, as evidenced by the fitting slope and correlation coefficient both approaching 1.00. The O_4 dSCDs calculated from the aerosol profiles of two resolutions based on the RTM were consistent, suggesting that SCITRIAN is capable of extracting similar information from aerosol profiles with resolutions of 100 m and 200 m.

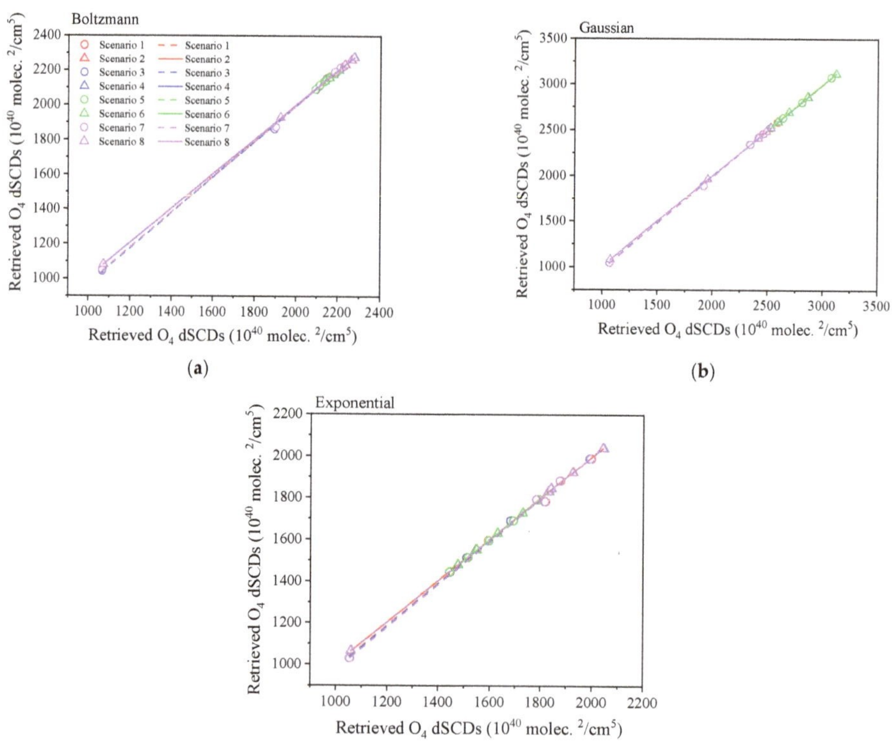

(**c**)

Figure 5. Correlation plots between the retrieved O_4 dSCDs by PriAM and the input O_4 dSCDs in (**a**) Boltzmann shape, (**b**) Gaussian shape, and (**c**) exponential shape. The caption of Figure 4 applies. Dotted lines and solid lines represent the results obtained from 100 m and 200 m resolution inputs, respectively.

Table 3. The correlation coefficient (R), slopes, and intercepts of the linear regressions for O_4 dSCDs simulated by the PriAM algorithm and the input O_4 dSCDs in the eight scenarios. The scenarios correspond to Figure 4.

Shape	Parameters	Scenarios							
		1	2	3	4	5	6	7	8
Boltzmann	Slope	1.02	0.9	1.02	0.99	0.97	1.00	1.02	0.99
	Intercept	−46.50	16.02	−54.86	13.69	60.41	4.91	−45.07	14.14
	R	1.000	1.000	1.000	1.000	0.999	1.000	1.000	1.000
Gaussian	Slope	0.99	1.00	1.02	0.99	1.00	0.99	1.02	0.99
	Intercept	21.16	12.99	−41.43	13.88	−8.61	34.46	−50.73	13.43
	R	1.000	1.000	1.000	1.000	1.000	1.000	1.000	1.000
Exponential	Slope	1.01	0.99	1.01	0.99	0.99	0.98	1.02	0.99
	Intercept	−27.62	13.44	−31.38	15.69	16.28	28.91	−50.72	14.33
	R	0.999	1.000	0.999	1.000	1.000	1.000	0.999	1.000

In MAX-DOAS profile inversion, the retrieval error encompasses the smoothing error (S_s), measurement error (S_m), and residual error (S_r). The total error (S_t) of the inversion is calculated using the error transfer formula: $S_t = \sqrt{S_S^2 + S_m^2 + S_r^2}$. The retrieval can be regarded as an estimate of the true state that has been smoothed by the averaging kernel. Hence, the error contribution due to smoothing is termed the smoothing error. The error caused by the measurement of trace gas is referred to as the measurement error. The error from the value function is labeled the residual error. The retrieval errors for the aerosol profiles at different resolutions and EA settings are depicted in Figure 6. The S_m, S_s, and S_t remain largely constant for both resolutions, whereas there are slight variations among the four EA settings. The S_r changed due to the variations in resolutions of the input aerosol profiles and the number of EAs. The S_r of a 100 m resolution input was significantly higher than that of a 200 m input. This is why the retrieved aerosol profile from a 200 m resolution input was closer to the true value compared to that from a 100 m resolution input. The increase in S_r suggested that the value function could not be effectively minimized during the nonlinear iterative calculation. This demonstrates that the nonlinear solving process needs optimization to improve the accuracy of the inversion algorithm under high-resolution aerosol profiles. The lowest S_r value from the 100 m resolution input was recorded for Setting III with 6 EAs, whereas, for the 200 m resolution input, it was observed for Setting IV with 6 EAs. This aligns precisely with the retrieved results of the aerosol profile.

The gain in information on the atmospheric state can be quantified through the OEM retrieval, which is represented by the AVK matrix. Therefore, the information and sensitivity obtained by PriAM inversion were assessed under different vertical resolutions of the aerosol profile as input. In Figure 7, the retrieved envelopes of the AVK are presented for three different shapes. The corresponding values of H_m and DOFS are provided in Table 4. The retrieved envelopes of the AVK, DOFS, and H_m from altitude resolutions of 100 m and 200 m as input were found to be in agreement. This suggests that the information obtained from both resolution profiles as input used in the PriAM retrieval process is consistent. Additionally, a decrease in the number of EAs resulted in a reduction in the retrieved envelopes of the AVK and DOFS, indicating that the maximum information gain from the measurements was achieved at 10 EAs. Therefore, despite the minimum deviation in the aerosol profile found at 6 EA retrieval, the EA setting with 10 EAs is still recommended, considering the small deviation and the uncertainty in the profile.

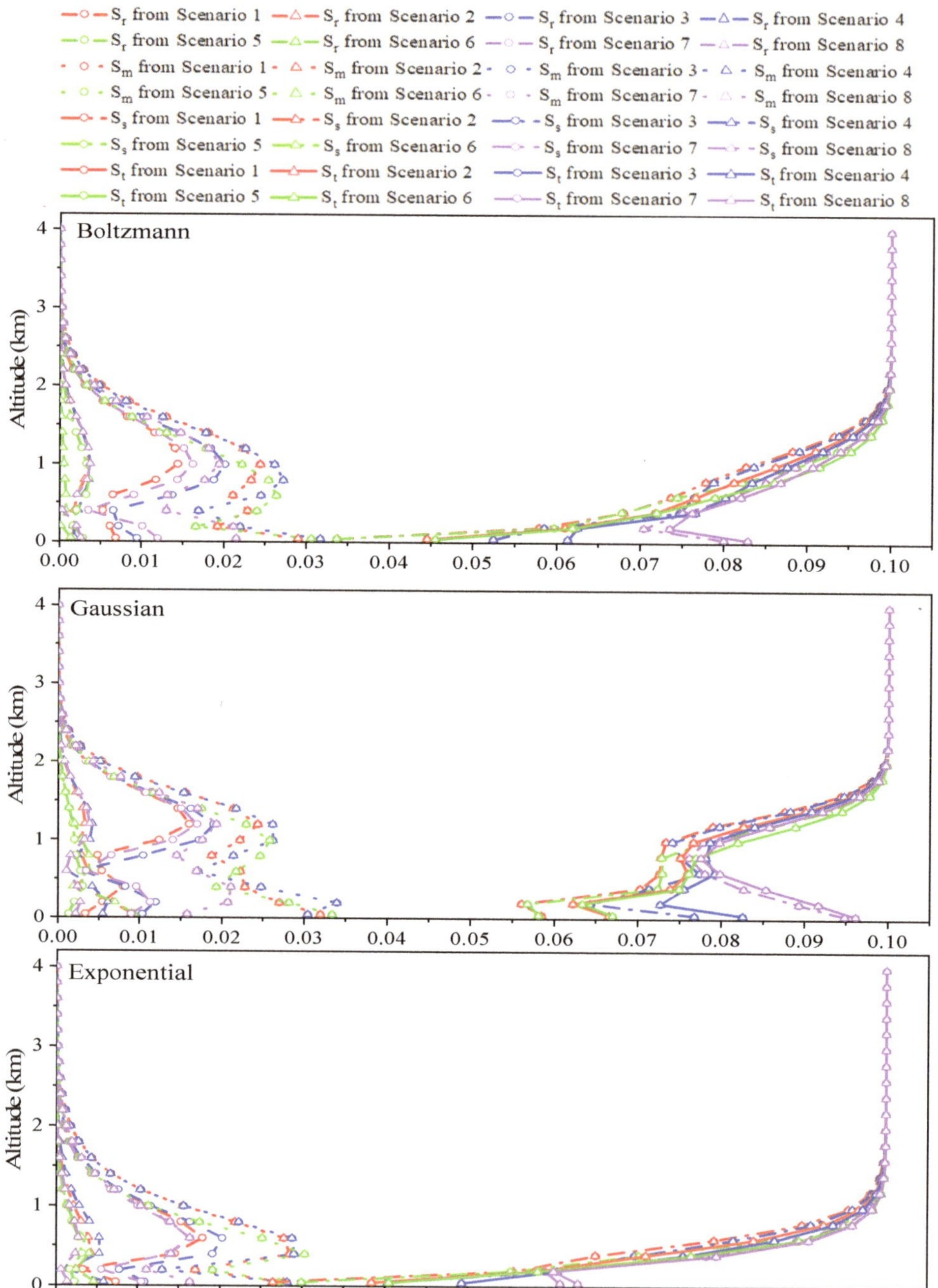

Figure 6. The retrieval error of the aerosol profile by PriAM under input profiles of two resolutions and four settings of EAs. The caption of Figure 4 applies.

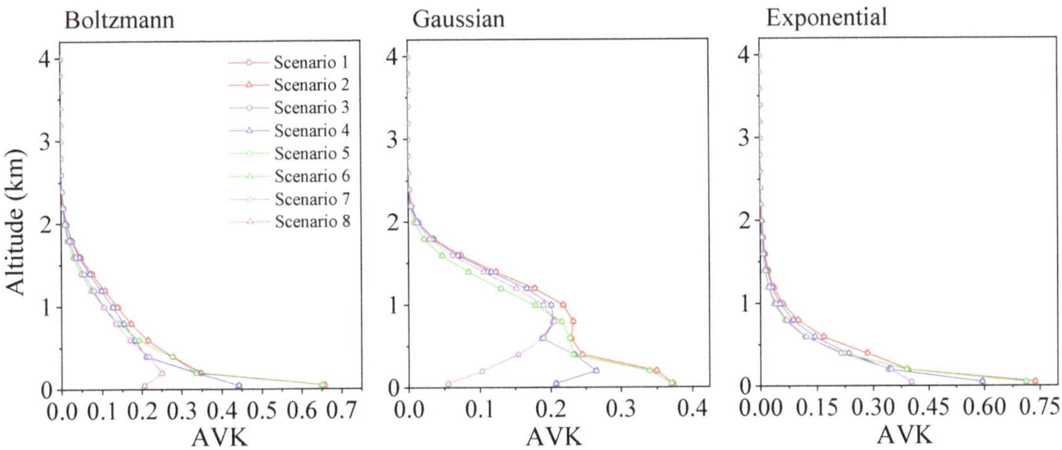

Figure 7. The retrieved envelopes of the AVK, H_m, and DOFS by PriAM under different scenarios in the Boltzmann shape (**Left column**), Gaussian shape (**Middle column**), and exponential shape (**Right column**). The caption of Figure 4 applies.

Table 4. The retrieved DOFS and H_m for three aerosol shapes from the eight scenarios. The scenarios correspond to Figure 4.

Shape	Parameters	Scenarios							
		1	2	3	4	5	6	7	8
Boltzmann	DOFS	1.95	1.95	1.58	1.58	1.73	1.73	1.10	1.10
	H_m	1.4	1.4	1.6	1.6	1.2	1.2	1.6	1.6
Gaussian	DOFS	1.95	1.95	1.55	1.55	1.73	1.73	1.06	1.06
	H_m	1.6	1.6	1.8	1.8	1.6	1.6	1.8	1.8
Exponential	DOFS	1.66	1.66	1.38	1.38	1.46	1.46	1.05	1.05
	H_m	0.8	0.8	0.8	0.8	0.6	0.6	0.8	0.8

3.1.2. The Aerosol Profile Retrieval in MAX-DOAS Observations

In real MAX-DOAS measurements, the determination of gas concentration in the atmosphere relies on observations at various elevation angles. To ensure the accuracy of the simulation, different EAs were matched with corresponding observation accuracies. The impact of EAs on the retrieval profile's accuracy was investigated, with Hefei's MAX-DOAS observations using 10 EAs (1, 2, 3, 4, 5, 6, 8, 15, 30, and 90°) and a comparative study with 6 EAs in the simulation. The aerosol profile for Hefei on 4 September 2020 was retrieved using the PriAM algorithm presenting results at 8:00 LT (local time) in Figure 8. The input profile at a 200 m resolution with AOD of 0.3, as detailed in Table 2, served as the universal *a priori* aerosol profile in the PriAM retrieval. Results indicated that aerosol extinctions below 1.5 km retrieved by 6 EAs and 10 EAs differed significantly from the Lidar aerosol profile, with 6 EAs showing higher values than 10 EAs. Among the EA settings, the aerosol profile retrieved from Setting IV with 6 EAs closely matched the Lidar profile, consistent with the simulation results in Section 3.1.1. Overall, the aerosol results from MAX-DOAS at 8:00 LT tended to be underestimated compared to Lidar observations.

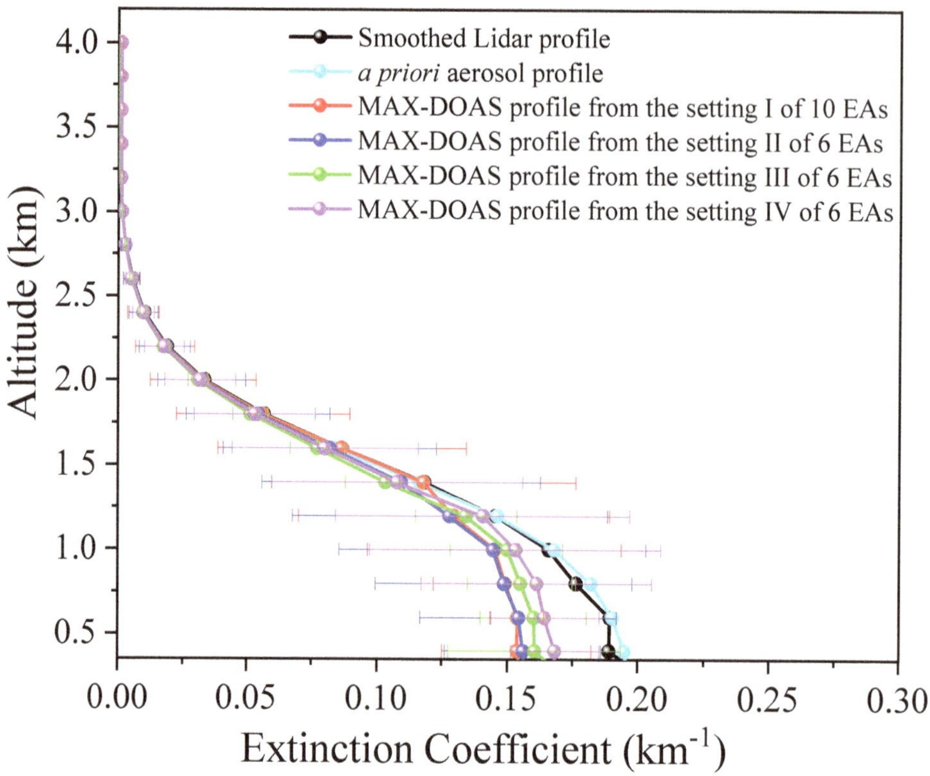

Figure 8. The comparison of aerosol profiles measured by Lidar (smoothed according to Equation (3)) and retrieved using PriAM under different settings of EAs at 8:00 LT on 4 September 2020. The colors refer to aerosol profiles under different scenarios shown on the top right. Error bars on the retrieved aerosol profile are retrieval errors. EA settings according to Table S1.

The retrieved envelopes of the AVK, H_m, and DOFS at 8:00 LT under different EA settings are shown in Figure 9. The DOFS obtained from the inversions using 10 EAs were higher than that from inversions using 6 EAs, indicating that the information obtained from 10 EAs was greater. The retrieved H_m from Setting III with 6 EAs was the lowest compared to other EA settings, suggesting that the retrieval result from settings with only low EAs is sensitive to pollution at low altitudes. AODs retrieved by MAX-DOAS in four scenarios were compared with Lidar-measured AODs (Figure 10), with Lidar AODs being the result of AVK smoothing based on the preprocessed aerosol profile. Retrieved AODs at 8:00 LT from four scenarios were lower than Lidar values. AODs retrieved from 6 EAs were closer to Lidar AODs than those from 10 EAs, although the differences between AODs from 10 EAs and 6 EAs were negligible. In general, retrieved AODs under different EAs were close, demonstrating consistent results. The retrieved DOFS from different EAs indicated that employing a greater number of EAs can obtain more information from observations, particularly from a number below a 5° elevation angle, mainly because the lowest elevations contain most information with respect to gas profile shapes [41].

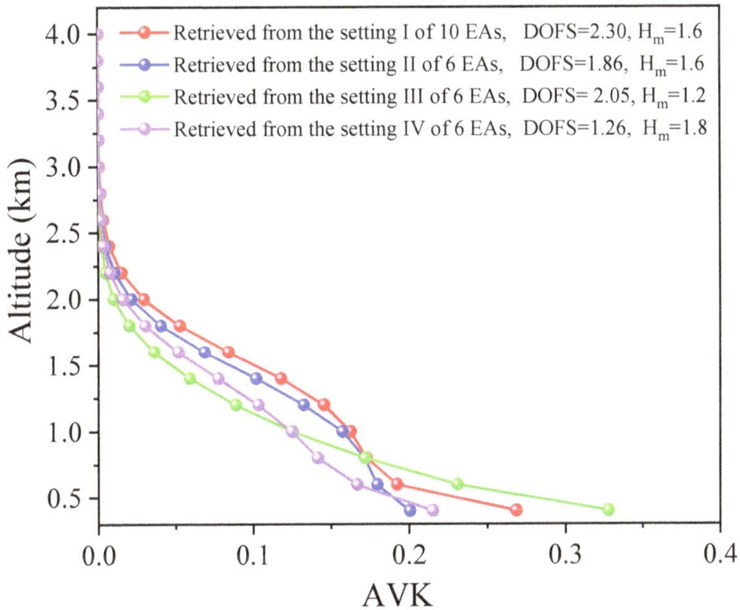

Figure 9. The retrieved envelopes of the AVK, H_m, and DOFS using PriAM under different settings of EAs at 8:00 LT on 4 September 2020. EA settings according to Table S1.

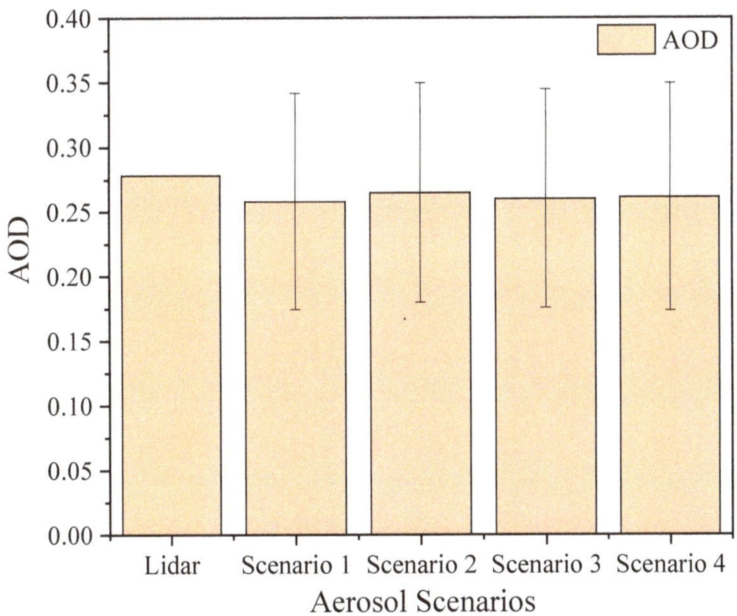

Figure 10. The comparison of the AODs measured by Lidar and retrieved by MAX−DOAS from different scenarios at 8:00 LT on 4 September 2020. (1) Scenario 1: retrieved by MAX−DOAS based on Setting I with 10 EAs; (2) Scenario 2: retrieved by MAX−DOAS based on Setting II with 6 EAs; (3) Scenario 3: retrieved by MAX−DOAS based on Setting III with 6 EAs; (4) Scenario 4: retrieved by MAX−DOAS based on Setting IV with 6 EAs. EA settings according to Table S1.

3.1.3. Lidar Results as the Input Aerosol Profile

In the previous investigation, discrepancies were observed between the aerosol profiles retrieved from MAX-DOAS observations and the actual Lidar aerosol profiles. This disparity could be attributed to insufficient O_4 dSCDs information gathered by MAX-DOAS, leading to inconsistencies between MAX-DOAS inversion and Lidar observations. To explore this, Lidar aerosol profiles were utilized as input profiles in the RTM to calculate O_4 dSCDs. Subsequently, the aerosol profiles were retrieved using the calculated O_4 dSCDs by PriAM and compared with input Lidar profiles. Lidar aerosol profiles at 8:00 LT on 4 September 2020, at different vertical resolutions (i.e., 100 m and 200 m), were selected for simulation. Figure 11 illustrates a comparison between the input and retrieved aerosol profiles from Lidar profiles used as input profiles and from MAX-DOAS measurements at 10 EAs. The aerosol profiles retrieved from Lidar profiles at 100 m and 200 m vertical resolution inputs closely resembled the Lidar aerosol profiles, validating the consistency with the simulation results in Section 3.1.1. However, the aerosol profiles retrieved by MAX-DOAS measurements differed from those with Lidar profiles as input, indicating underestimation. This disparity might be due to (1) the influence of the AVK smoothing [19], (2) the different location and viewing geometry of MAX-DOAS and Lidar [18,42,43], and (3) uncertainties associated with the Lidar profiles. The original Lidar aerosol profiles at 8:00 LT and the retrieved aerosol profiles are displayed in Figure 12. The aerosol profiles retrieved by MAX-DOAS measurements demonstrated consistency with the original Lidar profile, albeit exhibiting significant differences in concentration, particularly above 1.0 km. These were close to the smoothed Lidar profile at altitudes exceeding 1.4 km. Moreover, the difference between the MAX-DOAS inversion results and the Lidar results can largely be explained by the smoothing effects. Figure 13 illustrates the correlation between the retrieved and input aerosol profiles. The correlation coefficient and slope between the Lidar aerosol profile and the retrieved aerosol profile from Lidar at 100 m and 200 m altitude resolution inputs were close to 1.0. The correlation coefficient between the MAX-DOAS measurements and the Lidar results surpassed 0.99, while the slope exhibited noticeable deviation.

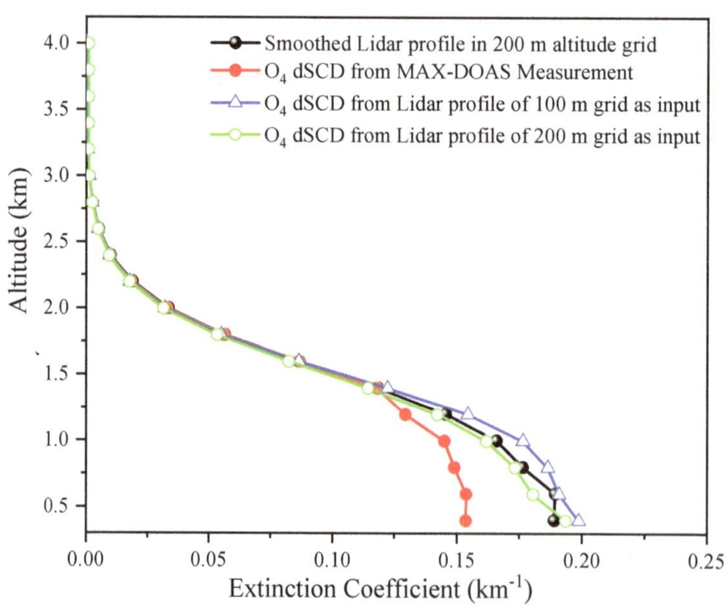

Figure 11. The comparison of aerosol profiles measured by Lidar (smoothed according to Equation (3)) and retrieved from different O_4 dSCDs scenarios at 8:00 LT on 4 September 2020. The colors refer to the O_4 dSCDs scenarios shown in the top right corner.

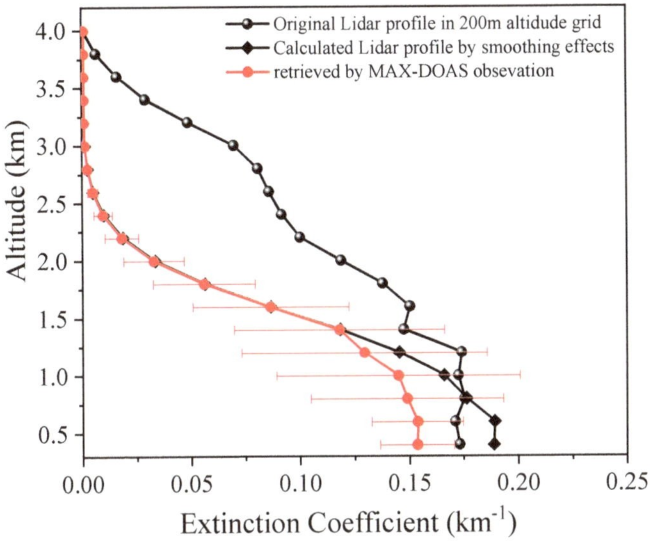

Figure 12. Aerosol profiles derived from the original Lidar observation, the smoothed Lidar profile, and MAX-DOAS measurements at 8:00 LT on 4 September 2020. The colors refer to the aerosol profiles shown in the top right corner. Error bars on the retrieved aerosol profile are retrieval errors.

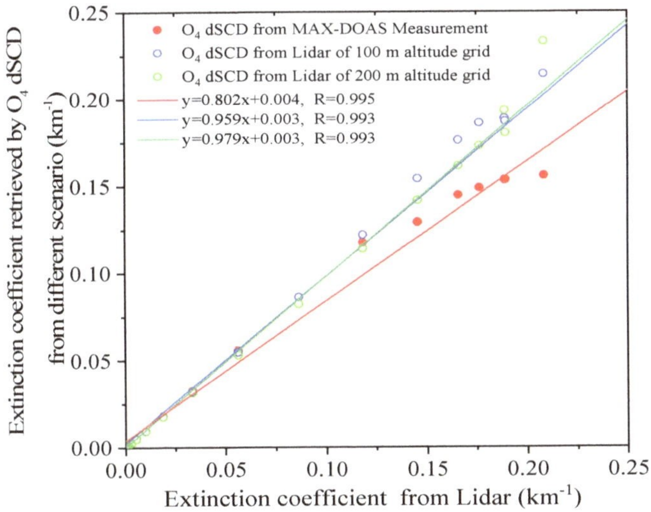

Figure 13. Correlation plots between the retrieved and the input aerosol extinction at 8:00 LT on 4 September 2020. The colors and symbol refer to the retrieved aerosol extinction from different O_4 dSCDs scenarios shown in the top left.

To investigate whether the difference in aerosol profile retrievals was caused by the difference in O_4 dSCDs, the O_4 dSCDs simulated from Lidar profiles at two resolutions and those observed by MAX-DOAS were compared (Figure 14). The O_4 dSCDs calculated by the actual Lidar profile were lower than those from MAX-DOAS observations. The correlation coefficient of the O_4 dSCDs between the MAX-DOAS observations and Lidar simulation was higher than 0.98, with a slope greater than 0.83. The retrieval results based on the Lidar profiles at different resolutions were consistent with the simulation results in

Section 3.1.1, further validating the conclusions of Section 3.1.1. That is, the inversion result from 100 m input overestimates the true value, while that from 200 m input is the opposite. The retrieved AODs from the three scenarios were compared with the smooth Lidar AOD (Figure 15). The AOD retrieved from MAX-DOAS observations was significantly underestimated compared with the AOD from the Lidar. The AOD retrieved by O_4 dSCD based on the actual Lidar profile simulation was also underestimated compared to the AOD from the Lidar, but the value was close. This shows that the accuracy of the aerosol information obtained by MAX-DOAS observations needs improvement, but the MAX-DOAS inversion algorithm can extract aerosol information from O_4 dSCDs effectively.

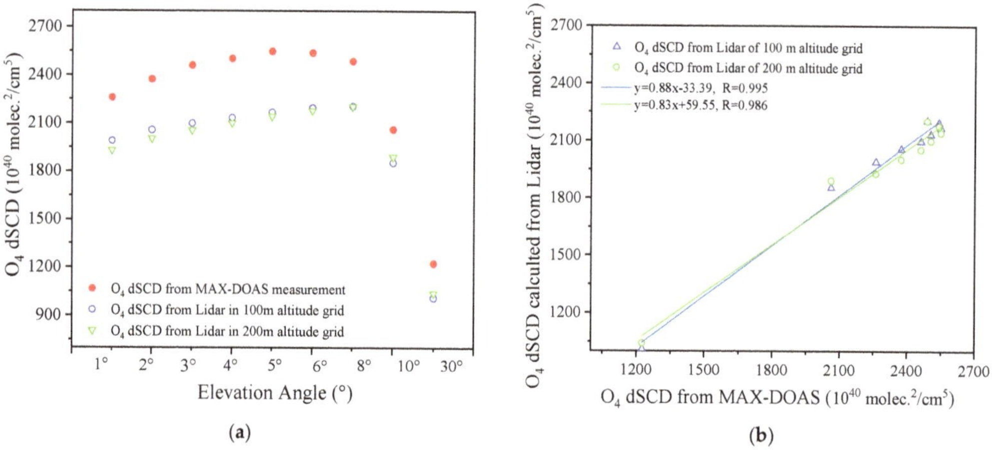

Figure 14. (**a**) The comparison and (**b**) correlation plots between the O_4 dSCDs obtained from the MAX-DOAS measurements and the Lidar simulation at 8:00 LT on 4 September 2020.

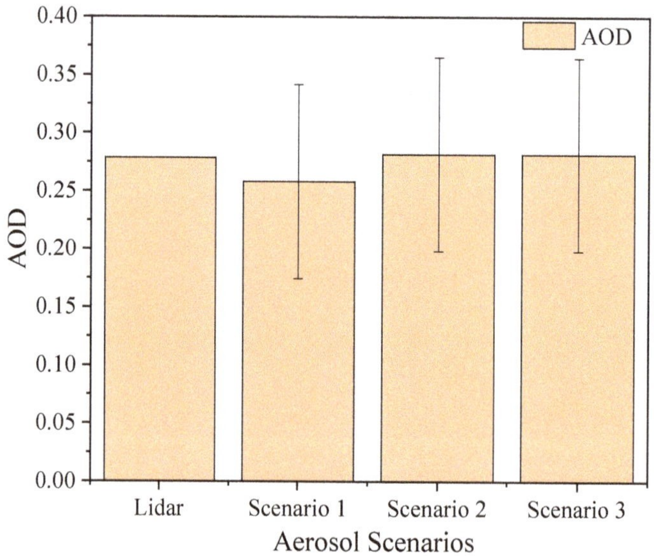

Figure 15. The comparison of AODs measured by Lidar and retrieved from different O_4 dSCDs scenarios at 8:00 LT on 4 September 2020. (1) Scenario 1: O_4 dSCDs obtained by MAX-DOAS measurement; (2) Scenario 2: O_4 dSCDs calculated using Lidar profile at a 100 m vertical resolution input; (3) Scenario 3: O_4 dSCDs calculated using Lidar profile at a 200 m vertical resolution input.

3.2. NO_2 Results

3.2.1. Effect of Vertical Resolution and Elevation Angles on NO_2 Profile Retrieval

The three selected NO_2 profiles were utilized for trace gas simulation under an AOD of 0.3. The NO_2 VCD was set to 1.0×10^{16} molec. cm^{-2}. Vertical resolutions of 100 m and 200 m were applied to the NO_2 profile. Additionally, the effect of EAs on NO_2 profile retrieval by MAX-DOAS was investigated. During the calculation of NO_2 dSCDs based on the RTM, the input aerosol profile, following a Boltzmann shape, with a vertical resolution of 200 m, served as the aerosol *a priori* profile. In comparison to aerosol inversion results, the retrieved NO_2 profiles based on the NO_2 profile with two vertical resolutions as input exhibited no significant overestimation or underestimation (Figure 16). The influence of vertical resolutions on NO_2 profile inversion appeared to be notably less pronounced than that observed in aerosol profile inversion. Similarly, and in line with the findings of aerosol inversion, the different EA settings had little impact on NO_2 profile retrievals. In summary, the impact of the input NO_2 profile's vertical resolution and EA settings on NO_2 profile retrievals can be considered negligible.

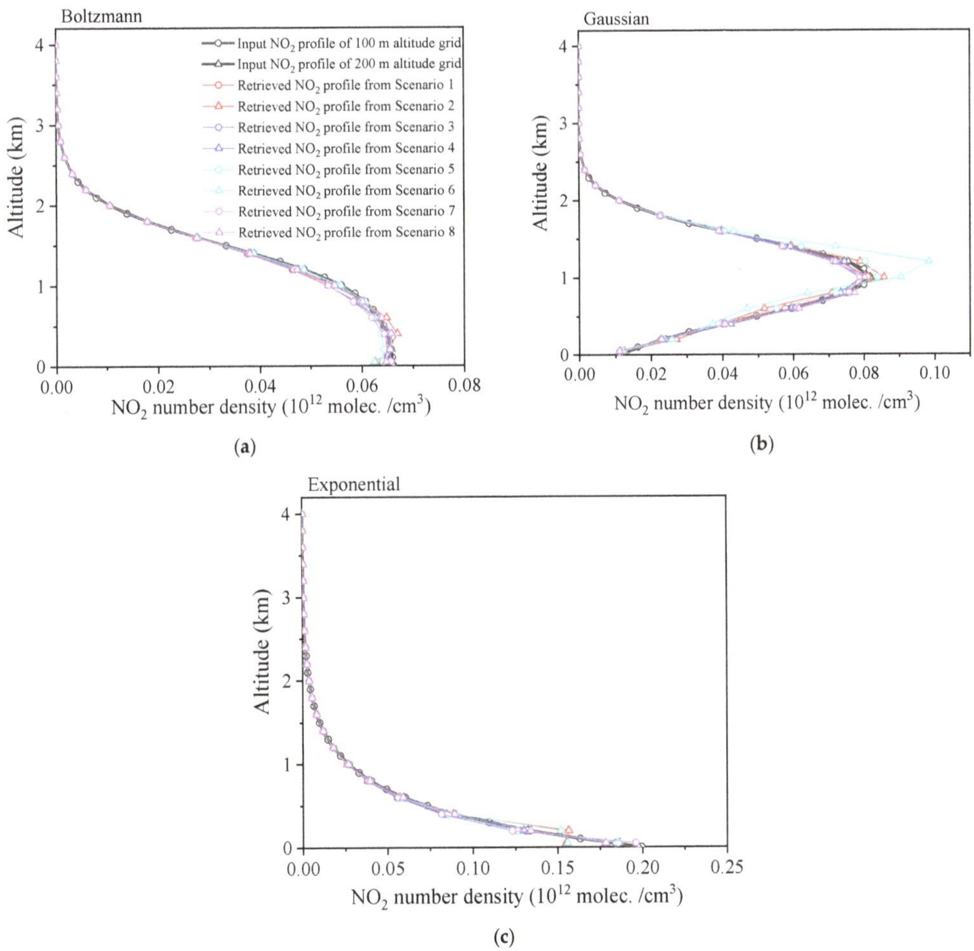

Figure 16. The comparison of the retrieved NO_2 profiles by PriAM and the corresponding input profiles in (**a**) Boltzmann shape, (**b**) Gaussian shape, and (**c**) exponential shape. The caption of Figure 4 applies.

The retrieved envelopes of the AVK, H_m, and DOFS under different scenarios were further compared (Figure 17 and Table 5). Under the same EA conditions, the retrieved envelopes of the AVK, H_m, and DOFS using input profiles with different altitude resolutions were essentially consistent. Similarly, the retrieved envelopes of the AVK, H_m, and DOFS at the same VCD scenarios exhibited consistency, indicating that AOD had minimal influence on NO_2 profile retrieval. The retrieved DOFS decreased as the number of EAs decreased, suggesting a reduction in the NO_2 information obtained from the observations as the number of EAs decreased.

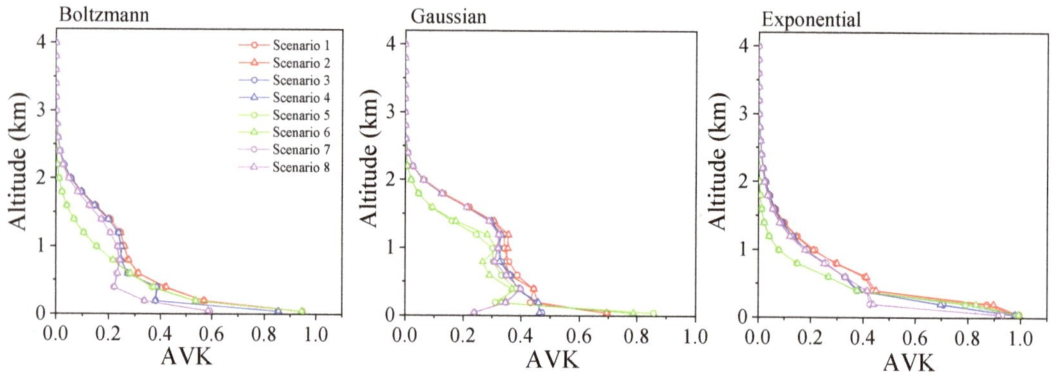

Figure 17. The retrieved envelopes of the AVK, H_m, and DOFS by PriAM in NO_2 profile retrievals in Boltzmann shape (**Left column**), Gaussian shape (**Middle column**), and exponential shape (**Right column**). The caption of Figure 4 applies.

Table 5. The retrieved DOFS and H_m for three NO_2 shapes from the eight scenarios. The scenarios correspond to Figure 4.

Shape	Parameters	Scenarios							
		1	2	3	4	5	6	7	8
Boltzmann	DOFS	3.16	3.17	2.74	2.74	2.52	2.52	2.11	2.11
	H_m	1.8	1.8	1.8	1.8	1.2	1.2	1.8	1.8
Gaussian	DOFS	2.91	2.91	2.45	2.46	2.17	2.16	1.97	1.97
	H_m	1.8	1.8	2.0	2.0	1.6	1.6	1.6	1.6
Exponential	DOFS	2.89	2.91	2.52	2.52	2.37	2.41	2.03	2.03
	H_m	1.4	1.2	1.2	1.2	0.8	0.8	1.2	1.2

3.2.2. NO_2 Profile Retrieval in MAX-DOAS Observations

The study focused on evaluating the impact of EAs on NO_2 profile retrieval through MAX-DOAS observations. The *a priori* aerosol profile utilized in the retrieval process was derived from a total of 10 EAs. For NO_2 profile retrieval, the MAX-DOAS data collected at 8:00 LT on 4 September 2020 were employed. Figure 18 illustrates the influence of different EA settings on NO_2 profile retrieval, revealing slight variations, primarily within the altitude range of 400 m to 1.6 km. Specifically, the NO_2 number density retrieved with 10 EAs was marginally higher than that obtained with 6 EAs. Comparisons were made between NO_2 profiles retrieved from the *a priori* aerosol profiles corresponding to different EAs settings in the NO_2 inversion process (Figure 18). Minor differences were observed, particularly below an altitude of 2.0 km, where profiles retrieved with 6 EAs exhibited slightly higher values compared to those with 10 EAs. Nevertheless, these differences were deemed insignificant. Additionally, the NO_2 profiles retrieved from NO_2 and aerosols of the corresponding EAs were higher than those from aerosol profiles with 10 EAs.

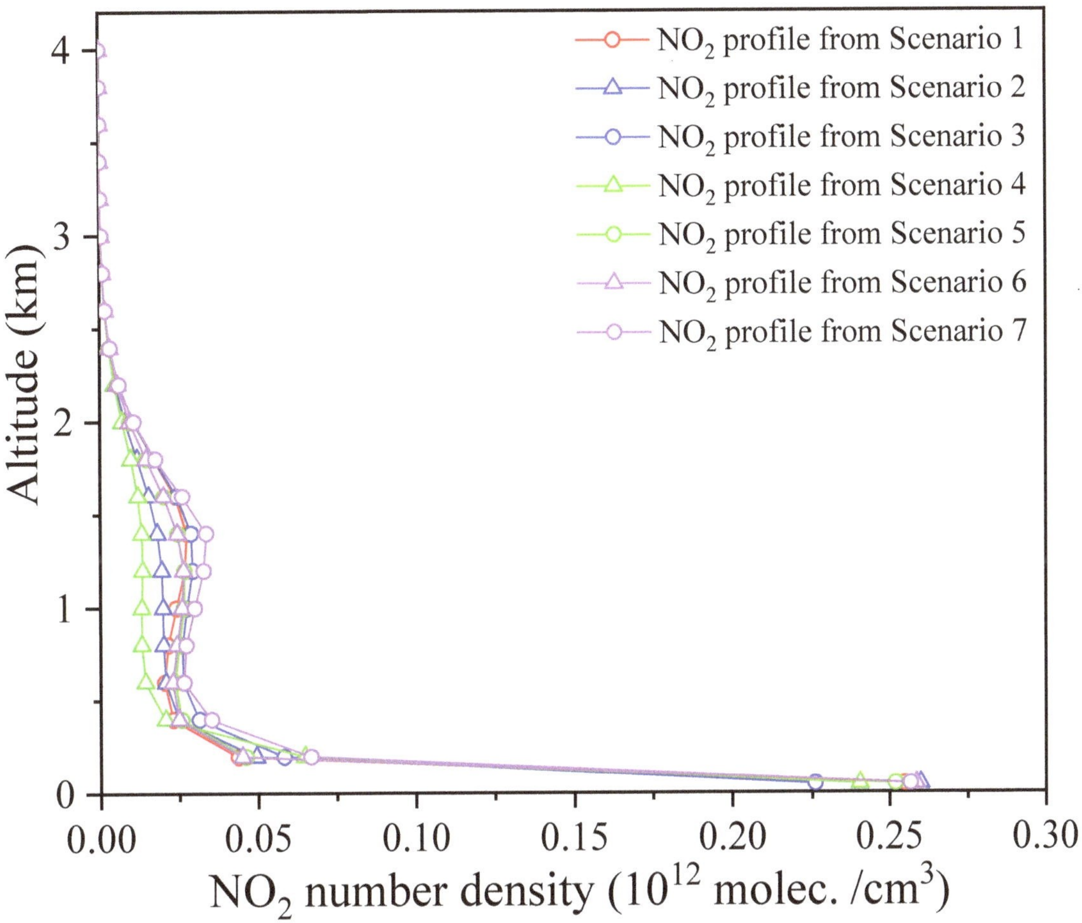

Figure 18. The comparison of the NO_2 profiles retrieved by PriAM under different scenarios of the *a priori* aerosol profile at 8:00 LT on 4 September 2020. Colors and symbols refer to the retrieved NO_2 profiles in different scenarios shown in the top right. (1) Scenario 1: retrieved from both NO_2 and aerosols of Setting I with 10 EAs; (2) Scenario 2: retrieved from NO_2 of Setting II with 6 EAs and aerosol of 10 EAs; (3) Scenario 3: retrieved from both NO_2 and aerosols of Setting II with 6 EAs; (4) Scenario 4: retrieved from NO_2 of Setting III with 6 EAs and aerosol of 10 EAs; (5) Scenario 5: retrieved from both NO_2 and aerosols of Setting III with 6 EAs; (6) Scenario 6: retrieved from NO_2 of Setting IV with 6 EAs and aerosol of 10 EAs; (7) Scenario 7: retrieved from both NO_2 and aerosols of Setting IV with 6 EAs. EA settings according to Table S1.

Analyzing the retrieved envelopes of the AVK, H_m, and DOFS under various scenarios revealed consistent DOFS values from NO_2 with the same EA settings (Figure 19). This indicates that the PriAM algorithm's ability to acquire gas information remains unaffected by the aerosol state. NO_2 VCDs retrieved from seven different scenarios are presented in Figure 20, showing a relative difference of less than 3%. Hence, the impact of EAs on NO_2 VCD retrievals can be considered negligible.

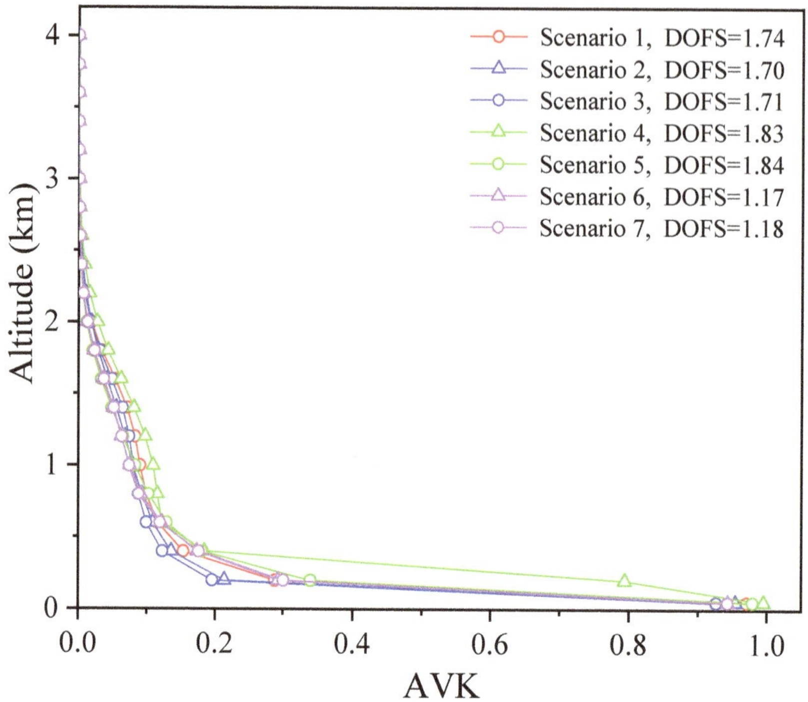

Figure 19. The retrieved envelopes of the AVK and DOFS corresponding to Figure 18. The retrieved DOFS is shown in the top right. The caption of Figure 18 applies.

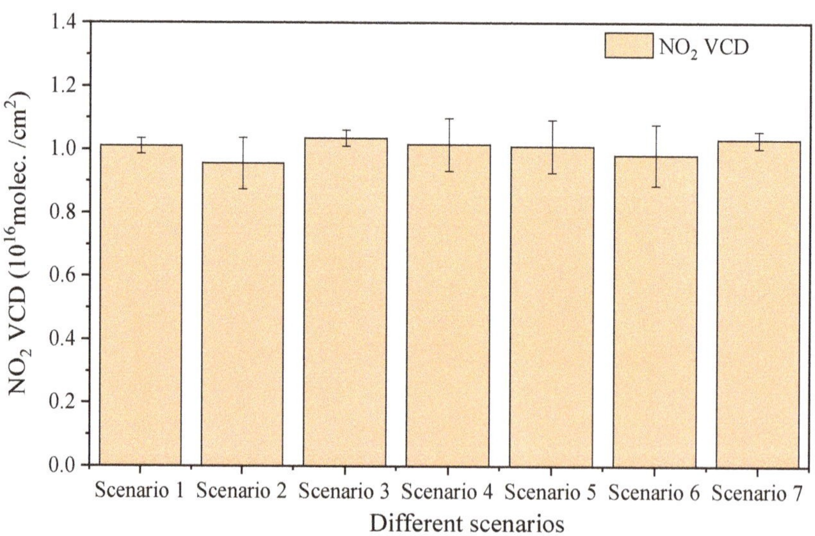

Figure 20. The comparison of NO_2 VCDs retrieved from different scenarios at 8:00 LT on 4 September 2020. The scenarios correspond to Figure 18.

To assess the accuracy of NO_2 profiles retrieved by PriAM, the proximity of simulated NO_2 dSCDs to those measured by MAX-DOAS was crucial. Figure 21 demonstrates the comparison of NO_2 dSCDs simulated by PriAM under seven different scenarios with those measured by MAX-DOAS. The correlation coefficients exceeded 0.97, with a slope of 1 ± 0.15, indicating that the inversion algorithms effectively utilized information from MAX-DOAS observations.

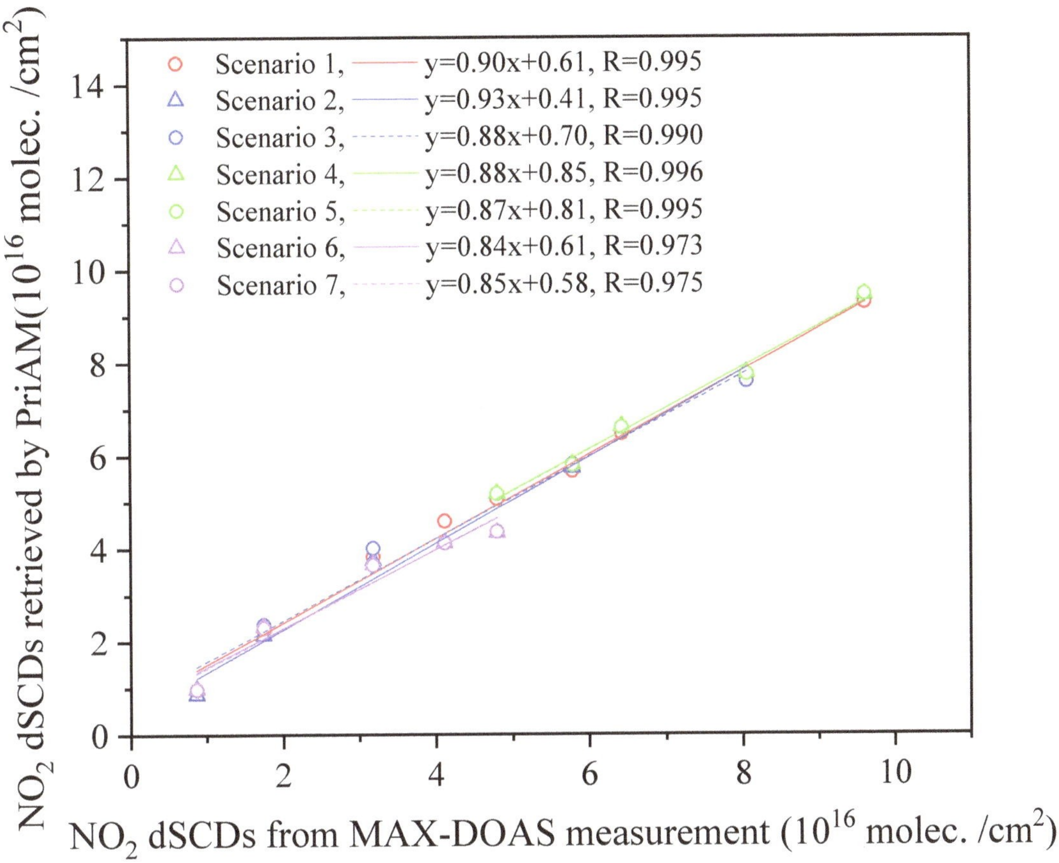

Figure 21. Correlation plots between the measured and the simulated NO_2 dSCDs at 8:00 LT on 4 September 2020. Colors and symbols refer to the NO_2 dSCDs from different scenarios shown in the top left. The scenarios correspond to Figure 18.

3.2.3. Lidar Profiles as Aerosol Profiles in NO_2 Profile Retrievals

To verify the impact of vertical resolutions in aerosol profiles on NO_2 profile retrievals, Lidar profiles with vertical resolutions of 50 m and 200 m were employed as aerosol information in the PriAM algorithm for NO_2 profiles retrieval. The same MAX-DOAS data used in Section 3.2.2 were utilized for NO_2 profile retrieval. NO_2 profiles were retrieved by PriAM based on aerosol profiles obtained from both MAX-DOAS retrievals and Lidar measurements (Figure 22). The results indicated that the effect of vertical resolutions for the Lidar profiles as aerosol information on NO_2 profile retrieval was insignificant. This finding aligns with previous reports [22], confirming that the retrieved NO_2 profiles were consistent based on the input aerosol profile and the retrieved aerosol profile.

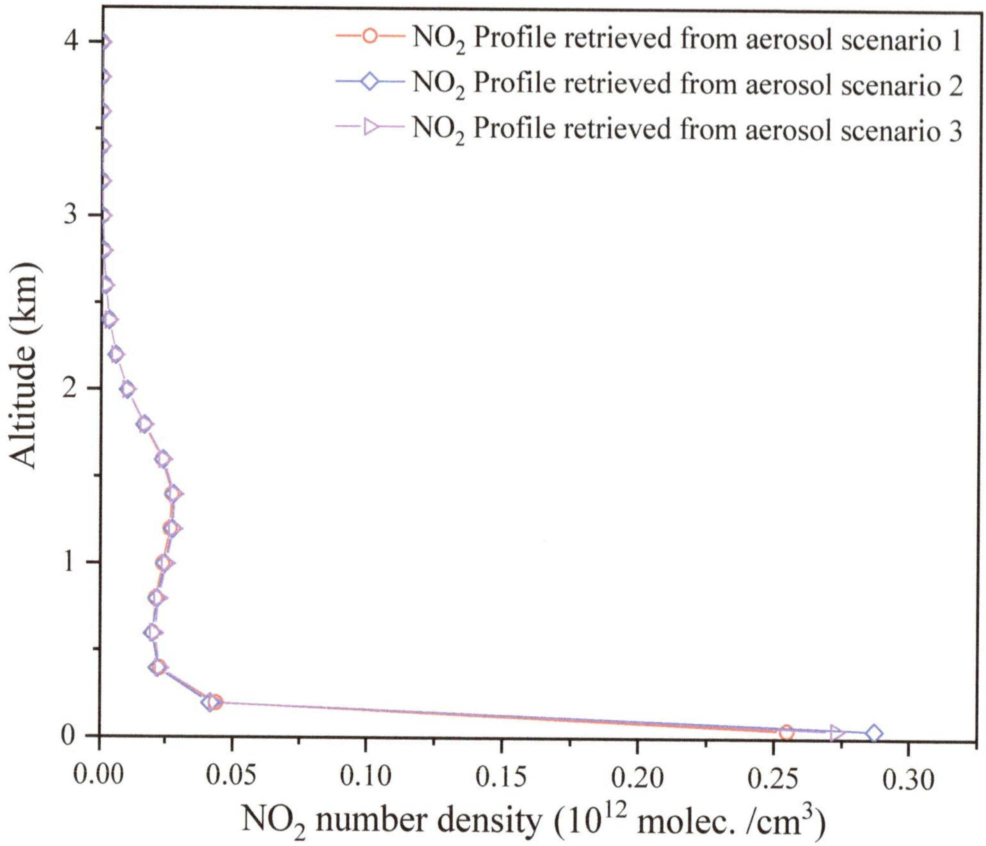

Figure 22. The comparison of NO_2 profiles retrieved by PriAM under different aerosol profile scenarios at 8:00 LT on 4 September 2020. The colors refer to the retrieved NO_2 profile in different scenarios shown in the top right. (1) Aerosol scenario 1: retrieved aerosol profile from MAX-DOAS observations; (2) Aerosol scenario 2: Smoothed Lidar aerosol profile with a 200 m vertical resolution; (3) Aerosol scenario 3: Smoothed Lidar aerosol profile with a 50 m vertical resolution.

3.3. Discussion

While the verification relied on the results from the MAX-DOAS observation at 8:00 LT on 4 September 2020, encompassing only one day, it is crucial to note that diverse scenarios were explored for aerosol and NO_2 retrievals through synthetic data simulation. The findings from the actual measurements aligned with those derived from synthetic data simulation, although it is essential to acknowledge that synthetic measurements represent an idealized scenario and disparities between true and retrieved states may be more pronounced in actual atmospheric measurements [18]. On one hand, real-world atmospheric inhomogeneities, particularly clouds in the observation EAs, can exert a significant influence on MAX-DOAS measurements. On the other hand, in actual atmospheric retrieval, while the conclusion regarding the impact of EAs and vertical resolutions on inversion remains consistent with the simulation results, the relative deviation between the retrieval and true values depends on the shape and concentration of the gas profile and the *a priori* profile. Nonetheless, the vertical resolution of the input profile has a relatively insignificant effect on the shape of the retrieved profile. Specifically, when dealing with particularly high uplifted layers, the retrieval results using Setting III with 6 EAs exhibit a notable underestimation of the height and concentration of the uplifted layer. A study by Vlemmix et al. (2011) highlighted that accurate retrieval of NO_2 is achievable in a homogeneously mixed boundary layer; however, accurate measurement and precise aerosol extinction information are imperative for retrieving NO_2 in the free troposphere [41]. Considering the DOFS obtained from both simulation and measurement, it is advisable to prioritize Setting I with 10 EAs.

4. Conclusions

The study focused on exploring the impact of vertical resolution and elevation angles on NO_2 and aerosol profile retrievals through a combination of simulation and measurements. In the simulation, three shapes—Boltzmann, Gaussian, and exponential—were employed for aerosol and NO_2 profile retrievals. The vertical resolutions of the input profiles were set at 100 and 200 m, while fixed values for AOD and NO_2 VCD were 0.3 and 1.0×10^{16} molec. cm^{-2}, respectively. The aerosol and NO_2 profiles were measured via MAX-DOAS and Lidar observations at the Science Island site in Hefei on 4 September 2020. Four different EA settings were employed: Setting I with 10 EAs used in MAX-DOAS observations (i.e., 1, 2, 3, 4, 5, 6, 8, 15, 30, and 90°), Setting II with 6 EAs (i.e., 2, 4, 8, 15, 30, and 90°), Setting III with 6 EAs (i.e., 1, 2, 3, 4, 5, and 90°), and Setting IV with 6 EAs (i.e., 5, 6, 8, 15, 30, and 90°).

Firstly, the simulation investigated the influence of vertical resolutions in the input aerosol profiles and EAs on aerosol profile retrieval using the RTM. Compared to the input aerosol profile, the retrieved aerosol profiles from the two vertical resolution inputs showed slight deviations. The Boltzmann shape's scale height and the concentration value at the peak of the Gaussian shape profile retrieved from a 100 m input showed a slight overestimation. Notably, the vertical resolution of the input profile minimally affected the shape of the retrieved aerosol profile. Discrepancies between aerosol profiles from MAX-DOAS measurements and Lidar observations were attributed to AVK smoothing, differences in location and viewing geometry, and uncertainties in Lidar profiles. In the practical inversion of MAX-DOAS observations, aerosol extinction retrieval below 1.5 km for different EAs deviated slightly from the input. The higher DOFS from 10 EAs in aerosol retrieval suggested that using more EAs could enhance information acquisition.

Secondly, the study delved into the impact of vertical resolution of aerosol and NO_2 profiles on NO_2 profile retrieval. NO_2 profiles retrieved by PriAM under input NO_2 profiles with different vertical resolutions exhibited minimal differences. However, NO_2 profiles below 2.0 km, using corresponding EAs for both NO_2 and aerosols, were higher compared to those from aerosol with 10 EAs. The impact of the Lidar profile as aerosol information on NO_2 profile retrieval was negligible, similar to previous simulation. Additionally, the influence of EAs on NO_2 profile retrieval was less pronounced than for aerosol profile retrieval.

Supplementary Materials: The following supporting information can be downloaded at: https://www.mdpi.com/article/10.3390/rs15225431/s1, Table S1: Parameter settings used in the RTM in the simulation [22,44].

Author Contributions: Conceptualization, X.T. and P.X.; methodology, X.T. and M.C.; software, X.T.; validation, J.X., A.L. and P.X.; formal analysis, X.T.; investigation, B.R. and M.C.; resources, J.Z. and Z.W.; data curation, X.T., T.Z. and G.F.; writing—original draft preparation, X.T.; writing—review and editing, P.X. and J.X.; supervision, A.L.; project administration, X.T. and P.X.; funding acquisition, X.T. and W.L. All authors have read and agreed to the published version of the manuscript.

Funding: This research was funded by the National Natural Science Foundation of China (Grant Nos. 42105132, U19A2044) and the National Key Research and Development Program of China (No. 2022YFC3700303).

Data Availability Statement: Underlying research data are available upon request from the first author according to the reader's needs (xtian@ahu.edu.cn).

Acknowledgments: The authors would also like to thank BIRA-IASB, who developed the software packages and code for the SCIAMACHY radiative transfer model (SCIATRAN) and QDOAS.

Conflicts of Interest: The authors declare that they have no conflict of interest.

References

1. Hönninger, G.; von Friedeburg, C.; Platt, U. Multi axis differential optical absorption spectroscopy (MAX-DOAS). *Atmos. Chem. Phys.* **2004**, *4*, 231–254. [CrossRef]
2. Wagner, T.; Dix, B.; von Friedeburg, C.; Friess, U.; Sanghavi, S.; Sinreich, R.; Platt, U. MAX-DOAS O4 measurements: A new technique to derive information on atmospheric aerosols—Principles and information content. *J. Geophys. Res. Atmos.* **2004**, *109*. [CrossRef]
3. Heckel, A.; Richter, A.; Tarsu, T.; Wittrock, F.; Hak, C.; Pundt, I.; Junkermann, W.; Burrows, J.P. MAX-DOAS measurements of formaldehyde in the Po-Valley. *Atmos. Chem. Phy.* **2005**, *5*, 909–918. [CrossRef]
4. Friess, U.; Monks, P.S.; Remedios, J.J.; Rozanov, A.; Sinreich, R.; Wagner, T.; Platt, U. MAX-DOAS O4 measurements: A new technique to derive information on atmospheric aerosols: 2. Modeling studies. *J. Geophys. Res.-Atmos.* **2006**, *111*, D14. [CrossRef]
5. Platt, U.; Stutz, J. *Differential Optical Absorption Spectroscopy, Principles and Applications*; Springer: Berlin/Heidelberg, Germany, 2008; pp. 314–330.
6. Irie, H.; Kanaya, Y.; Akimoto, H.; Iwabuchi, H.; Shimizu, A.; Aoki, K. First retrieval of tropospheric aerosol profiles using MAX-DOAS and comparison with lidar and sky radiometer measurements. *Atmos. Chem. Phys.* **2008**, *8*, 341–350. [CrossRef]
7. Tian, X.; Xu, J.; Xie, P.H.; Li, A.; Hu, Z.K.; Li, X.M.; Ren, B.; Wu, Z.Y. Retrieving Tropospheric Vertical Distribution in HCHO by Multi-Axis Differential Optical Absorption Spectroscopy. *Spectrosc. Spect. Anal.* **2019**, *39*, 2325–2331.
8. Ma, J.Z.; Dörner, S.; Donner, S.; Jin, J.L.; Cheng, S.Y.; Guo, J.R.; Zhang, Z.F.; Wang, J.Q.; Liu, P.; Zhang, G.Q.; et al. MAX-DOAS measurements of NO_2, SO_2, HCHO, and BrO at the Mt. Waliguan WMO GAW global baseline station in the Tibetan Plateau. *Atmos. Chem. Phys.* **2020**, *20*, 6973–6990. [CrossRef]
9. Ma, J.Z.; Beirle, S.; Jin, J.L.; Shaiganfar, R.; Yan, P.; Wagner, T. Tropospheric NO_2 vertical column densities over Beijing: Results of the first three years of ground-based MAX-DOAS measurements (2008–2011) and satellite validation. *Atmos. Chem. Phys.* **2013**, *13*, 1547–1567. [CrossRef]
10. Javed, Z.; Tanvir, A.; Bilal, M.; Su, W.J.; Xia, C.Z.; Rehman, A.; Zhang, Y.Y.; Sandhu, O.; Xing, C.Z.; Ji, X.; et al. Recommendations for HCHO and SO_2 Retrieval Settings from MAX-DOAS Observations under Different Meteorological Conditions. *Remote Sens.* **2021**, *13*, 2244. [CrossRef]
11. Hendrick, F.; Müller, J.-F.; Clémer, K.; Wang, P.; De Mazière, M.; Fayt, C.; Gielen, C.; Hermans, C.; Ma, J.Z.; Pinardi, G.; et al. Four years of ground-based MAX-DOAS observations of HONO and NO_2 in the Beijing area. *Atmos. Chem. Phys.* **2014**, *14*, 765–781. [CrossRef]
12. Kanaya, Y.; Irie, H.; Takashima, H.; Iwabuchi, H.; Akimoto, H.; Sudo, K.; Gu, M.; Chong, J.; Kim, Y.J.; Lee, H.; et al. Long-term MAX-DOAS network observations of NO_2 in Russia and Asia (MADRAS) during the period 2007–2012: Instrumentation, elucidation of climatology, and comparisons with OMI satellite observations and global model simulations. *Atmos. Chem. Phys.* **2014**, *14*, 7909–7927. [CrossRef]
13. Wang, T.; Wang, P.; Theys, N.; Tong, D.; Hendrick, F.; Zhang, Q.; Van Roozendael, M. Spatial and temporal changes in SO_2 regimes over China in the recent decade and the driving mechanism. *Atmos. Chem. Phys.* **2018**, *18*, 18063–18078. [CrossRef]
14. Wang, Y.; Lampel, J.; Xie, P.; Beirle, S.; Li, A.; Wu, D.; Wagner, T. Ground-based MAX-DOAS observations of tropospheric aerosols, NO_2, SO_2 and HCHO in Wuxi, China, from 2011 to 2014. *Atmos. Chem. Phys.* **2017**, *17*, 2189–2215. [CrossRef]
15. Luo, Y.H.; Dou, S.K.; Fan, G.; Huang, S.; Si, F.; Zhou, H.; Wang, Y.; Pei, C.; Tang, F.; Yang, D.; et al. Vertical distributions of tropospheric formaldehyde, nitrogen dioxide, ozone and aerosol in southern China by ground-based MAX-DOAS and LIDAR measurements during PRIDE-GBA 2018 campaign. *Atmos. Environ.* **2020**, *226*, 117384. [CrossRef]
16. Zhang, J.; Wang, S.; Guo, Y.L.; Zhang, R.F.; Qin, X.F.; Huang, K.; Wang, D.F.; Fu, Q.Y.; Wang, J.; Zhou, B. Aerosol vertical profile retrieved from ground-based MAX-DOAS observation and characteristic distribution during wintertime in Shanghai, China. *Atmos. Environ.* **2018**, *192*, 193–205. [CrossRef]
17. Rodgers, C.D. *Inverse Methods for Atmospheric Sounding—Theory and Practice, Series on Atmospheric, Oceanic and Planetary Physics—Volume 2*; World Scientific Publishing Co., Pte. Ltd.: Singapore, 2000. [CrossRef]
18. Frieß, U.; Beirle, S.; Alvarado Bonilla, L.; Bösch, T.; Friedrich, M.M.; Hendrick, F.; Piters, A.; Richter, A.; van Roozendael, M.; Rozanov, V.V.; et al. Intercomparison of MAX-DOAS vertical profile retrieval algorithms: Studies using synthetic data. *Atmos. Meas. Tech.* **2019**, *12*, 2155–2181. [CrossRef]
19. Tirpitz, J.-L.; Frieß, U.; Hendrick, F.; Alberti, C.; Allaart, M.; Apituley, A.; Bais, A.; Beirle, S.; Berkhout, S.; Bognar, K.; et al. Intercomparison of MAX-DOAS vertical profile retrieval algorithms: Studies on field data from the CINDI-2 campaign. *Atmos. Meas. Tech.* **2021**, *14*, 1–35. [CrossRef]

20. Vlemmix, T.; Hendrick, F.; Pinardi, G.; De Smedt, I.; Fayt, C.; Hermans, C.; Piters, A.; Wang, P.; Levelt, P.; Van Roozendael, M. MAX-DOAS observations of aerosols, formaldehyde and nitrogen dioxide in the Beijing area: Comparison of two profile retrieval approaches. *Atmos. Meas. Tech.* **2015**, *8*, 941–963. [CrossRef]
21. Hartl, A.; Wenig, M.O. Regularisation model study for the least-squares retrieval of aerosol extinction time series from UV/VIS MAX-DOAS observations for a ground layer profile parameterisation. *Atmos. Meas. Tech.* **2013**, *6*, 1959–1980. [CrossRef]
22. Tian, X.; Wang, Y.; Beirle, S.; Xie, P.; Wagner, T.; Xu, J.; Li, A.; Dörner, S.; Ren, B.; Li, X. Technical note: Evaluation of profile retrievals of aerosols and trace gases for MAX-DOAS measurements under different aerosol scenarios based on radiative transfer simulations. *Atmos. Chem. Phys.* **2021**, *21*, 12867–12894. [CrossRef]
23. Wang, Y.; Apituley, A.; Bais, A.; Beirle, S.; Benavent, N.; Borovski, A.; Bruchkouski, I.; Chan, K.L.; Donner, S.; Drosoglou, T.; et al. Inter-comparison of MAX-DOAS measurements of tropospheric HONO slant column densities and vertical profiles during the CINDI-2 campaign. *Atmos. Meas. Tech.* **2020**, *13*, 5087–5116. [CrossRef]
24. Frieß, U.; Baltink, H.K.; Beirle, S.; Clémer, K.; Hendrick, F.; Henzing, B.; Irie, H.; de Leeuw, G.; Li, A.; Moerman, M.M.; et al. Intercomparison of aerosol extinction profiles retrieved from MAX-DOAS measurements. *Atmos. Meas. Tech.* **2016**, *9*, 3205–3222. [CrossRef]
25. Irie, H.; Takashima, H.; Kanaya, Y.; Boersma, K.F.; Gast, L.; Wittrock, F.; Brunner, D.; Zhou, Y.; Van Roozendael, M. Eight-component retrievals from ground-based MAX-DOAS observations. *Atmos. Meas. Tech.* **2011**, *4*, 1027–1044. [CrossRef]
26. Bösch, T.; Rozanov, V.; Richter, A.; Peters, E.; Rozanov, A.; Wittrock, F.; Merlaud, A.; Lampel, J.; Schmitt, S.; de Haij, M.; et al. BOREAS—A new MAX-DOAS profile retrieval algorithm for aerosols and trace gases. *Atmos. Meas. Tech.* **2018**, *11*, 6833–6859. [CrossRef]
27. Ren, B.; Xie, P.; Xu, J.; Li, A.; Qin, M.; Hu, R.; Zhang, T.; Fan, G.; Tian, X.; Zhu, W.; et al. Vertical characteristics of NO_2 and HCHO, and the ozone formation regimes in Hefei, China. *Sci. Total Environ.* **2022**, *823*, 153425. [CrossRef]
28. Wang, X.; Zhang, T.; Xiang, Y.; Lv, L.; Fan, G.; Ou, J. Investigation of atmospheric ozone during summer and autumn in Guangdong Province with a lidar network. *Sci. Total Environ.* **2021**, *751*, 141740. [CrossRef]
29. Fan, G.; Zhang, T.; Fu, Y.; Dong, Y.; Liu, W. Temporal and spatial distribution characteristics of ozone based on differential absorption lidar in Beijing. *Chin. J. Lasers* **2014**, *41*, 1014003.
30. Kreher, K.; Van Roozendael, M.; Hendrick, F.; Apituley, A.; Dimitropoulou, E.; Friess, U.; Richter, A.; Wagner, T.; Lampel, J.; Abuhassan, N.; et al. Intercomparison of NO_2, O_4, O_3 and HCHO slant column measurements by MAX-DOAS and zenith-sky UV-visible spectrometers during CINDI-2. *Atmos. Meas. Tech.* **2020**, *13*, 2169–2208. [CrossRef]
31. Vandaele, A.C.; Hermans, C.; Simon, P.C.; Carleer, M.; Colin, R.; Fally, S.; Merienne, M.F.; Jenouvrier, A.; Coquart, B. Measurements of the NO_2 absorption cross-section from 42,000 cm^{-1} to 10,000 cm^{-1} (238–1000 nm) at 220 K and 294 K. *J. Quant. Spectrosc. Radiat. Transf.* **1998**, *59*, 171–184. [CrossRef]
32. Thalman, R.; Volkamer, R. Temperature dependent absorption cross-sections of O2–O2 collision pairs between 340 and 630 nm and at atmospherically relevant pressure. *Phys. Chem. Chem. Phys.* **2013**, *15*, 15371–15381. [CrossRef]
33. Serdyuchenko, A.; Gorshelev, V.; Weber, M.; Chehade, W.; Burrows, J.P. High spectral resolution ozone absorption cross-sections—Part 2: Temperature dependence. *Atmos. Meas. Tech.* **2014**, *7*, 625–636. [CrossRef]
34. Meller, R.; Moortgat, G.K. Temperature dependence of the absorption cross sections of formaldehyde between 223 and 323 K in the wavelength range 225–375 nm. *J. Geophys. Res -Atmos.* **2000**, *105*, 7089–7101. [CrossRef]
35. Fleischmann, O.C.; Hartmann, M.; Burrows, J.P.; Orphal, J. New ultraviolet absorption cross-sections of BrO at atmospheric temperatures measured by time-windowing Fourier transform spectroscopy. *J. Photochem. Photobiol. A Chem.* **2004**, *168*, 117–132. [CrossRef]
36. Wagner, T.; Beirle, S.; Deutschmann, T. Three-dimensional simulation of the Ring effect in observations of scattered sun light using Monte Carlo radiative transfer models. *Atmos. Meas. Tech.* **2009**, *2*, 113–124. [CrossRef]
37. Wang, Y.; Li, A.; Xie, P.H.; Chen, H.; Mou, F.S.; Xu, J.; Wu, F.C.; Zeng, Y.; Liu, J.G.; Liu, W.Q. Measuring tropospheric vertical distribution and vertical column density of NO_2 by multi-axis differential optical absorption spectroscopy. *Acta Phys. Sin.-Ch. Ed.* **2013**, *62*, 20. [CrossRef]
38. Wang, Y.; Beirle, S.; Hendrick, F.; Hilboll, A.; Jin, J.; Kyuberis, A.A.; Lampel, J.; Li, A.; Luo, Y.; Lodi, L.; et al. MAX-DOAS measurements of HONO slant column densities during the MAD-CAT campaign: Inter-comparison, sensitivity studies on spectral analysis settings, and error budget. *Atmos. Meas. Tech.* **2017**, *10*, 3719–3742. [CrossRef]
39. Rozanov, A.; Rozanov, V.; Buchwitz, M.; Kokhanovsky, A.; Burrows, J.P. SCIATRAN 2.0—A new radiative transfer model for geophysical applications in the 175–2400 nm spectral region. *Adv. Space Res.* **2005**, *36*, 1015–1019. [CrossRef]
40. Liu, Q.; Ding, W.D.; Xie, L.Q.; Zhang, J.; Zhu, J.; Xia, X.A.; Liu, D.Y.; Yuan, R.; Fu, Y. Aerosol properties over an urban site in central East China derived from ground sun-photometer measurements. *Sci. China Earth Sci.* **2017**, *60*, 297–314. [CrossRef]
41. Vlemmix, T.; Piters, A.J.M.; Berkhout, A.J.C.; Gast, L.F.L.; Wang, P.; Levelt, P.F. Ability of the MAX-DOAS method to derive profile information for NO: Can the boundary layer and free troposphere be separated? *Atmos. Meas. Tech.* **2011**, *4*, 2659–2684. [CrossRef]
42. Gratsea, M.; Bösch, T.; Kokkalis, P.; Richter, A.; Vrekoussis, M.; Kazadzis, S.; Tsekeri, A.; Papayannis, A.; Mylonaki, M.; Amiridis, V.; et al. Retrieval and evaluation of tropospheric-aerosol extinction profiles using multi-axis differential optical absorption spectroscopy (MAX-DOAS) measurements over Athens, Greece. *Atmos. Meas. Tech.* **2021**, *14*, 749–767. [CrossRef]

43. Karagkiozidis, D.; Friedrich, M.M.; Beirle, S.; Bais, A.; Hendrick, F.; Voudouri, K.A.; Fountoulakis, I.; Karanikolas, A.; Tzoumaka, P.; Van Roozendael, M.; et al. Retrieval of tropospheric aerosol, NO_2, and HCHO vertical profiles from MAX-DOAS observations over Thessaloniki, Greece: Intercomparison and validation of two inversion algorithms. *Atmos. Meas. Tech.* **2022**, *15*, 1269–1301. [CrossRef]
44. NASA. *US Standard Atmosphere, Office, U.S.*; Government Printing: Washington, DC, USA, 1976.

remote sensing

MDPI

Article

Automatic Identification and Segmentation of Diffuse Aurora from Untrimmed All-Sky Auroral Videos

Qian Wang [1,2,3], Peiqi Hao [1] and Han Pan [4,*]

1 School of Telecommunication and Information Engineering, Xi'an University of Posts & Telecommunications, Xi'an 710121, China
2 Xi'an Key Laboratory of Image Processing Technology and Applications for Public Security, Xi'an 710121, China
3 International Joint Research Center for Wireless Communication and Information Processing Technology of Shaanxi Province, Xi'an 710121, China
4 School of Aeronautics and Astronautics, Shanghai Jiao Tong University, Shanghai 200240, China
* Correspondence: panhan@sjtu.edu.cn

Highlights

What are the main findings?

- An automated framework achieves 96.3% frame-wise accuracy, 87.7% Edit score, and F1@50 of 83.0% for diffuse aurora identification and temporal segmentation in long ASI videos.
- Large-scale statistical analysis of 358,560 all-sky images reveals a robust diurnal pattern of diffuse aurora occurrence, with a consistent peak around 07:00 UT and a minimum near 11:00 UT across multiple years.

What are the implications of the main findings?

- The results demonstrate that physics-informed, vision-based methods can reliably extract physically meaningful auroral statistics from massive remote sensing video archives without manual inspection.
- The derived occurrence statistics provide quantitative observational support for established diffuse aurora particle precipitation and drift mechanisms, enabling scalable studies of magnetosphere-ionosphere coupling.

Abstract

Diffuse aurora is a widespread and long-lasting auroral emission that plays an important role in diagnosing magnetosphere-ionosphere coupling and magnetospheric plasma transport. Despite its scientific significance, diffuse aurora remains challenging to identify automatically in all-sky imager (ASI) observations due to its weak optical intensity, indistinct boundaries, and gradual temporal evolution. These characteristics, together with frequent cloud contamination, limit the effectiveness of conventional keogram-based or morphology-driven detection approaches and hinder large-scale statistical analyses based on long-term optical datasets. In this study, we propose an automated framework for the identification and temporal segmentation of diffuse aurora from untrimmed all-sky auroral videos. The framework consists of a frame-level coarse identification module that combines weak morphological information with inter-frame temporal dynamics to detect candidate diffuse-auroral intervals, and a snippet-level segmentation module that dynamically aggregates temporal information to capture the characteristic gradual onset-plateau-decay evolution of diffuse aurora. Bidirectional temporal modeling is employed to improve boundary localization, while an adaptive mixture-of-experts mechanism reduces redundant temporal variations and enhances discriminative features relevant to diffuse emission. The proposed method is evaluated using multi-year 557.7 nm ASI observations

https://doi.org/10.3390/rs18030402

acquired at the Arctic Yellow River Station. Quantitative experiments demonstrate state-of-the-art performance, achieving 96.3% frame-wise accuracy and an Edit score of 87.7%. Case studies show that the method effectively distinguishes diffuse aurora from cloud-induced pseudo-diffuse structures and accurately resolves gradual transition boundaries that are ambiguous in keograms. Based on the automated identification results, statistical distributions of diffuse aurora occurrence, duration, and diurnal variation are derived from continuous observations spanning 2003–2009. The proposed framework enables robust and fully automated processing of large-scale all-sky auroral images, providing a practical tool for remote sensing-based auroral monitoring and supporting objective statistical studies of diffuse aurora and related magnetospheric processes.

Keywords: diffuse aurora; all-sky optical imaging; temporal evolution segmentation; physics-informed modeling; auroral morphology identification

1. Introduction

Aurora is a luminous phenomenon resulting from high-energy charged particles precipitating along magnetic field lines into the Earth's polar ionosphere, where collisions with atmospheric atoms and molecules generate characteristic emissions [1,2]. Aurora is a direct manifestation of solar-terrestrial interactions in the polar regions, particularly magnetosphere-ionosphere coupling, and serves as a key observational window for studying space weather events [3–8]. Ground-based all-sky auroral observations have revealed two major types of aurora based on morphological and dynamic characteristics: discrete aurora and diffuse aurora [9]. Discrete aurora is short-lived and exhibits well-defined luminous structures, including arcs, bands, curls, and rays [10–12]. In contrast, diffuse aurora generally displays weak, uniform brightness, indistinct morphology, long duration, slow drift, and periodic pulsations [13]. Different auroral types correspond to distinct magnetospheric processes, and studying them separately is essential for understanding auroral mechanisms.

Diffuse aurora is a persistent and gradually evolving phenomenon in the polar ionosphere, mainly produced by the continuous precipitation of magnetospheric electrons [14–17]. Unlike discrete aurora, which appears as transient and localized arcs, diffuse aurora exhibits smooth temporal transitions and broad spatial coverage [18–21]. Its identification and temporal characterization require large-scale data analysis over long observation sequences, motivating automated approaches for detecting gradual transitions and distinguishing them from discrete auroral forms.

Diffuse aurora remains challenging to identify and characterize in practice. Although particle spectrograms from satellites such as the Defense Meteorological Satellite Program (DMSP) can indicate precipitation events, they often fail to capture the full temporal and spatial evolution of diffuse aurora. As demonstrated in our YRS-DMSP case (Figure 1), the all-sky optical images clearly show the emergence of diffuse aurora at lower latitudes between 08:10 and 08:20 UT, whereas the corresponding particle spectrogram exhibits no distinct enhancement during the same interval. This discrepancy arises because the satellite, moving from higher to lower latitudes, observes only a narrow portion of the precipitation region and may miss early-stage developments visible in ground-based imagery. In addition, particle data are prone to local disturbances and instrumental uncertainties, such as around 08:13 UT, which can obscure the gradual transitions typical of diffuse aurora. In contrast, optical imagery provides a continuous, wide-field record of auroral morphology, making it a more reliable basis for identifying diffuse auroral evolution.

https://doi.org/10.3390/rs18030402

Therefore, leveraging computer vision to objectively detect and segment diffuse aurora from large-scale all-sky video archives is both scientifically justified and operationally essential.

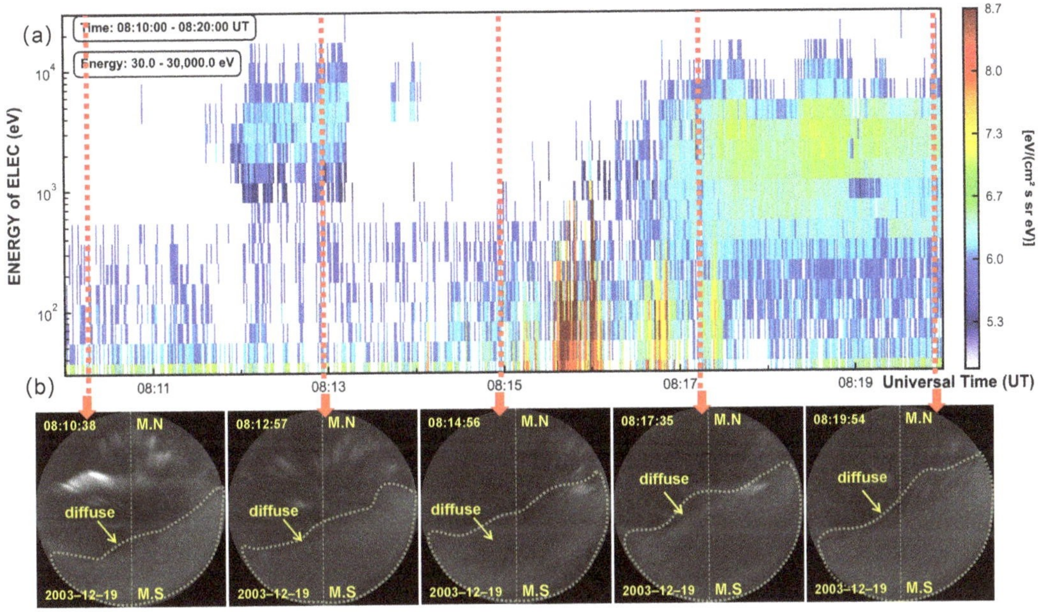

Figure 1. Joint YRS optical and DMSP in situ observations illustrating the correspondence between diffuse-aurora evolution and source-region particle precipitation (19 December 2003, 08:10–08:20 UT). (**a**) DMSP electron energy-flux spectrogram; vertical red dashed lines mark the sampling times of the co-temporal all-sky images in (**b**).

Identifying and segmenting diffuse aurora in untrimmed all-sky videos can be formulated as a temporal action segmentation problem, where the goal is to assign frame-wise labels to continuous sequences with gradual transitions. This formulation naturally fits the diffuse aurora scenario, which evolves continuously rather than through abrupt onsets. Existing video understanding methods can be broadly categorized into three groups: (i) frame-based models that infer temporal relations among per-frame features [22–28], (ii) two-stage frameworks that first extract frame features and then build long-range temporal context [29,30], and (iii) parallel models that simultaneously learn frame-level and sequence-level representations with bidirectional interaction [31].

Frame-based methods represent the foundational paradigm in temporal action segmentation. The MS-TCN (Multi-Stage Temporal Convolutional Network) [24] utilizes multi-stage dilated temporal convolutions to learn frame-level feature representations, achieving progressive feature refinement through its hierarchical architecture. However, due to the inherent limitations of convolutional receptive fields, this method struggles to capture long-range temporal dependencies beyond the receptive field and is prone to over-segmentation artifacts. ASFormer [26] replaces temporal convolutions with a Transformer architecture to enhance long-range temporal modeling capability. Nevertheless, owing to the high computational complexity of standard self-attention mechanisms, this approach adopts windowed temporal attention, which still constrains its capacity to model global temporal relationships across extended sequences. Two-stage methods attempt to decouple frame-level and action-level representation learning. UVAST (Unified Video Action Segmentation Transformer) [29] adopts a sequence-to-sequence framework, where the first stage computes frame-level features and the second stage constructs action-level

https://doi.org/10.3390/rs18030402

representations based on these features to generate predictions. However, this method fuses the two-stage prediction outputs only at the final stage, failing to fully exploit the complementary information between frame-level and action-level features for iterative optimization. Parallel methods represent a more recent paradigm that enables continuous interaction between different representation levels. FACTM (Frame-Action Cross-attention Temporal Modeling) [31] jointly learns morphological representations and temporal dynamics through bidirectional interaction between the frame branch and the action branch. Compared to two-stage methods, this parallel architecture enables continuous mutual optimization of frame-action features across multiple processing stages.

Given that diffuse aurora exhibits weak spatial contrast, slow temporal evolution, and quasi-periodic modulations, we adopt the parallel paradigm. This parallel modeling strategy enables mutual reinforcement between spatial cues (useful for suppressing clouds, stars, and background noise) and temporal dynamics, which is essential for capturing the gradual onset, development, and decay of diffuse auroral emissions.

Building on this perspective, we develop an automatic framework for identifying and segmenting diffuse aurora from continuous all-sky image sequences. The framework is grounded in the physical nature of diffuse aurora: its spatial morphology is weak and unstructured, while its temporal evolution is gradual and often intertwined with transitions to or from discrete aurora. To address these challenges, our method performs collaborative modeling between spatial morphology and temporal dynamics. The intra-frame branch focuses on morphological cues—enhancing sensitivity to faint, diffuse emissions and suppressing background contamination from clouds and stars—whereas the inter-frame branch captures long-term temporal continuity, slow drifts, and quasi-periodic modulations characteristic of diffuse aurora. The bidirectional interaction between these two representations allows the system to recognize diffuse auroral activity even under complex observing conditions, improving the determination of onset and termination phases in a physically interpretable way. The main contributions of this study are as follows:

(1) Physics-informed video segmentation of diffuse aurora. We design a segmentation and identification framework guided by the physical characteristics of diffuse aurora. Expert-driven feature selection enhances sensitivity to weak auroral structures, while snippet-level aggregation captures gradual temporal transitions. These designs enable reliable localization and classification of diffuse-auroral intervals in long untrimmed ASI videos. Its performance is shown in Figure 2.

(2) Scalable detection and statistical characterization. By applying the trained model to long-term video archives, we automatically extract diffuse auroral episodes across multiple years, enabling large-sample analyses of their occurrence rate, duration, and temporal variations—providing new opportunities for data-driven auroral climatology and for linking optical observations with magnetospheric particle dynamics.

(3) Improved operational utility for auroral monitoring. The proposed framework reduces reliance on manual labeling and subjective interpretation, offering a reproducible and objective means of monitoring diffuse aurora at scale—potentially supporting coordinated studies with in situ satellite measurements such as DMSP or THEMIS.

https://doi.org/10.3390/rs18030402

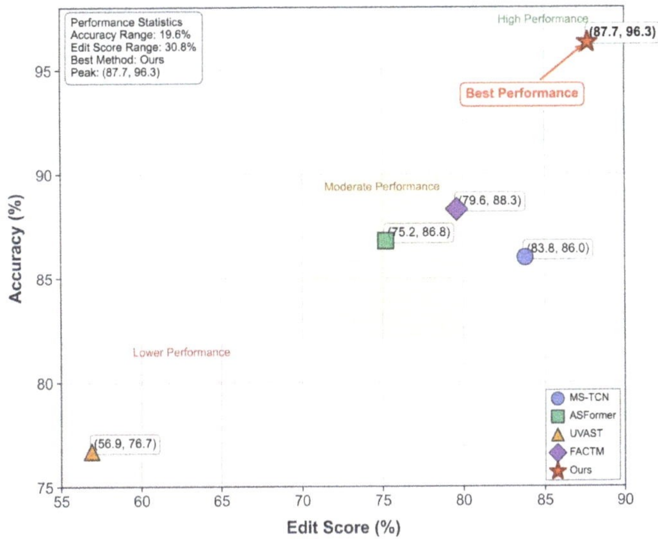

Figure 2. Comparison of the proposed method with existing methods in identifying and delineating diffuse auroral activities. The evaluation is performed on long-duration all-sky auroral image sequences using frame-wise accuracy (Acc) and Edit score. The proposed method attains the highest performance (96.3% Acc and 87.7% Edit), indicating its enhanced capability in tracing the spatially extended, weakly structured emissions and gradual temporal evolution of diffuse aurora, as well as in resolving their indistinct morphological boundaries.

2. Observations and Dataset

2.1. All-Sky Imaging System

The auroral data employed in this study were acquired by the All-Sky Imager (ASI) system at the Arctic Yellow River Station (YRS) in Ny-Ålesund, Svalbard. The station is located at geographic coordinates 78.92°N, 11.93°E, with a corrected geomagnetic latitude of 76.24° and magnetic local time MLT = UT + 3 h. As shown in Figure 3a, the YRS optical system consists of three identical ASI cameras capturing auroras at wavelengths of 427.8 nm, 557.7 nm, and 630.0 nm during the polar night period (November through February). Each frame is acquired with a 7 s exposure time at 10 s intervals between consecutive frames [32]. Figure 3b presents the internal structure of the CCD camera employed in the ASI system, highlighting the sensor configuration and key optical components, including the fish-eye lens, relay lens, filter lens, and focus lens.

Previous long-term observational studies have shown that diffuse aurora is most prominently and consistently observed in the 557.7 nm band, whereas discrete aurora typically exhibits strong emissions across multiple wavelength bands [13]. As a result, the green-line channel provides the highest contrast and signal-to-noise ratio for diffuse auroral structures, while effectively suppressing contamination from bright discrete arcs, airglow, and background illumination. By restricting the dataset to 557.7 nm observations, we impose a wavelength-level physical prior at the data selection stage. This constraint does not rely on geomagnetic indices or external parameters, but ensures that the visual manifestations of diffuse aurora are maximally preserved, thereby providing a reliable and consistent foundation for subsequent morphology- and dynamics-based analysis.

The raw observational data undergo the following preprocessing pipeline [33]: dark current subtraction, intensity stretching (16-bit to 8-bit grayscale conversion), rotation align-

https://doi.org/10.3390/rs18030402

ment to magnetic north, edge noise removal, and cropping of highly distorted circular peripheral regions, ultimately yielding 440 × 440 pixel images suitable for subsequent analysis.

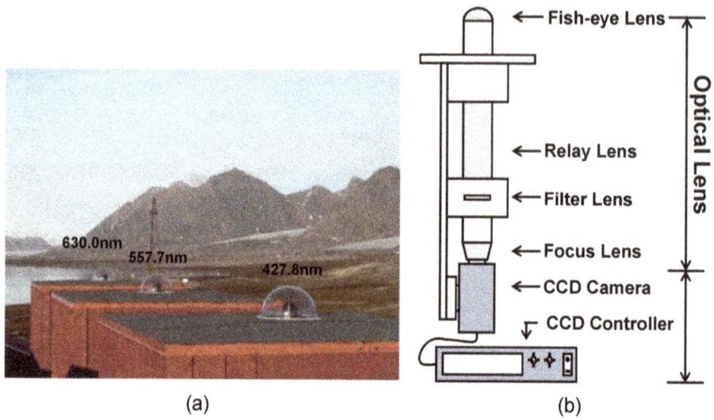

Figure 3. Observational data and construction of diffuse aurora samples. (**a**) Geographic location of the Arctic YRS, Ny-Ålesund, Svalbard, and the optical system equipped with three all-sky imagers operating at 427.8, 557.7, and 630.0 nm; (**b**) Internal structure of the CCD camera used in the all-sky imagers.

2.2. *Diffuse Aurora Dataset*

To support the development of the automated diffuse aurora identification and segmentation framework, we constructed an annotated dataset based on all-sky auroral images at 557.7 nm wavelength from the YRS optical system.

2.2.1. Aurora Dataset 1 (AD1)

AD1 comprises 40 long sequences, each containing approximately 1000 frames (totaling 40,002 images), collected from December 2003 to December 2007. This dataset is used for model training and quantitative evaluation. The 40 sequences were not randomly selected but were systematically curated according to the following criteria to ensure representativeness and diversity: (1) Temporal coverage: The selected sequences span multiple observing seasons from December 2003 to December 2007. (2) Diversity of observing conditions: The sequence selection intentionally includes the following representative scenarios: pure diffuse aurora events under clear sky backgrounds, observing scenes with cloud contamination, complex scenarios where diffuse aurora coexists with discrete aurora, and different morphological subtypes of diffuse aurora (veiling, patchy, pulsating, etc.). (3) Annotation feasibility: Priority was given to sequences with relatively clear boundaries and high inter-expert annotation consistency to ensure training label quality. Ground-truth frame-level labels were manually annotated by experienced auroral physicists following a consistent protocol according to the morphological and dynamical definitions of diffuse aurora [9]: (i) identifying the onset, mature, and decay phases of each diffuse auroral event; (ii) ensuring temporal precision within ±1–2 frames (10–20 s); (iii) merging ambiguous micro-gaps to maintain physically meaningful event continuity. Inter-annotator disagreements were resolved through joint review and discussion. In each sequence, frame-by-frame annotations were classified into two categories: diffuse aurora and others. The diffuse aurora category is characterized by weak structural morphological features such as uniform luminosity and diffuse boundaries, along with dynamic features including slow quasi-periodic pulsations and gradual evolution patterns. As shown in Figure 4a–c, three typical subtypes of diffuse aurora are presented. The "others" category

https://doi.org/10.3390/rs18030402

encompasses discrete auroral structures and background/cloud contamination. Discrete auroral structures, as shown in Figure 4d, exhibit sharp emission boundaries and rapid temporal variability. Cloud motion, depicted in Figure 4e, displays weak structural features superficially similar to diffuse aurora but demonstrates distinct dynamic characteristics of random advection and rapid texture deformation.

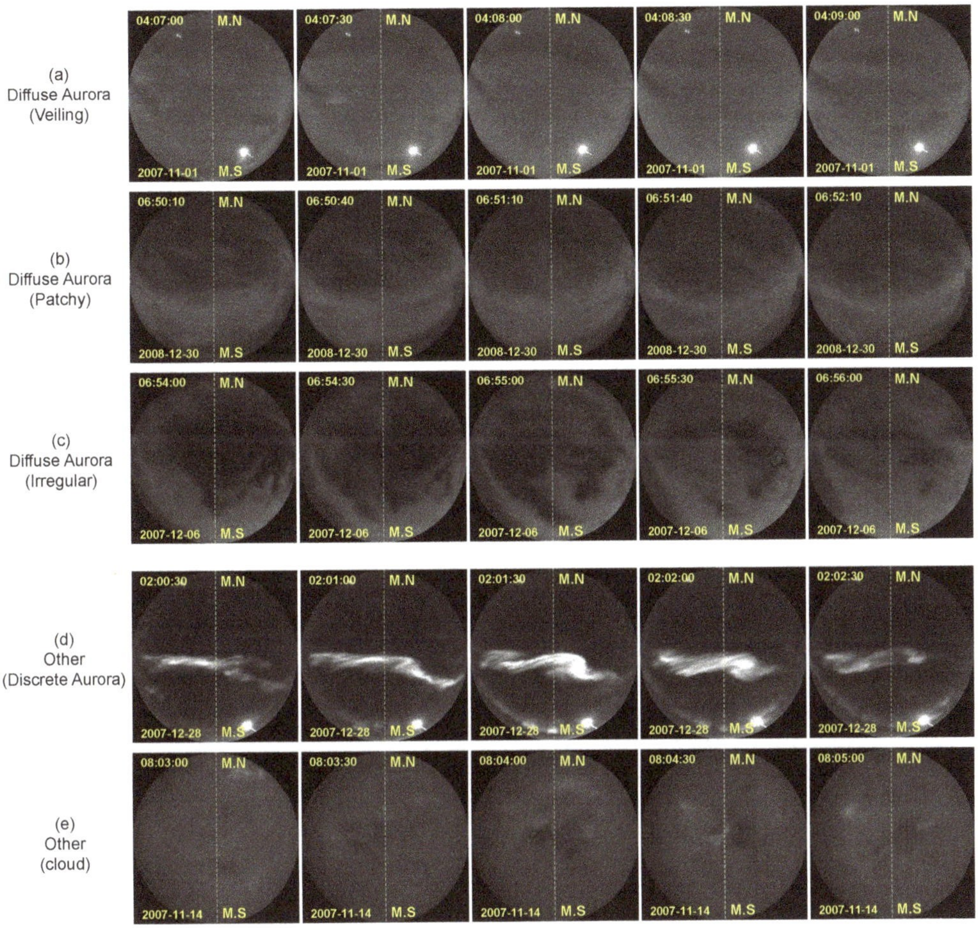

Figure 4. Representative temporal evolution sequences of different auroral and non-auroral phenomena used for dataset construction. (**a**) Diffuse Aurora (Veiling); (**b**) Diffuse Aurora (Patchy); (**c**) Diffuse Aurora (Irregular); (**d**) Other (Discrete Aurora); (**e**) Other (cloud).

2.2.2. Geomagnetic Conditions of AD1

We performed a statistical analysis of the geomagnetic conditions corresponding to the AD1 dataset. The Kp index (Planetarische Kennziffer) is a standardized measure of global geomagnetic activity that quantifies disturbances in the horizontal component of Earth's magnetic field [34]. The Kp index has a direct physical connection to diffuse aurora occurrence through its influence on plasmasphere configuration and cold plasma distribution, which in turn affects the generation efficiency of plasma waves responsible for scattering electrons into the loss cone [35,36].

Figure 5a presents the histogram of sample distribution as a function of Kp index, and Figure 5b illustrates the proportions of samples under different geomagnetic activity

https://doi.org/10.3390/rs18030402

levels through a pie chart. The statistical results indicate that the majority of samples in the dataset (61.8%) were acquired under geomagnetically quiet conditions (Kp = 0–2), with the remainder mainly occurring during moderate activity levels (Kp = 3–4, 35.7%). Only a small fraction of samples (2.5%) correspond to geomagnetically active conditions (Kp $\geq$ 5).

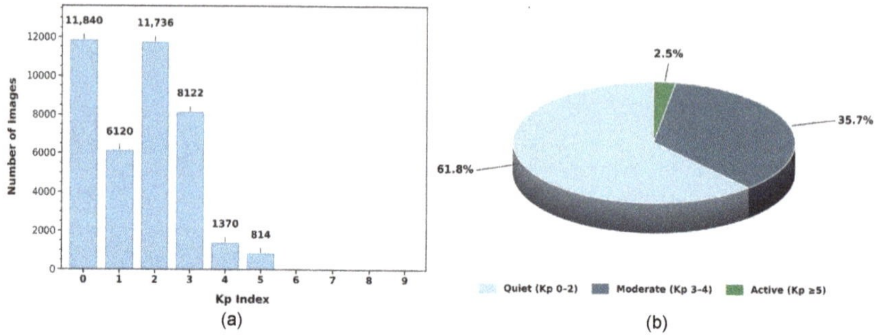

Figure 5. Geomagnetic condition statistics of the AD1 dataset. (**a**) Histogram showing the distribution of samples as a function of Kp index; (**b**) pie chart illustrating the proportion of samples under different geomagnetic activity levels (quiet: Kp 0–2, moderate: Kp 3–4, active: Kp $\geq$ 5).

This outcome is physically meaningful rather than coincidental. Diffuse aurora is more readily observable during quiet periods because: (1) an extended plasmasphere provides favorable cold plasma conditions for efficient wave–particle scattering; (2) reduced discrete auroral activity makes weaker diffuse emissions easier to identify; (3) the auroral oval contracts to higher latitudes, providing clearer contrast between diffuse aurora boundaries and the background. These physical considerations informed our dataset construction and support the representativeness of AD1 for diffuse aurora studies.

2.2.3. Aurora Dataset 2 (AD2)

AD2 comprises continuous ASI observations collected from October 2008 to December 2009, totaling 358,560 images. This dataset is used for statistical analysis of diffuse auroral events based on automatically identified results. Specifically, observations covering 48 days (October to December 2008) and 35 days (October to December 2009) were selected for statistical analysis of diffuse auroral events based on automatic identification results. The dataset is partitioned into training, validation, and test subsets with no temporal overlap to prevent scene-level data leakage.

3. Methods

3.1. Overview Architecture

We adopt an enhanced FACT-based architecture [31] tailored to the specific challenges of diffuse-aurora event detection from long untrimmed all-sky video sequences. As shown in Figure 6, our framework consists of three major components: input processing, Frame-level Coarse Identification Module (FCIM), and Snippet-level Gradual-transition Segmentation Module (SGSM).

In the input processing stage, raw auroral video sequences are first passed through feature extraction and preprocessing pipelines to produce frame-level embeddings and initial action tokens. Let F_0 denote the initial frame features pre-extracted by an Inflated 3D ConvNet (I3D) [37] and let A_0 denote the action tokens initialized as $A_0 = 0$. F_0 is fed into a Multi-Stage Temporal Convolutional Network (MSTCN) [24] to capture multi-scale temporal dependencies:

https://doi.org/10.3390/rs18030402

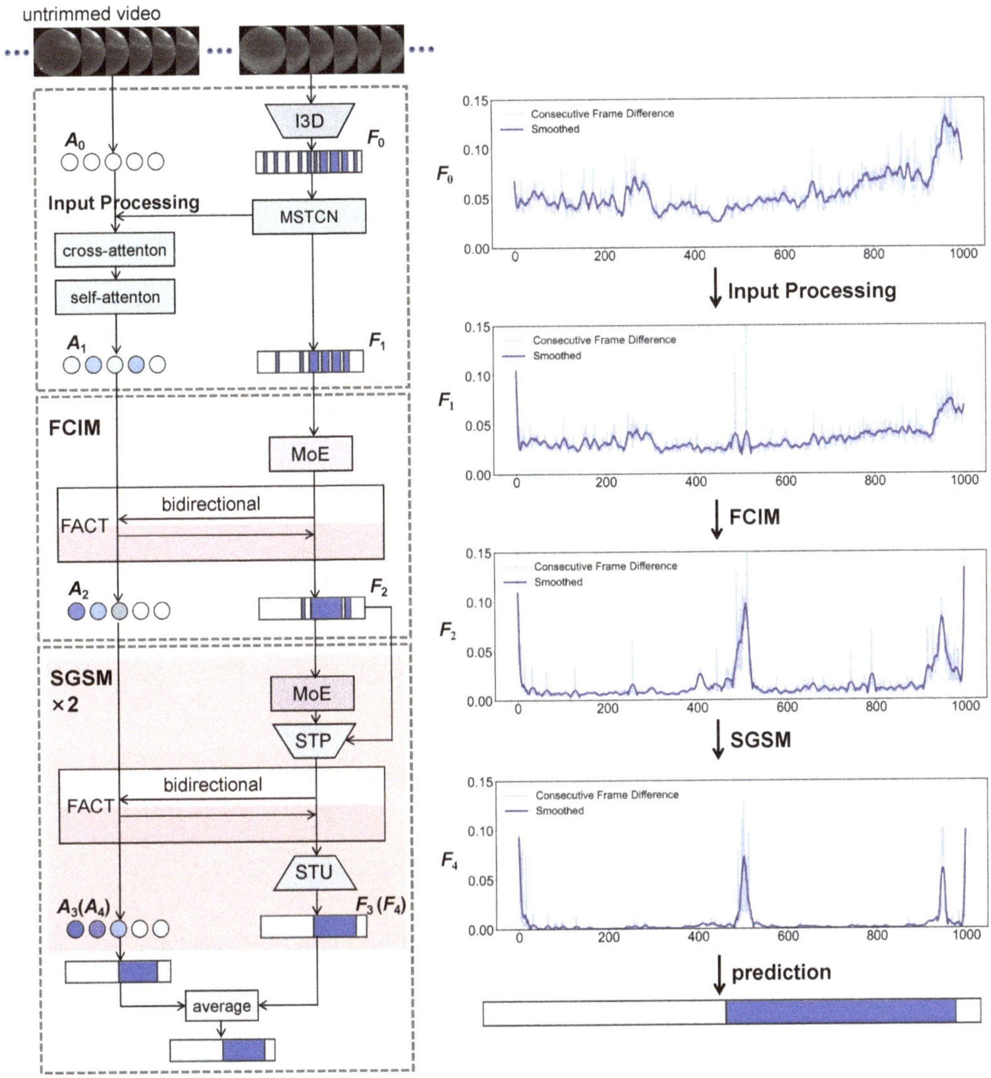

Figure 6. Framework for automatic diffuse aurora identification and segmentation, consisting of input processing, Frame-level Coarse Identification Module (FCIM), and Snippet-level Gradual-Transition Segmentation Module (SGSM).

In this FCIM stage, before feeding frame features into FACT, we pass them through a Mixture-of-Experts (MoE) module that is specially configured for auroral imagery. This MoE enhancement enables the FACT layer to better distinguish diffuse-aurora characteristics (e.g., weak morphology, slow drift, varied background) when performing coarse interval identification.

This architectural design builds on FACT's original capability for frame-action cross-attention temporal modeling, but adds aurora-specific logic (MoE gating, snippet-pooling with GRU context, adaptive sampling) to improve detection sensitivity to the gradual transitions and weak structural signatures of diffuse aurora. The right panel of Figure 6 visualizes the inter-frame feature differences extracted after each stage. From the initial input to the final stage, the difference patterns gradually evolve from scattered fluctuations

https://doi.org/10.3390/rs18030402

to clear and localized peaks. This visualization reflects how the hierarchical processing enhances temporal distinctiveness—frame-to-frame variations that were previously indistinct become increasingly concentrated around transition intervals. Such progression intuitively explains how the network transforms raw sequential data into structured temporal cues suitable for diffuse-aurora segmentation.

3.2. Frame-Level Coarse-Grained Identification

The Frame-level Coarse-grained Identification Module (FCIM) is designed for the coarse identification of candidate diffuse-aurora segments in long, untrimmed all-sky video sequences. At its core, FCIM adopts the Frame-Action Cross-Attention Temporal (FACT) modeling [31] to jointly learn frame-wise morphological representations and action-level temporal dynamics. Unlike conventional frame-only architectures, FACT maintains two interactive branches: a frame branch that encodes spatial morphology within each frame, and an action-token branch that abstracts temporal semantics across frames. Through alternating cross-attention operations, the two branches co-learn complementary cues: the frame branch refines its local structure with temporal context from the tokens, while the token branch acquires temporally coherent action descriptors from the evolving frame features. This bidirectional information exchange allows the model to understand not only what the auroral pattern looks like at each instant, but also how it evolves through time.

Such a mechanism is particularly suitable for diffuse aurora, whose spatial appearance often resembles thin cloud layers in still images but diverges dramatically once motion is considered. While clouds exhibit random advection and rapid texture deformation, diffuse aurora evolve through wave-like intensity modulations and gradual drifts governed by magnetospheric dynamics. FACT's explicit disentanglement between per-frame morphology and inter-frame temporal tokens provides a natural way to capture these subtle but physically meaningful transitions. The action tokens summarize long-range temporal dependencies—such as slow brightenings or fading arcs—whereas the frame features retain fine-scale spatial structures (diffuse boundaries, filamentary texture). Consequently, the FCIM can separate true auroral motion from background cloud or brightness fluctuations, yielding a robust coarse-level identification of auroral activity segments.

Before entering the FACT block, the frame features are first processed by a Mixture-of-Experts (MoE) [38,39] layer tailored for auroral data, as shown in Figure 6. Auroral imagery presents strong variability in brightness, noise level, and morphological type (uniform glow, patchy bands, or pulsating structures). The MoE introduces multiple lightweight expert networks, each specializing in different feature statistics, while a gating network dynamically assigns each frame or sequence to the most suitable experts. By integrating MoE-enhanced frame features with action tokens through FACT, the FCIM learns complementary morphological and dynamical representations. This enables coarse-grained localization of candidate diffuse aurora segments while filtering out irrelevant background and redundant temporal patterns, laying the foundation for segmentation.

3.3. Snippet-Level Gradual-Transition Segmentation

The SGSM takes the frame features and action tokens produced by the previous stage as input and explicitly models the gradual onset–plateau–decay behavior of diffuse aurora by operating on temporally aggregated snippets rather than isolated frames. Conceptually, SGSM answers the question: when auroral intensity changes slowly over many frames, how can we form temporally coherent representations that expose true gradual transitions while suppressing noise? Figure 7 illustrates an example of diffuse aurora with gradual transitions, highlighting the challenge of precisely determining event start and end frames

https://doi.org/10.3390/rs18030402

from visual inspection alone. Such gradual transitions, if processed frame-by-frame, can lead to noisy predictions or misclassification of the early and late stages of the event.

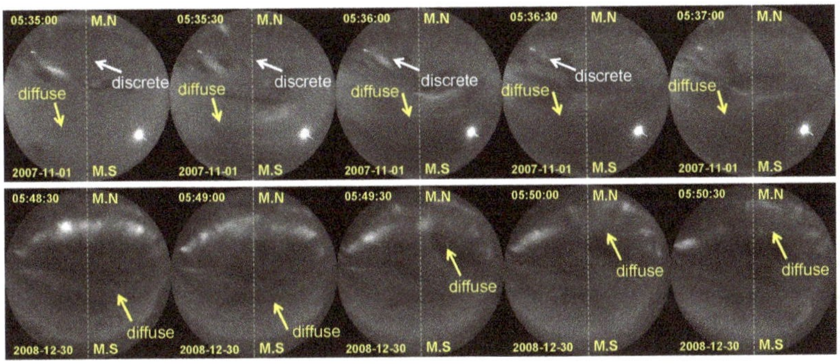

Figure 7. All-sky aurora images obtained at 30 s sampling frequency during the (**a**) onset and (**b**) decay phases of diffuse aurora.

The module proceeds through two successive refinement iterations (as shown in the lower-left of Figure 6). Each iteration contains MoE enhancement, snippet-based pooling, contextual modeling, FACT-based refinement, and temporal restoration. Let (A_2, F_2) denote the action tokens and frame features output from the previous stage. We write $F_2 = \{f_t\}_{t=1}^{T}$ for the frame feature sequence and $A_2 = \{a_k\}_{k=1}^{K}$ for the action tokens. The incoming frame features are first passed through the MoE. Using preliminary frame labels from the previous stage, consecutive frames with identical coarse labels are grouped into snippets. Each snippet represents a short, semantically consistent interval obtained through average pooling along the temporal axis. This step effectively downsamples the long frame sequence into a shorter sequence of snippet-level descriptors, reducing redundancy and emphasizing gradual transitions rather than frame-to-frame noise. To capture how diffuse aurora evolves across time, the snippet sequence is processed by a Bi-directional GRU (BiGRU) [40], which integrates temporal dependencies from both past and future snippets. This allows the model to interpret smooth intensity changes or fading boundaries that are not obvious when viewed frame-by-frame. After contextual enhancement, snippet-level features are fed into a FACT block for cross-attention between frame features and action tokens. The snippet representations are then upsampled back to the frame-level resolution (by assigning each snippet's refined descriptor to its constituent frames). The process is illustrated in Figure 8.

The final classification is used to achieve the final prediction. A linear classifier maps frame features to per-frame logits $L_f(t) = W_f f_t + b_f$. The token-level logits $L_a(k) = W_a a_k + b_a$ are aligned to the frame timeline by the FACT attention alignment matrix $\mathcal{A}$ (the softmax attention weights computed in the action $\rightarrow$ frame pass). If $\mathcal{A}$ has shape $K \times T$, token logits are projected to frame space by:

$$L_{a\rightarrow f}(t) = \sum_{k=1}^{K} A_{tk} L_a(k), \tag{1}$$

producing token-informed, frame-aligned logits. The two sources are combined by a small fusion weight λ:

$$P_{\text{final}}(t) = \lambda \cdot \text{Softmax}(L_a(k)) + (1-\lambda) \cdot \text{Softmax}(L_{a\rightarrow f}(t)). \tag{2}$$

The per-frame label is the argmax of $P_{\text{final}}(t)$. This fusion preserves both local, morphology-driven cues and longer-range, token-level context.

https://doi.org/10.3390/rs18030402

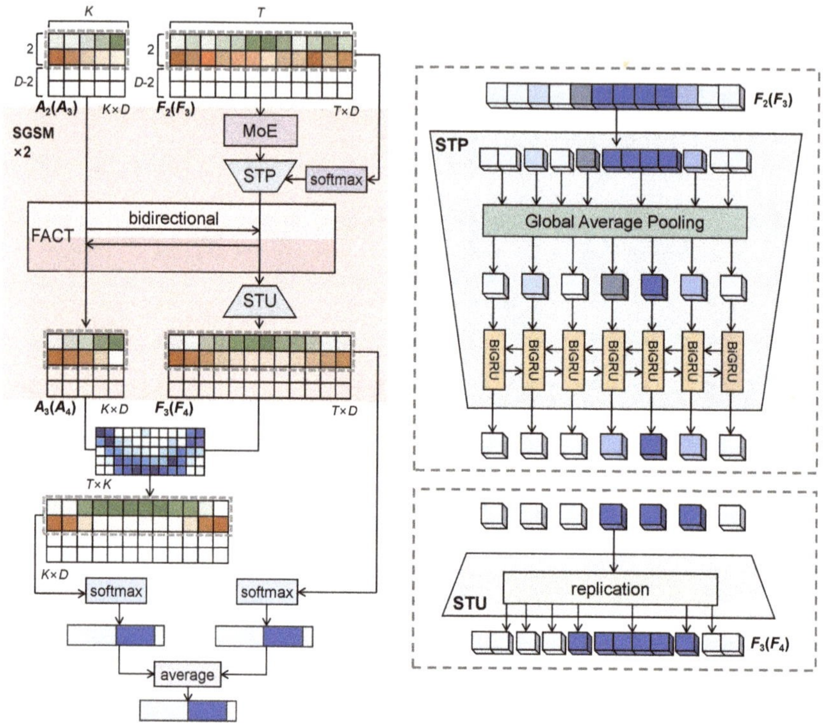

Figure 8. Illustration of snippet-based pooling and BiGRU processing in SGSM. Similar frames (indicated by similar color) are grouped into snippets through snippet-based pooling and processed as unified temporal units.

4. Results

4.1. Experiment Data and Setting

4.1.1. Implementation Details

All experiments are implemented in Python 3.8.20 and PyTorch 2.4.1 with CUDA 11.8. Models are trained for 1,000 epochs using a batch size of 2. Optimization follows the Adam optimizer [41] with an initial learning rate of 0.001, weight decay of 0.0001, and gradient clipping at a maximum norm of 10.0. The learning rate decays by a factor of 0.1 every 20 epochs. The loss function combines cross-entropy loss [42] with matching loss [24] to enforce temporal consistency. The FACT backbone adopts 512-dimensional hidden representations with 6 attention layers, each configured with 8 self-/cross-attention heads, and 0.3 dropout probability. The MoE module is configured with 6 experts, activating k = 2 experts per forward pass, and is applied in both FCIM and SGSM modules. The action-token pathway uses 5 action tokens with frame positional encoding.

4.1.2. Evaluation Metrics

To ensure consistent evaluation of both identification and segmentation performance, three standard temporal action segmentation metrics [29,43] are employed: (1) Frame-wise Accuracy (Acc), (2) Segmental F1 score (F1@{10, 25, 50}), and (3) Edit distance score, providing complementary assessments of per-frame correctness, boundary localization, and temporal structural consistency [44].

https://doi.org/10.3390/rs18030402

4.2. Comparison with Existing Methods

We compare our method with representative temporal action segmentation approaches for diffuse aurora identification. Existing methods can be broadly grouped into three categories: (1) frame-based methods, which learn temporal relations from frame-level features; (2) two-stage methods, which separately construct frame-level and action-level representations; and (3) parallel methods, which jointly refine frame and action features through iterative interaction.

Table 1 presents quantitative comparisons between our method and representative approaches from each category: MS-TCN [24], ASFormer [26], UVAST [29], and FACTM [31], which represent the most competitive frame-based, two-stage, and parallel methods, respectively. Our method achieves superior performance across all evaluation metrics, attaining 96.3 ± 0.4% accuracy, 87.7 ± 0.2% edit distance score, and F1@{10, 25, 50} scores of 86.8 ± 0.5%, 86.6 ± 0.5%, and 83.0 ± 0.4%, respectively. Notably, our method exhibits the smallest standard deviations across all metrics among the evaluated approaches, indicating consistent and stable performance over multiple independent runs. In contrast, the two-stage method UVAST demonstrates considerably higher variance (e.g., Edit score standard deviation of 5.3), suggesting less stable training convergence. The low variability of our method can be attributed to the explicit modeling of gradual temporal evolution through the SGSM, which provides robust regularization against stochastic training fluctuations.

Table 1. Performance comparison of our method with current mainstream action segmentation approaches on the diffuse aurora identification and segmentation task. Results are reported as mean ± standard deviation over five independent runs. The best results are highlighted in bold.

	Method	Acc	Edit	F1@{10, 25, 50}		
Frame-based	MS-TCN	86.0 ± 0.2	83.8 ± 0.6	70.3 ± 1.0	70.0 ± 0.8	59.5 ± 0.7
	AsFormer	86.8 ± 0.8	75.2 ± 1.2	78.6 ± 0.7	75.3 ± 0.7	70.5 ± 0.7
Two-stage	UVAST	76.7 ± 2.2	56.9 ± 5.3	59.1 ± 2.8	52.3 ± 2.0	33.0 ± 1.8
Parallel	FACTM	88.3 ± 0.7	79.6 ± 0.4	79.9 ± 2.0	79.9 ± 2.0	71.9 ± 0.3
	Ours	**96.3 ± 0.4**	**87.7 ± 0.2**	**86.8 ± 0.5**	**86.6 ± 0.5**	**83.0 ± 0.4**

Frame-based methods (MS-TCN, ASFormer) mainly capture local temporal dependencies and rely heavily on frame-level morphology, making them unsuitable for modeling the long, slowly evolving nature of diffuse aurora. Two-stage methods (UVAST) fuse features only at the final stage, leading to weak interaction between morphological and temporal cues. Parallel methods (FACTM) improve temporal reasoning through bidirectional refinement, yet they do not explicitly model the onset-plateau-decay evolution of diffuse aurora and struggle with long-duration temporal redundancy. Our method addresses these limitations through a two-stage coarse-to-fine design. The FCIM stage integrates intra-frame morphology and inter-frame dynamics to produce robust coarse segmentation. The SGSM stage further models the characteristic gradual evolution of diffuse aurora using snippet-level temporal aggregation and a BiGRU, enabling accurate onset and decay boundary refinement. Additionally, an MoE-based adaptive enhancement mechanism suppresses redundant temporal information and strengthens discriminative representations. This combination yields the most accurate and stable identification of diffuse aurora events among all evaluated methods.

To further validate the effectiveness of our method in diffuse aurora segmentation tasks, we conducted comprehensive qualitative analysis experiments. Figure 9 presents a comparative visualization of segmentation results obtained by applying our method and four mainstream methods to the same untrimmed all-sky auroral video (total length

https://doi.org/10.3390/rs18030402

of 1000 frames). As observed in the figure, the frame-based methods MS-TCN and AS-Former both exhibit minor segmentation errors at auroral event boundaries, manifested as misclassification between true positive (TP) and true negative (TN) regions, as well as confusion between false positive (FP) and false negative (FN) instances. The two-stage method UVAST demonstrates severe over-segmentation phenomena, generating numerous erroneous boundaries across multiple temporal intervals, which results in continuous diffuse aurora events being incorrectly partitioned into separate events. The parallel method FACTM reduces local segmentation errors through its frame-action bidirectional information exchange mechanism; however, due to the intrinsic gradual evolution characteristics of diffuse aurora, its boundary localization accuracy remains insufficient during the onset and decay phases. In contrast, the segmentation results of our method exhibit high consistency with the ground truth annotations, achieving the most precise temporal segmentation of diffuse aurora events.

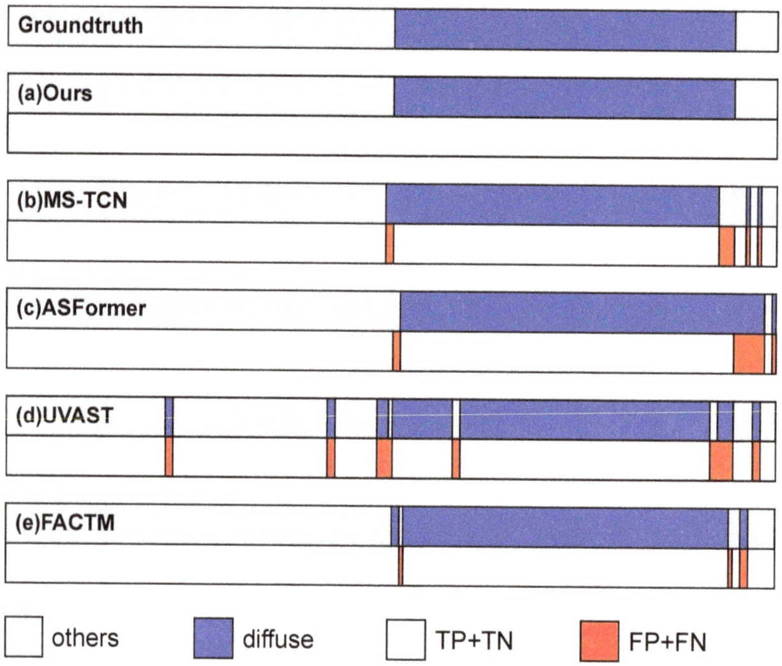

Figure 9. Qualitative comparison of identification and segmentation results for diffuse aurora using different methods. (**a**) Our proposed method; (**b**) MS-TCN; (**c**) ASFormer; (**d**) UVAST; (**e**) FACTM.

Figure 10 further visualizes the frame-to-frame feature differences extracted by each method, where the vertical axis represents the normalized consecutive inter-frame feature difference intensity. Analysis reveals that our method generates clear and sharp feature difference peaks at true diffuse aurora event boundaries, indicating its capability to effectively capture critical temporal transition points of diffuse aurora phenomena. In comparison, MS-TCN exhibits relatively smooth inter-frame difference signals with low sensitivity to boundary features, leading to ambiguous temporal localization. Although ASFormer shows feature difference responses at certain boundaries, the overall signal contains substantial noise components that may induce misclassifications. UVAST's feature difference distribution is unstable, with numerous spurious peaks throughout the sequence, which fundamentally explains the underlying cause of its severe over-segmentation phenomenon. While the FACTM method displays inter-frame difference responses at major boundaries,

https://doi.org/10.3390/rs18030402

the peak prominence is insufficient, and considerable interference signals exist in non-boundary regions, constraining its boundary localization accuracy. In summary, through explicit modeling of the gradual evolution patterns of diffuse aurora and adaptive feature enhancement mechanisms, our method achieves precise characterization of event boundaries at the feature representation level, thereby demonstrating superior segmentation performance in both qualitative and quantitative evaluations.

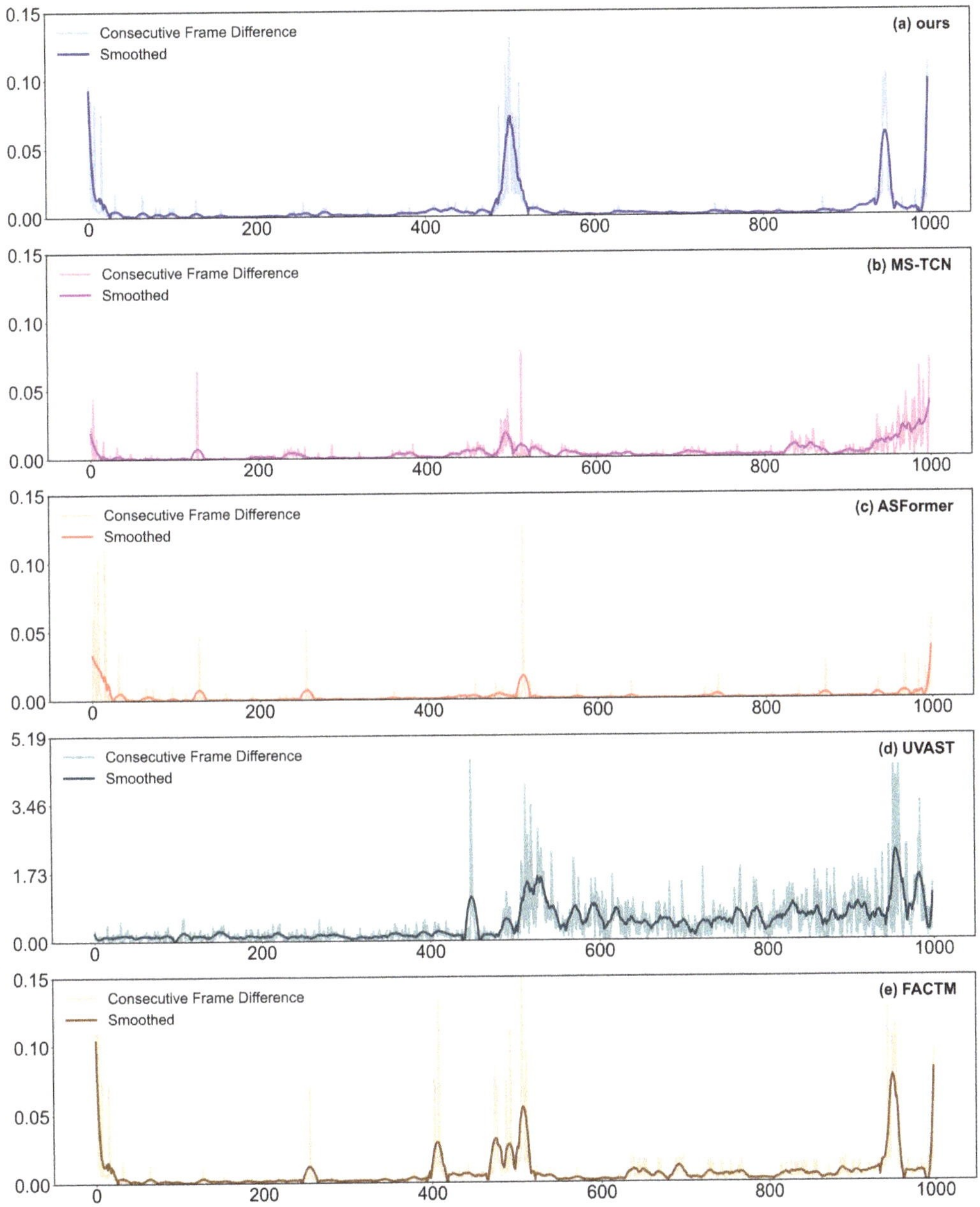

Figure 10. Comparative of inter-frame differences for the identification and segmentation of diffuse auroras using different methods. (**a**) Our proposed method; (**b**) MS-TCN; (**c**) ASFormer; (**d**) UVAST; (**e**) FACTM.

https://doi.org/10.3390/rs18030402

4.3. Ablation and Sensitivity Analysis

4.3.1. Effectiveness of Modules

To validate the effectiveness of each module in our method and their contributions to overall performance, we conducted systematic ablation experiments. The experimental design first establishes a baseline model that directly processes initial video frame features extracted by the I3D network to learn preliminary representations of action tokens and frame features. Building upon this baseline, we progressively integrate the FCIM and SGSM to evaluate their individual effects and synergistic contributions to model performance.

The quantitative results of the ablation experiments are presented in Table 2. When the FCIM and SGSM are independently integrated into the baseline model, all major evaluation metrics achieve significant improvements. Specifically, the integration of the FCIM alone yields the most substantial improvement in Acc, elevating it from the baseline's 80.1% to 90.1%, whereas the addition of the SGSM alone produces a more pronounced enhancement in the Edit metric, improving it from the baseline's 70.6% to 78.6%. This differential improvement stems from the distinct functional emphases of the two modules: the FCIM employs an iterative bidirectional information exchange mechanism that addresses the weak structural morphological characteristics of diffuse aurora by supplementing morphological cues with rich dynamic features, capturing inter-frame correlation features across the entire temporal range, and thereby significantly enhancing overall recognition accuracy; in contrast, the SGSM focuses on fine-grained characterization of the onset and decay phases of diffuse aurora, introducing snippet descriptor aggregation and BiGRU temporal context modeling mechanisms to enhance snippet-level semantic guidance specifically tailored to the gradual evolution characteristics of diffuse aurora, making it particularly suitable for gradual boundary refinement tasks and thus exhibiting superior performance in Edit. When both the FCIM and SGSM are simultaneously integrated into the network, both Acc and Edit scores achieve further improvements, reaching 89.6% and 80.7%, respectively; however, the F1@0.50 score remains at a relatively low level, indicating that the model has not yet achieved optimal stability in boundary localization.

Table 2. Ablation study on the contributions of the proposed FCIM and SGSM. Results are reported as mean ± standard deviation over five independent runs. "✓" and "–" indicate the presence and absence of the MoE module, respectively. The best results are highlighted in bold.

Baseline	FCIM	SGSM1	SGSM2	Acc	Edit	F1@{10, 25, 50}		
✓	–	–	–	80.1 ± 1.9	70.6 ± 2.5	66.4 ± 2.9	63.4 ± 2.4	52.7 ± 3.7
✓	✓	–	–	90.1 ± 0.2	78.6 ± 1.1	79.4 ± 0.9	76.4 ± 0.9	62.9 ± 1.0
✓	–	✓	–	89.6 ± 0.5	80.7 ± 1.6	82.1 ± 1.4	78.6 ± 1.4	66.7 ± 1.2
✓	✓	✓	–	95.0 ± 0.7	84.6 ± 0.4	85.2 ± 1.0	85.2 ± 1.0	78.5 ± 1.0
✓	✓	✓	✓	**96.3 ± 0.4**	**87.7 ± 0.2**	**86.8 ± 0.5**	**86.6 ± 0.5**	**83.0 ± 0.4**

Our method achieves optimal performance across all evaluation metrics by simultaneously integrating the FCIM with the two-step cascaded SGSM, with Acc, Edit, and F1@{10, 25, 50} reaching 96.3%, 87.7%, 86.8%, 86.6%, and 83.0%, respectively. This architectural design first achieves deep interaction between morphological and dynamic features in the FCIM to generate initial candidate segmentation results, then explicitly models the gradual evolution behavior of diffuse aurora through a two-step cascaded refinement process in the SGSM, providing fine-grained processing of onset and decay phase boundaries, and ultimately producing accurate and physically interpretable temporal segmentation results.

Figure 11 further illustrates a comparison of diffuse aurora identification and segmentation results in the same video sequence when only the FCIM is added (Baseline + FCIM) versus when both the FCIM and SGSM are fully integrated (Baseline + FCIM + SGSM).

https://doi.org/10.3390/rs18030402

We selected a video containing two representative snippets for visualization analysis—corresponding to a pulsating aurora scenario and a complex scenario where diffuse aurora coexists with discrete auroral structures. The results demonstrate that because the FCIM emphasizes feature modeling across the entire temporal range, the frame-by-frame processing approach generates considerable noisy predictions when confronted with highly similar consecutive frames, leading to spurious segmentation phenomena in ambiguous scenarios. In contrast, the complete method effectively suppresses the influence of temporal noise information through the application of BiGRU processing to snippet-level descriptors in the SGSM, while simultaneously enhancing the representation capability for gradual transition features, thereby achieving accurate identification of gradual onset and decay boundaries. The visualization results clearly show that our method can accurately capture the true boundaries of diffuse aurora events, whereas the configuration using only FCIM produces boundary offset errors of four frames and two frames, validating the critical role of the SGSM in boundary refinement tasks.

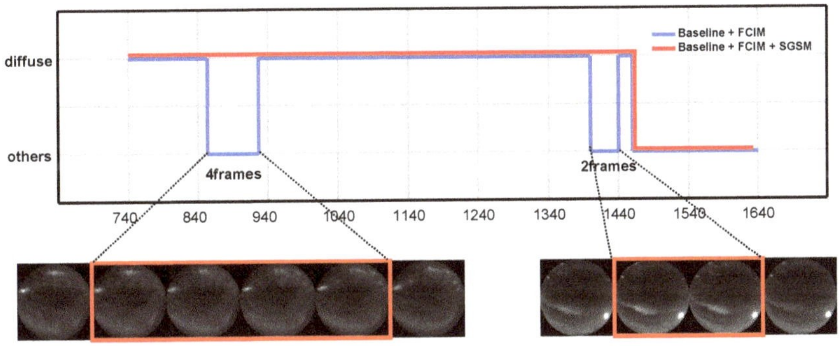

Figure 11. Comparative analysis of segmentation results between FCIM-only and FCIM + SGSM integrated models. The red boxes highlight ASI frames corresponding to time intervals where the two models produce different detection results.

4.3.2. Effectiveness of MoE

This section delves into the effectiveness of the MoE mechanism across different modules and its contribution to model performance. Specifically, we add and remove the MoE mechanism in the FCIM and SGSM separately, and systematically evaluate the architectural design that achieves optimal performance by comparing experimental results of different configuration combinations.

As shown in Table 3, compared to the baseline configuration without the MoE mechanism, when MoE is independently integrated into either the FCIM or SGSM alone, model performance metrics improve but have not yet reached optimal performance. Specifically, introducing MoE only in the FCIM achieves 95.2% Acc and 85.1% Edit, whereas introducing MoE only in the SGSM achieves 95.9% Acc and 86.7% Edit. The MoE mechanism, based on the intrinsic characteristics of diffuse aurora's extended duration and slow evolution, adaptively selects and integrates specialized expert pathways that focus on morphological feature extraction, motion feature capture, or background noise suppression through gating networks, thereby achieving enhancement of discriminative features and suppression of redundant information. However, introducing the MoE mechanism in only a single module cannot adequately address the feature redundancy problem generated at different stages during the learning process, limiting further performance improvement. Experimental results demonstrate that simultaneously integrating the MoE mechanism in both the FCIM and SGSM achieves optimal performance metrics. This complete configuration effectively addresses the core challenges of weak structural morphological features of diffuse au-

https://doi.org/10.3390/rs18030402

rora, highly redundant inter-frame features caused by gradual onset and decay processes, and difficulties in boundary localization, achieving precise identification and temporal segmentation of diffuse aurora events in untrimmed all-sky auroral video sequences through adaptive focus on key frame features.

Table 3. Ablation study on the integration strategies of the Mixture of Experts (MoE) mechanism within the FCIM and SGSM. Results are reported as mean ± standard deviation over five independent runs. "✓" and "–" indicate the presence and absence of the MoE module, respectively. The best results are highlighted in bold.

MoE in FCIM	MoE in SGSM	Acc	Edit	F1@{10, 25, 50}		
–	–	93.2 ± 1.3	84.9 ± 2.8	84.2 ± 3.3	84.2 ± 3.3	79.6 ± 2.1
✓	–	95.2 ± 0.2	85.1 ± 0.3	85.2 ± 0.9	85.2 ± 0.9	81.5 ± 0.2
–	✓	95.9 ± 0.3	86.7 ± 0.3	85.8 ± 0.7	85.8 ± 0.7	82.5 ± 0.6
✓	✓	**96.3 ± 0.4**	**87.7 ± 0.2**	**86.8 ± 0.5**	**86.6 ± 0.5**	**83.0 ± 0.4**

Figure 12 presents the performance of the MoE mechanism across different module configurations as a three-dimensional performance surface, clearly illustrating the trends of Acc and Edit as functions of the MoE integration status in the FCIM and SGSM. The visualization results intuitively demonstrate that when both the FCIM and SGSM employ the MoE mechanism, model performance reaches a global optimum, achieving significant improvement compared to the configuration without MoE. The gradient distribution of the performance surface reveals the synergistic enhancement effect of the MoE mechanism in the two modules: introducing MoE in either module alone can bring performance improvement, but only the dual-module synergistic configuration can fully exploit the comprehensive advantages of adaptive feature selection and multi-scale redundancy suppression.

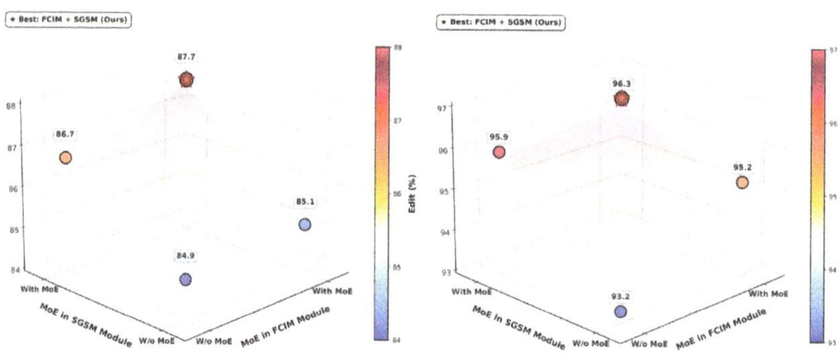

Figure 12. Three-dimensional performance visualization of MoE mechanism integration across different modules. Colored circles denote different MoE configurations, where the color indicates the corresponding performance value as shown by the colorbar. The star marker highlights the best configuration (FCIM + SSGM).

To further intuitively demonstrate the feature enhancement performance of the MoE mechanism in focusing on key frames and suppressing redundant information, we conducted a comparative analysis of the temporal evolution characteristics of frame-by-frame features in untrimmed all-sky auroral video sequences before and after applying the MoE mechanism. Figure 13a,b show frame-level feature analysis results under two different auroral activity scenarios, where the blue and red curves represent the feature evolution states before and after applying the MoE mechanism, respectively. Each set of visualization

https://doi.org/10.3390/rs18030402

results contains two subplots: the upper plot shows normalized inter-frame feature difference analysis, characterizing the magnitude of feature changes between consecutive frames; the lower plot shows cosine similarity analysis of inter-frame features, quantifying the consistency degree of feature representations between adjacent frames. The results indicate that before applying the MoE mechanism, inter-frame features exhibit highly complex and unstable temporal evolution patterns, primarily attributable to the intrinsic weak structural morphological characteristics of diffuse aurora, background noise interference, and slowly gradual evolution characteristics, resulting in blurred event boundary signals that are difficult to precisely localize. After enhancement by the MoE mechanism, frame-level features dynamically select and weight-fuse specialized expert pathways through gating networks, effectively suppressing redundant temporal information and background noise interference while adaptively focusing on more discriminative key frame features, thereby achieving precise characterization of diffuse aurora event boundaries at the feature representation level. Specifically, this manifests as at event boundaries, inter-frame difference signals exhibit sharper peak responses, while during stable phases within events, feature similarity maintains higher stability with significantly reduced noise fluctuations. This feature enhancement mechanism fundamentally improves the model's representation capability for the gradual transition characteristics of diffuse aurora, establishing a solid foundation for achieving high-precision temporal segmentation.

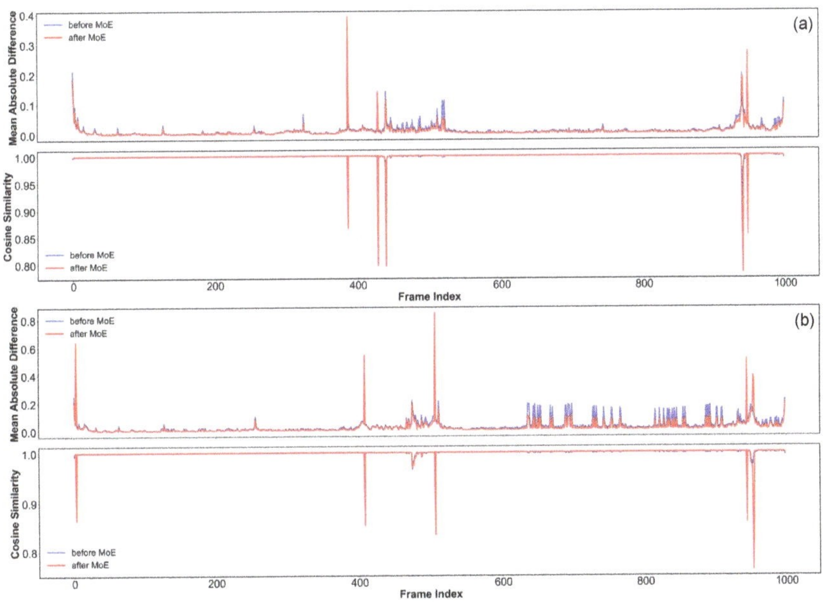

Figure 13. Comparison of per-frame feature stability before and after the application of the MoE. (**a**,**b**) show two example sequences.

4.4. Case Studies of Challenging Diffuse Aurora Events

Diffuse aurora presents several characteristics that make automated identification and segmentation particularly challenging: its luminosity is weak and often spatially uniform, its boundaries are fuzzy and slowly evolving, and its temporal behavior can include quasi-periodic modulations such as pulsating patches. These properties not only weaken the contrast within individual frames, but also cause substantial ambiguity when diffuse structures coexist with bright discrete aurora or when cloud motion introduces similar large-scale smooth patterns in all-sky images.

https://doi.org/10.3390/rs18030402

To demonstrate the robustness of our method under these conditions, Figure 14 presents a set of representative challenging cases. (a) Intra-class diversity: Diffuse aurora itself exhibits substantial variability in brightness, spatial extent, and temporal evolution—including classical diffuse sheets and pulsating diffuse structures. Our method successfully identifies these diverse patterns, capturing both stable and rapidly modulated diffuse emissions. (b) Hard negative cases: Diffuse aurora is easily confused with weak discrete arcs or cloud-covered scenes, both of which produce smooth, low-contrast signatures that resemble diffuse luminosity in single-frame observations. The model effectively suppresses these hard negatives by leveraging temporal evolution cues and expert-guided feature modulation, yielding correct rejections in these ambiguous intervals. (c) Overlapping phenomena: When diffuse aurora coexists with bright discrete structures, the diffuse component may appear extremely subtle and is prone to omission by traditional keogram-based inspection. Our method still retrieves the diffuse intervals by consolidating long-range temporal continuity and weak structural cues, producing consistent identification even when diffuse signatures are visually dominated by discrete emissions. These case studies highlight the method's capability to handle complex real-world auroral scenes that challenge both manual keogram inspection and conventional vision-based approaches, demonstrating its practical utility for large-scale auroral video analysis.

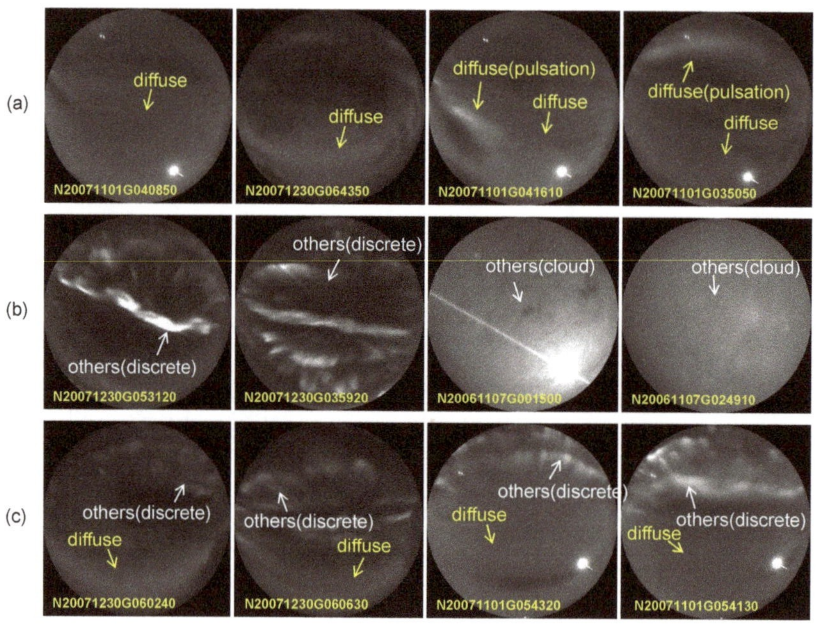

Figure 14. Representative challenging cases for diffuse-aurora identification. (**a**) Large intra-class variation in brightness, extent, and motion, all correctly identified. (**b**) False-positive-prone scenes with weak discrete arcs or cloud cover, where our method remains stable. (**c**) Coexisting diffuse and discrete aurora, where weak diffuse emission is still successfully detected.

A further challenge of diffuse aurora identification lies in its gradual temporal evolution: the onset is typically weak and spatially inconspicuous, and the transition between growth, plateau, and decay often lacks sharp visual boundaries. To illustrate how our method performs under this condition, we present two representative sequences, each arranged in a consistent diagnostic layout: from top to bottom, the keogram for the same time interval (geomagnetic north–south column stacked in time), the automated identification

https://doi.org/10.3390/rs18030402

and segmentation generated by our framework, and a small set of all-sky frames at physically meaningful moments (onset, plateau, decay, or ambiguous background conditions). The keogram offers a compact view of large-scale temporal structure, the segmentation band beneath it shows the frame-wise labels produced by our model, and the selected image snapshots make the physical interpretation explicit.

Figure 15 shows the results for the 28 December 2007 sequence (06:00–10:00 UT). In the early stages of diffuse-emission development, the luminous patch exhibits both small spatial extent and low radiance, producing only subtle traces in the keogram. As the event evolves, remnants of discrete structures intermittently appear and complicate the temporal profile, making human inspection of keograms prone to imprecise onset and termination markings. In contrast, the proposed method leverages both the morphological signature of diffuse emission (broad, uniform, and slowly varying bands) and its temporal behavior (extended duration without rapid structural transitions). These two cues reinforce one another, enabling our framework to produce an onset estimate that aligns with the physical scene observed in the image frames.

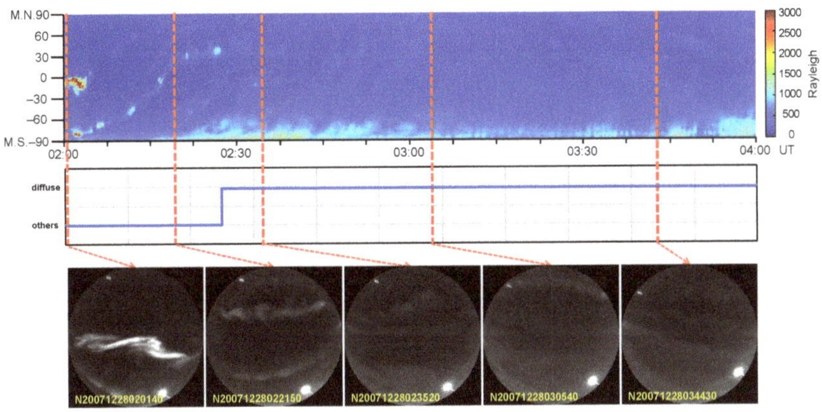

Figure 15. Results for 28 December 2007 (06:00–10:00 UT). The keogram (top panel) exhibits diffuse-like signatures; however, the proposed method correctly classifies the interval as non-diffuse. The blue curve in the middle panel denotes the automatically identified diffuse aurora label over time. Red dashed vertical lines indicate the time instants selected from the keogram, and the corresponding all-sky imager (ASI) frames are shown in the bottom panel, as highlighted by the red arrows. The ASI images confirm that the observed emissions are dominated by cloud cover rather than true diffuse aurora.

Figure 16 presents the identification and segmentation results for the all-sky auroral sequence recorded on 14 November 2007 (02:00–04:00 UT). A keogram-only inspection could easily misinterpret the entire interval as diffuse aurora, since the stacked intensity pattern resembles a weak, persistent band. However, our method labels the entire duration as *non-diffuse*. Inspection of the all-sky frames confirms that the scene is dominated by cloud cover rather than auroral emission. This case highlights a common pitfall of keogram-based screening: cloud-induced brightness variations may mimic the large-scale temporal smoothness of diffuse aurora, whereas our model, operating on full-frame spatial information and learned morphological cues, correctly rejects such false positives.

https://doi.org/10.3390/rs18030402

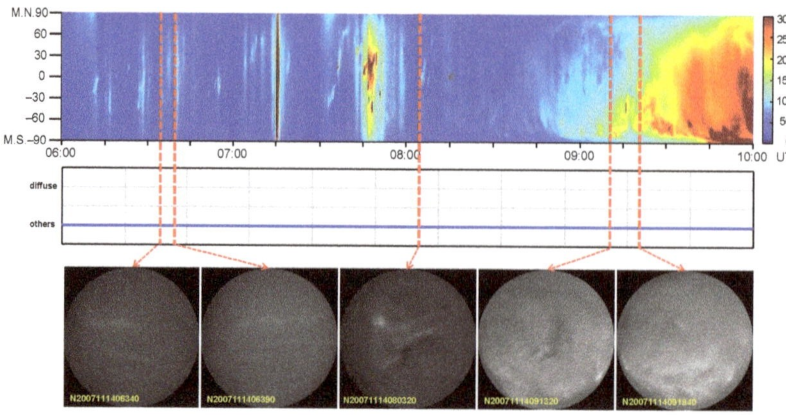

Figure 16. Diffuse-aurora identification on 14 November 2007 (02:00–04:00 UT). The blue curve denotes the automatically identified diffuse aurora label, and red dashed lines on keogram indicate the time instants of the displayed ASI frames.

4.5. Method Validation Across Different Geomagnetic Activity Levels

The occurrence of diffuse aurora is closely linked to geomagnetic activity, as its generation mechanisms—primarily involving wave-particle interactions with whistler-mode chorus waves and electron cyclotron harmonic (ECH) waves—are modulated by magnetospheric dynamics [7,15]. The Kp index (Planetarische Kennziffer) is a standardized measure of global geomagnetic activity that quantifies disturbances in the horizontal component of Earth's magnetic field [34,36]. This index is derived from weighted averages of observations from 13 mid-latitude geomagnetic observatories worldwide and represents a 3 h quasi-logarithmic local index. The Kp index has a direct physical connection to diffuse aurora occurrence through its influence on plasmasphere configuration and cold plasma distribution, which in turn affects the generation efficiency of plasma waves responsible for scattering electrons into the loss cone [35]. To evaluate the applicability of the proposed method under different geomagnetic conditions, we conducted systematic experiments across three representative geomagnetic activity states: geomagnetically quiet periods, moderate-to-active periods, and substorm intervals. These conditions were characterized using the Kp index and the SuperMAG SML index, respectively.

4.5.1. Case Study: Method Performance During Geomagnetically Quiet Conditions

Figure 17 presents an all-sky auroral observation sequence from YRS during 00:00–12:00 UT on 29 November 2006. The panels from top to bottom display: keogram showing auroral intensity evolution; classification results from the proposed method (diffuse aurora versus other types); Kp index variation represented as a color-coded bar chart, where yellow indicates geomagnetically quiet conditions and orange indicates moderate activity; and representative all-sky images at key moments. The Kp index evolution during this interval was 2.3 $\rightarrow$ 3.7 $\rightarrow$ 2.0 $\rightarrow$ 0.7, exhibiting a complete geomagnetic activity cycle transitioning from quiet conditions through moderate activity and returning to quiet conditions.

https://doi.org/10.3390/rs18030402

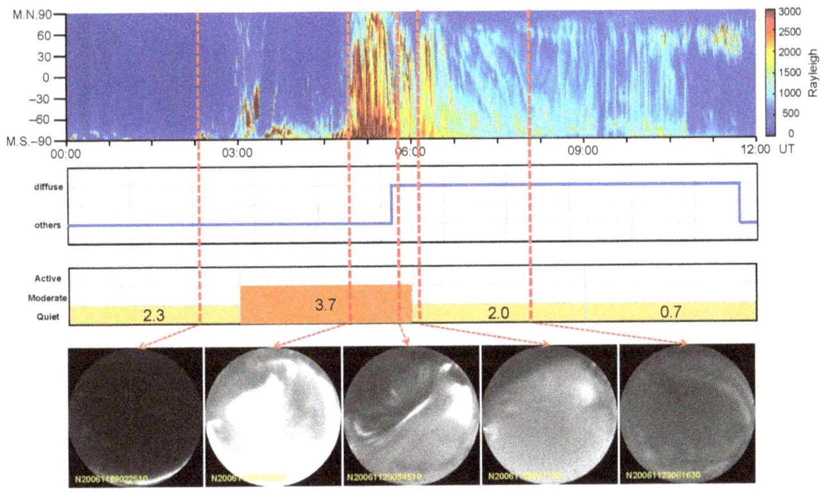

Figure 17. Method validation during geomagnetically quiet-to-moderate conditions: Case study of 29 November 2006 (00:00–12:00 UT). The yellow and orange intervals indicate periods of quiet and moderate geomagnetic conditions, respectively. The annotated numbers denote the Kp index. The blue curve denotes the automatically identified diffuse aurora label, and red dashed lines on keogram indicate the time instants of the displayed ASI frames.

Initial phase (Kp = 2.3, 00:00–03:00 UT): despite the quiet-level Kp index, the keogram during this phase indicates weak auroral activity, with no distinct diffuse aurora signatures visible in the all-sky images. The proposed method achieves accurate identification based on actual morphological and dynamical characteristics, demonstrating that it does not erroneously classify intervals as diffuse aurora solely due to low Kp values.

Moderate activity phase (Kp = 3.7, 03:00–06:00 UT): As the Kp index increases, magnetospheric disturbances intensify, resulting in significantly enhanced auroral intensity. The keogram reveals pronounced high-intensity regions, indicating the emergence of discrete auroral structures. The corresponding all-sky images show a transition from intense discrete auroral events to a mixed state where discrete emission structures coexist with diffuse aurora. This phase, characterized by the coexistence of discrete and diffuse auroral morphologies, poses challenges for automated detection algorithms, as strong discrete emissions may partially obscure or locally dominate the weaker diffuse background. Despite these challenges, the proposed method successfully identifies diffuse auroral occurrences prior to 05:45 UT. The method focuses on the slowly varying, spatially extended emission patterns characteristic of diffuse aurora, while being less sensitive to the rapidly evolving and highly localized structures associated with discrete aurora. As a result, diffuse auroral activity can still be detected during this mixed phase, even when discrete auroral emissions are present and visually dominant.

Pure diffuse aurora phase (Kp = 2.0 → 0.7, 06:00–12:00 UT): As geomagnetic activity weakens, discrete auroral structures gradually diminish, and diffuse aurora becomes the dominant emission type. The keogram and all-sky images show a transition toward spatially extended, low-contrast, and slowly evolving auroral emissions, which are characteristic of diffuse aurora. During this phase, the proposed method consistently identifies diffuse auroral activity and accurately delineates its temporal extent, demonstrating its ability to capture the gradual onset and decay processes typical under geomagnetically quiet conditions.

https://doi.org/10.3390/rs18030402

4.5.2. Case Study: Method Performance During Geomagnetically Active Conditions

Figure 18 presents an all-sky auroral observation sequence during 03:00–15:00 UT on 6 January 2005. The panels show: (a) a keogram illustrating the temporal evolution of auroral intensity; (b) classification results produced by the proposed method; (c) the corresponding Kp index variation; and (d) representative all-sky images at selected times.

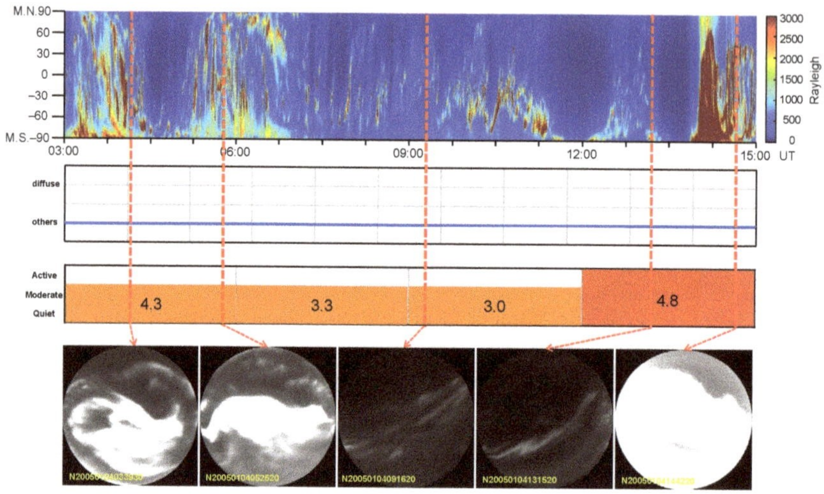

Figure 18. Method validation during geomagnetically active conditions: Case study of 6 January 2005 (03:00–15:00 UT). The orange and red intervals indicate periods of moderate and active geomagnetic conditions, respectively. The annotated numbers denote the Kp index, which decreases from 4.3 to 3.3 and 3.0 during the moderate phase and subsequently increases to 4.8 during the active phase. The blue curve denotes the automatically identified diffuse aurora label, and red dashed lines on keogram indicate the time instants of the displayed ASI frames.

During this interval, the Kp index varies from 4.3 to 3.3 and 3.0 and then increases again to 4.8, indicating persistently moderate to active geomagnetic conditions. Visual inspection of the all-sky images shows that the field of view is dominated by intense, spatially structured discrete auroral emissions. In several periods, extended diffuse-like glow regions are also visible, which may appear morphologically similar to diffuse aurora in terms of spatial smoothness and reduced edge contrast.

However, when examined in the temporal context provided by the keogram and image sequence, these diffuse-like regions are closely associated with rapid intensity fluctuations and structural reconfigurations of nearby discrete auroral forms, rather than exhibiting the gradual drift and quasi-periodic modulation commonly observed in diffuse aurora. The proposed method leverages joint morphological and temporal cues to distinguish such cases and does not identify diffuse aurora during this interval. This case illustrates the method's ability to avoid false positive detections of diffuse aurora under geomagnetically active conditions where discrete auroral activity dominates.

4.6. Case Study: Method Performance During Substorm Intervals

Figure 19 presents a typical moderate-intensity substorm event ($SML_{min} = -417$ nT) that occurred from 23:43 UT on 30 November 2007 to 02:08 UT on 1 December 2007. The panels from top to bottom display (a) SML index evolution with three substorm phases annotated; (b) keogram showing auroral intensity evolution; (c) classification results from the proposed method (diffuse aurora versus other types); and (d) representative all-sky images at key moments.

https://doi.org/10.3390/rs18030402

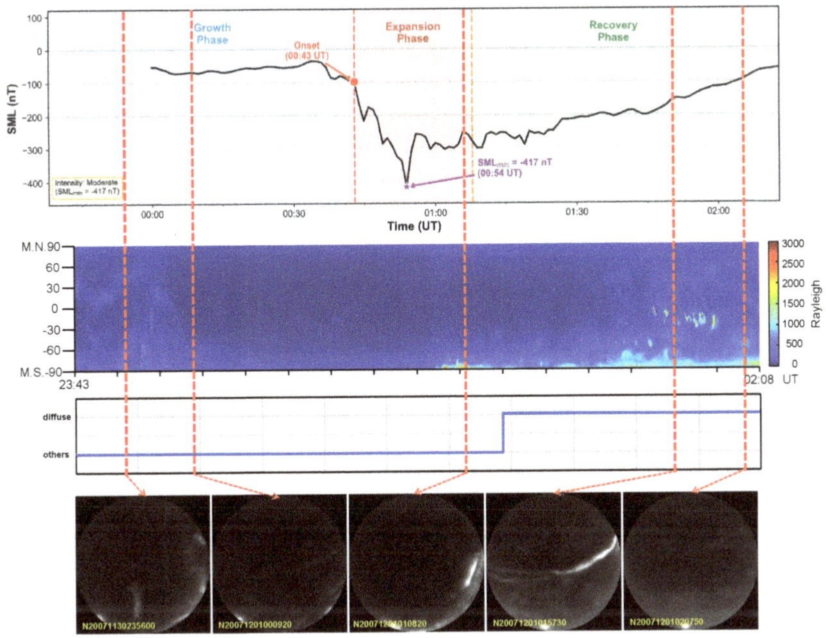

Figure 19. Method validation during a substorm event: Case study of 30 November–1 December 2007 (23:43–02:08 UT). The blue curve denotes the automatically identified diffuse aurora label, and red dashed lines on keogram indicate the time instants of the displayed ASI frames.

The all-sky images reveal the following characteristics across the substorm evolution: Growth-to-expansion phase transition (00:20–00:43 UT)—as wave-particle interactions intensify, discrete auroral structures begin to brighten. Expansion phase (00:43–01:00 UT): Bright discrete arc structures dominate the field of view, exhibiting rapid morphological changes. Recovery phase (01:00–02:08 UT): Although discrete arc structures persist initially, diffuse aurora gradually recovers and becomes the dominant type, displaying pronounced pulsation characteristics.

During the transition from the expansion phase to the recovery phase, the all-sky images show a gradual shift from dominant discrete aurora to a mixed state where discrete and diffuse aurora coexist. The gradual evolution of diffuse aurora combined with the influence of bright discrete auroral structures poses challenges for accurate onset localization of diffuse auroral events. The classification results demonstrate that the proposed method accurately captures the timing of diffuse aurora emergence, with the transition boundary precisely localized through the SGSM's snippet-level temporal aggregation and BiGRU contextual modeling.

During this event, YRS was located at approximately 03:43–05:08 MLT (post-midnight to dawn sector), positioned precisely in the region where injected electrons drift eastward following substorm onset. This observational location and temporal configuration are consistent with the characteristics described by [2,45], who reported that diffuse aurora occurrence peaks in the post-midnight sector approximately 30–40 min after substorm onset. This agreement validates the effectiveness of the proposed method in physically reasonable substorm–diffuse aurora association scenarios.

https://doi.org/10.3390/rs18030402

4.7. Statistical Study of Diffuse Auroral Events

To investigate the temporal distribution patterns of diffuse aurora, we applied the model trained on manually labeled all-sky aurora video data from 2003 to 2007 to automatically identify and segment unlabeled all-sky aurora data from 2008 and 2009. Specifically, we selected all-sky aurora observation data covering 48 days (October to December 2008) and 35 days (October to December 2009) for automatic identification and segmentation. To more intuitively present the temporal distribution patterns of diffuse aurora, we divided the 12 h period from 03:00 UT to 15:00 UT into intervals of 6 min each, creating a total of 120 intervals. Given that the temporal resolution of consecutive all-sky aurora images is 10 s, each time interval corresponds to 36 aurora frames. The number of frames identified as diffuse aurora in each interval was counted, and their occurrence probability across different time periods was calculated, thereby intuitively revealing their distribution patterns.

Figure 20 presents the temporal distribution statistics of diffuse aurora for 2008 and 2009. Figure 20 show the results for each year, respectively. The bars (colored from light blue to deep blue) reflect the number of frames in which diffuse aurora was identified within each time interval, with color intensity proportional to frame count, while the red curves illustrate the temporal variation trends of diffuse aurora occurrence rate. Statistical results indicate that diffuse auroras occur significantly more frequently in the morning than in the afternoon, with the peak appearing at approximately 07:00 UT and the minimum at approximately 11:00 UT This pattern is consistently observed in both years' data. This temporal distribution pattern is consistent with the particle precipitation physical mechanism of diffuse aurora. Diffuse aurora is primarily caused by hot electrons associated with substorm activity and injected into the central plasma sheet. After injection near midnight, these hot electrons predominantly drift toward the dawnside. During the drift process, a portion of the hot electrons is scattered into the loss cone via wave-particle interaction, subsequently precipitating and exciting diffuse aurora; meanwhile, the number density of the source electron population gradually diminishes. Since auroral emission intensity is proportional to the precipitating electron flux, the decline in electron number density in the source region directly leads to a reduction in the effective electron flux capable of producing visible aurora. Therefore, from midnight to noon, the number density of source electrons continuously decreases during the drift process. Although wave-particle interaction still causes some electron precipitation near noon (around 11:00 UT), the emission intensity has diminished to a level that is difficult to clearly identify by conventional observation equipment, resulting in the trough in diffuse aurora occurrence rate during this period. Conversely, during the morning hours of the drift process (around 07:00 UT), the source electron number density remains relatively high, capable of effectively exciting diffuse aurora over a larger spatial extent, thereby explaining the formation of the occurrence rate peak. The observed distribution pattern is consistent with the diffuse aurora behavior reported in previous studies [9], validating the effectiveness of the method presented in this work.

https://doi.org/10.3390/rs18030402

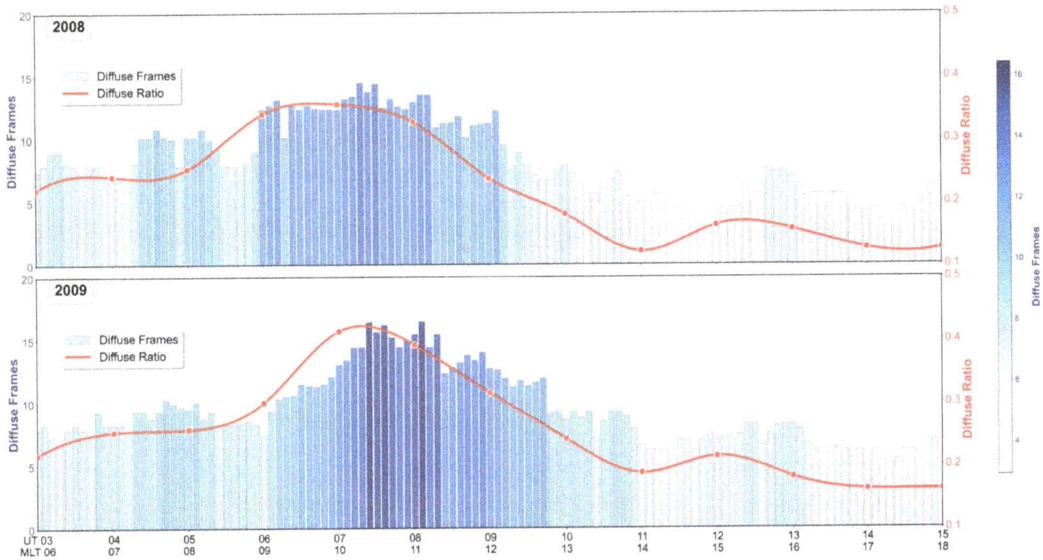

Figure 20. Occurrence distributions of diffuse aurora based on automatically identified results. The bars (colored from light blue to deep blue) show the number of frames in which diffuse aurora is identified within each time interval. The red curves illustrate the temporal variation of the diffuse aurora occurrence rate, while the red dots denote the values at exact hourly timestamps.

5. Discussions

This study demonstrates that diffuse aurora can be reliably identified and temporally segmented from long, untrimmed all-sky auroral image sequences by explicitly modeling its weak spatial morphology and gradual temporal evolution. The consistently high identification accuracy achieved across diverse observational conditions indicates that the proposed framework captures key phenomenological characteristics of diffuse aurora that have long challenged traditional analysis methods.

Previous studies have primarily relied on keograms, visual inspection, or heuristic morphological criteria to identify diffuse auroral intervals. While effective for qualitative analysis and selected case studies, such approaches are inherently limited when handling weak optical signatures, gradual onset and decay, or complex scenarios where diffuse aurora coexists with discrete auroral structures. In particular, keogram-based methods collapse two-dimensional spatial information into a single latitudinal slice, which can obscure subtle spatial–temporal patterns and introduce ambiguity. By contrast, our framework operates directly on full all-sky image sequences and enforces temporal continuity, enabling more robust discrimination and more precise localization of diffuse auroral onset and decay.

The observed performance gains can be interpreted in the context of the physical nature of diffuse aurora. Unlike discrete aurora, which is characterized by sharp spatial boundaries and rapid intensity variations, diffuse aurora reflects prolonged precipitation processes associated with magnetospheric transport and wave–particle interactions. Its optical manifestation is therefore weakly structured in space but coherent over extended time scales. The snippet-based temporal aggregation and bidirectional temporal modeling adopted in this work align naturally with the gradual evolution of diffuse aurora, providing a physically meaningful abstraction of its onset–development–decay behavior rather than a purely mathematical segmentation.

https://doi.org/10.3390/rs18030402

Statistical analysis shows that a substantial fraction of the diffuse aurora events in our dataset occur under geomagnetically quiet to mildly disturbed conditions (low to moderate Kp), which is consistent with established observations that such conditions are favorable for diffuse auroral precipitation. Notably, additional case studies demonstrate that the proposed method remains effective across a range of geomagnetic activity levels. Under moderately active conditions, the method can still identify diffuse aurora embedded within mixed auroral scenes, while under persistently active conditions dominated by intense discrete aurora, it avoids falsely labeling diffuse-like background emissions as genuine diffuse aurora. These results suggest that the method does not rely on geomagnetic indices as explicit inputs, but instead learns discriminative visual–temporal patterns that remain stable across varying geomagnetic environments.

Several limitations should nevertheless be acknowledged. The refinement stage depends on the quality of coarse candidate intervals generated during initial identification, which may lead to error propagation in extremely ambiguous cases. Optical observations are also affected by atmospheric conditions, cloud cover, and instrumental sensitivity, introducing unavoidable uncertainty. In addition, annotation ambiguity is non-negligible for weak or transitional auroral states, and finite temporal resolution may limit boundary precision during rapid transitions.

6. Conclusions

This study presents an automated framework for the identification and temporal segmentation of diffuse aurora from long, untrimmed all-sky auroral video sequences. By modeling diffuse aurora as a video segmentation problem and tailoring the framework to its weak morphology and gradual temporal evolution, the proposed method achieves reliable identification and accurate boundary localization under complex observational conditions.

Experimental results on multi-year 557.7 nm all-sky imaging data from the Arctic YRS demonstrate strong performance in both identification and segmentation, achieving 96.3% frame-wise accuracy and an Edit score of 87.7%. The method effectively handles challenging scenarios including cloud contamination, low-brightness emissions, and the coexistence of diffuse and discrete aurora.

Based on the automated results, large-scale statistical analyses of diffuse aurora occurrence and temporal distribution were conducted, revealing consistent diurnal patterns across multiple years. These results confirm the feasibility of using intelligent, vision-based approaches for systematic analysis of long-term auroral observations. The proposed framework provides a practical tool for efficient processing of extensive auroral image archives and supports future data-driven investigations of auroral dynamics and magnetosphere-ionosphere coupling.

Author Contributions: Conceptualization, Q.W., P.H. and H.P.; methodology, Q.W. and P.H.; software, P.H.; validation, P.H.; formal analysis, P.H. and H.P.; data curation, P.H.; writing—original draft preparation, P.H.; writing—review and editing, Q.W.; visualization, P.H.; supervision, Q.W. and H.P.; project administration, Q.W. and H.P.; funding acquisition, Q.W. All authors have read and agreed to the published version of the manuscript.

Funding: This work was supported by the National Natural Science Foundation of China (Grant No. 41874173), the Aviation Key Laboratory of Science and Technology on Aerospace Vehicle (Grant No. J2025-STAV-03-001), and the National Key Laboratory of Science and Technology on Space-Born Intelligent Information Processing (Grant No. TJ-04-23-03).

Data Availability Statement: Data will be made available on request.

Acknowledgments: The authors would like to thank the Polar Research Institute of China for providing the auroral data.

https://doi.org/10.3390/rs18030402

Conflicts of Interest: All authors declare that they have no conflicts of interest related to this work.

References

1. Newell, P.T.; Wing, S.; Meng, C.-I.; Sigillito, V. The Auroral Oval Position, Structure, and Intensity of Precipitation from 1984 Onward: An Automated On-Line Data Base. *J. Geophys. Res. Space Phys.* **1991**, *96*, 5877–5882.
2. Meng, C.-I.; Rycroft, M.J.; Frank, L.A. *Auroral Physics;* Cambridge University Press: Cambridge, UK, 1991.
3. Frey, H.U.; Phan, T.D.; Fuselier, S.A.; Mende, S.B. Continuous Magnetic Reconnection at Earth's Magnetopause. *Nature* **2003**, *426*, 533–537. [CrossRef] [PubMed]
4. Newell, P.T.; Sotirelis, T.; Wing, S. Diffuse, Monoenergetic, and Broadband Aurora: The Global Precipitation Budget. *J. Geophys. Res. Space Phys.* **2009**, *114*, A09207. [CrossRef]
5. Ni, B.; Liang, J.; Thorne, R.M.; Angelopoulos, V.; Horne, R.B.; Kubyshkina, M.; Spanswick, E.; Donovan, E.F.; Lummerzheim, D. Efficient Diffuse Auroral Electron Scattering by Electrostatic Electron Cyclotron Harmonic Waves in the Outer Magnetosphere: A Detailed Case Study. *J. Geophys. Res. Space Phys.* **2012**, *117*, A01218. [CrossRef]
6. Sandholt, P.E.; Farrugia, C.J. Does the Aurora Provide Evidence for the Occurrence of Antiparallel Magnetopause Reconnection? *J. Geophys. Res. Space Phys.* **2003**, *108*, 1466. [CrossRef]
7. Thorne, R.M.; Ni, B.; Tao, X.; Horne, R.B.; Meredith, N.P. Scattering by Chorus Waves as the Dominant Cause of Diffuse Auroral Precipitation. *Nature* **2010**, *467*, 943–946. [CrossRef]
8. Ni, B.; Thorne, R.M.; Zhang, X.; Bortnik, J.; Pu, Z.; Xie, L.; Hu, Z.-J.; Han, D.; Shi, R.; Zhou, C.; et al. Origins of the Earth's Diffuse Auroral Precipitation. *Space Sci. Rev.* **2016**, *200*, 205–259. [CrossRef]
9. Han, D.-S. Dayside Diffuse Aurora and the Cold-Plasma Structuring: A Brief Review. *Front. Astron. Space Sci.* **2021**, *8*, 725677. [CrossRef]
10. Feldstein, Y.I.; Vorobjev, V.G.; Zverev, V.L.; Förster, M. Investigations of the Auroral Luminosity Distribution and the Dynamics of Discrete Auroral Forms in a Historical Retrospective. *Hist. Geo Space Sci.* **2014**, *5*, 81–134. [CrossRef]
11. Sandholt, P.E.; Farrugia, C.J.; Moen, J.; Noraberg, Ø.; Lybekk, B.; Sten, T.; Hansen, T. A Classification of Dayside Auroral Forms and Activities as a Function of Interplanetary Magnetic Field Orientation. *J. Geophys. Res. Space Phys.* **1998**, *103*, 23325–23345. [CrossRef]
12. Wang, Q.; Bai, W.; Zhang, W.; Shi, J. Automatically Sketching Auroral Skeleton Structure in All-Sky Image for Measuring Aurora Arcs. *J. Geophys. Res. Space Phys.* **2024**, *129*, e2023JA031778. [CrossRef]
13. Han, D.-S.; Chen, X.-C.; Liu, J.-J.; Qiu, Q.; Keika, K.; Hu, Z.-J.; Liu, J.-M.; Hu, H.-Q.; Yang, H.-G. An Extensive Survey of Dayside Diffuse Aurora Based on Optical Observations at Yellow River Station. *J. Geophys. Res. Space Phys.* **2015**, *120*, 7447–7465. [CrossRef]
14. Lou, Y.; Cao, X.; Ni, B.; Tu, W.; Gu, X.; Fu, S.; Xiang, Z.; Ma, X. Diffuse Auroral Electron Scattering by Electrostatic Electron Cyclotron Harmonic Waves in the Dayside Magnetosphere. *Geophys. Res. Lett.* **2021**, *48*, e2020GL092208. [CrossRef]
15. Ni, B.; Thorne, R.M.; Horne, R.B.; Meredith, N.P.; Shprits, Y.Y.; Chen, L.; Li, W. Resonant Scattering of Plasma Sheet Electrons Leading to Diffuse Auroral Precipitation: 1. Evaluation for Electrostatic Electron Cyclotron Harmonic Waves. *J. Geophys. Res. Space Phys.* **2011**, *116*, A04218. [CrossRef]
16. Nishimura, Y.; Lessard, M.R.; Katoh, Y.; Miyoshi, Y.; Grono, E.; Partamies, N.; Sivadas, N.; Hosokawa, K.; Fukizawa, M.; Samara, M.; et al. Diffuse and Pulsating Aurora. *Space Sci. Rev.* **2020**, *216*, 4. [CrossRef]
17. Frahm, R.A.; Winningham, J.D.; Sharber, J.R.; Link, R.; Crowley, G.; Gaines, E.E.; Chenette, D.L.; Anderson, B.J.; Potemra, T.A. The Diffuse Aurora: A Significant Source of Ionization in the Middle Atmosphere. *J. Geophys. Res. Atmos.* **1997**, *102*, 28203–28214. [CrossRef]
18. Wang, Q.; Jing, S.; Yang, R.; Liu, Z.; Tang, Y.; Pan, H. Loosen Attention: Integrating Localized Channel and Coarse Spatial Attention for Enhanced Analysis of Complex Aurora Images. *Eng. Appl. Artif. Intell.* **2025**, *160*, 111907. [CrossRef]
19. Wang, Q.; Shi, J.; Liu, J.; Fan, J. Weakly Supervised Vortex Detection for Studying Correlation Between Multiscale Auroral Events. *IEEE Geosci. Remote Sens. Lett.* **2025**, *22*, 7500105. [CrossRef]
20. Yang, Q.; Wang, J.; Su, H.; Xing, Z. Automatic Recognition and Localization of Poleward Moving Auroral Forms (PMAFs) from All-Sky Auroral Videos. *Earth Space Sci.* **2023**, *10*, e2023EA002843. [CrossRef]
21. Tang, Y.; Niu, C.; Dong, M.; Ren, S.; Liang, J. Poleward Moving Aurora Recognition with Deep Convolutional Networks. In Proceedings of the Chinese Conference on Pattern Recognition and Computer Vision (PRCV), Xi'an, China, 8–11 November 2019; pp. 551–560.
22. Ahn, H.; Lee, D. Refining Action Segmentation with Hierarchical Video Representations. In Proceedings of the IEEE/CVF International Conference on Computer Vision (ICCV), Virtual, 11–17 October 2021; pp. 16302–16310.
23. Aziere, N.; Todorovic, S. Multistage Temporal Convolution Transformer for Action Segmentation. *Image Vis. Comput.* **2022**, *128*, 104567. [CrossRef]

https://doi.org/10.3390/rs18030402

24. Farha, Y.A.; Gall, J. MS-TCN: Multi-Stage Temporal Convolutional Network for Action Segmentation. In Proceedings of the IEEE/CVF Conference on Computer Vision and Pattern Recognition (CVPR), Long Beach, CA, USA, 15–20 June 2019; pp. 3575–3584.
25. Singhania, D.; Rahaman, R.; Yao, A. Coarse to Fine Multi-Resolution Temporal Convolutional Network. *arXiv* **2021**, arXiv:2105.10859. [CrossRef]
26. Yi, F.; Wen, H.; Jiang, T. ASFormer: Transformer for Action Segmentation. *arXiv* **2021**, arXiv:2110.08568. [CrossRef]
27. Liu, Y.; Huo, J.; Peng, J.; Sparks, R.; Dasgupta, P.; Granados, A.; Ourselin, S. SKIT: A Fast Key Information Video Transformer for Online Surgical Phase Recognition. In Proceedings of the IEEE/CVF International Conference on Computer Vision (ICCV), Paris, France, 1–6 October 2023; pp. 21074–21084.
28. Yu, F.; Koltun, V. Multi-Scale Context Aggregation by Dilated Convolutions. *arXiv* **2015**, arXiv:1511.07122.
29. Behrmann, N.; Golestaneh, S.A.; Kolter, Z.; Gall, J.; Noroozi, M. Unified Fully and Timestamp Supervised Temporal Action Segmentation via Sequence to Sequence Translation. In Proceedings of the European Conference on Computer Vision (ECCV), Tel Aviv, Israel, 23–27 October 2022; pp. 52–68.
30. Jiang, B.; Jin, Y.; Tan, Z.; Mu, Y. Video Action Segmentation via Contextually Refined Temporal Keypoints. In Proceedings of the IEEE/CVF International Conference on Computer Vision (ICCV), Paris, France, 1–6 October 2023; pp. 13836–13845.
31. Lu, Z.; Elhamifar, E. FACT: Frame-Action Cross-Attention Temporal Modeling for Efficient Action Segmentation. In Proceedings of the IEEE/CVF Conference on Computer Vision and Pattern Recognition (CVPR), Seattle, WA, USA, 17–21 June 2024; pp. 18175–18185.
32. Hu, Z.-J.; Yang, H.; Huang, D.; Araki, T.; Sato, N.; Taguchi, M.; Seran, E.; Hu, H.; Liu, R.; Zhang, B.; et al. Synoptic Distribution of Dayside Aurora: Multiple-Wavelength All-Sky Observation at Yellow River Station in Ny-Ålesund, Svalbard. *J. Atmos. Sol.-Terr. Phys.* **2009**, *71*, 794–804. [CrossRef]
33. Wang, Q.; Liang, J.; Hu, Z.-J.; Hu, H.-H.; Zhao, H.; Hu, H.-Q.; Gao, X.; Yang, H. Spatial Texture Based Automatic Classification of Dayside Aurora in All-Sky Images. *J. Atmos. Sol.-Terr. Phys.* **2010**, *72*, 498–508. [CrossRef]
34. Matzka, J.; Stolle, C.; Yamazaki, Y.; Bronkalla, O.; Morschhauser, A. The Geomagnetic Kp Index and Derived Indices of Geomagnetic Activity. *Space Weather* **2021**, *19*, e2020SW002641. [CrossRef]
35. Meredith, N.P.; Horne, R.B.; Thorne, R.M.; Summers, D.; Anderson, R.R. Substorm Dependence of Plasmaspheric Hiss. *J. Geophys. Res. Space Phys.* **2004**, *109*, A06209. [CrossRef]
36. Thomsen, M.F. Why Kp Is Such a Good Measure of Magnetospheric Convection. *Space Weather* **2004**, *2*, S11004. [CrossRef]
37. Carreira, J.; Zisserman, A. Quo Vadis, Action Recognition? A New Model and the Kinetics Dataset. In Proceedings of the IEEE/CVF Conference on Computer Vision and Pattern Recognition (CVPR), Honolulu, HI, USA, 21–26 July 2017; pp. 6299–6308.
38. Niu, Z.; Zhong, G.; Yue, G.; Wang, L.-N.; Yu, H.; Ling, X.; Dong, J. Recurrent Attention Unit: A New Gated Recurrent Unit for Long-Term Memory of Important Parts in Sequential Data. *Neurocomputing* **2023**, *517*, 1–9. [CrossRef]
39. Zhang, Y.; Wu, R.; Dascalu, S.M.; Harris, F.C., Jr. A Novel Extreme Adaptive GRU for Multivariate Time Series Forecasting. *Sci. Rep.* **2024**, *14*, 2991. [CrossRef]
40. Cho, K.; Van Merriënboer, B.; Gulcehre, C.; Bahdanau, D.; Bougares, F.; Schwenk, H.; Bengio, Y. Learning Phrase Representations Using RNN Encoder-Decoder for Statistical Machine Translation. In Proceedings of the Conference on Empirical Methods in Natural Language Processing (EMNLP), Doha, Qatar, 25–29 October 2014; pp. 1724–1734.
41. Kingma, D.P. Adam: A Method for Stochastic Optimization. *arXiv* **2014**, arXiv:1412.6980.
42. Goodfellow, I. *Deep Learning*; MIT Press: Cambridge, MA, USA, 2016.
43. Lea, C.; Flynn, M.D.; Vidal, R.; Reiter, A.; Hager, G.D. Temporal Convolutional Networks for Action Segmentation and Detection. In Proceedings of the IEEE/CVF Conference on Computer Vision and Pattern Recognition (CVPR), Honolulu, HI, USA, 21–26 July 2017; pp. 156–165.
44. Caba Heilbron, F.; Escorcia, V.; Ghanem, B.; Carlos Niebles, J. ActivityNet: A Large-Scale Video Benchmark for Human Activity Understanding. In Proceedings of the IEEE/CVF Conference on Computer Vision and Pattern Recognition (CVPR), Boston, MA, USA, 7–12 June 2015; pp. 961–970.
45. Newell, P.T.; Gjerloev, J.W. Evaluation of SuperMAG Auroral Electrojet Indices as Indicators of Substorms and Auroral Power. *J. Geophys. Res. Space Phys.* **2011**, *116*, A12211. [CrossRef]

https://doi.org/10.3390/rs18030402

MDPI AG
Grosspeteranlage 5
4052 Basel
Switzerland
Tel.: +41 61 683 77 34

Remote Sensing Editorial Office
E-mail: remotesensing@mdpi.com
www.mdpi.com/journal/remotesensing

www.ingramcontent.com/pod-product-compliance
Lightning Source LLC
Chambersburg PA
CBHW040824250926
51850CB00007BA/1251
* 9 7 8 3 7 2 5 8 6 9 8 0 0 *